SITE SURVEYING

SECOND EDITION

John Muskett

BEng, CEng, MICE
School of Civil Engineering and Building
Bolton Institute of Higher Education

Blackwell
Science

Blackwell Science Ltd, a Blackwell Publishing company
Editorial offices:
Blackwell Science Ltd, 9600 Garsington Road, Oxford OX4 2DQ, UK
Tel: +44 (0) 1865 776868
Blackwell Publishing Inc., 350 Main Street, Malden, MA 02148-5020, USA
Tel: +1 781 388 8250
Blackwell Science Asia Pty Ltd, 550 Swanston Street, Carlton, Victoria 3053, Australia
Tel: +61 (0)3 8359 1011

First published 1988 by BSP Professional Books, a division of Blackwell Scientific Publications Ltd
Reprinted 1991, 1993
Second edition published 1995 by Blackwell Science Ltd
Reprinted 2000, 2002, 2003, 2005

Library of Congress Cataloging-in-Publication Data is available

ISBN-10: 0-632-03848-9
ISBN-13: 978-0632-03848-0

A catalogue record for this title is available from the British Library

Set by SNP Best-set Typesetter Ltd., Hong Kong
Printed and bound by Marston Book Services, Oxford, Great Britain

The publisher's policy is to use permanent paper from mills that operate a sustainable forestry policy, and which has been manufactured from pulp processed using acid-free and elementary chlorine-free practices. Furthermore, the publisher ensures that the text paper and cover board used have met acceptable environmental accreditation standards.

For further information on Blackwell Publishing, visit our website:
www.thatconstructionsite.com

Contents

Preface xi

Acknowledgements xiii

Dedication xv

1 Introduction 1
- 1.1 Background 1
- 1.2 Site Surveys 2
- 1.3 Elements of Site Surveys 3
- 1.4 Reconnaissance 6
- 1.5 Booking 6
- 1.6 Good Practice 6
- 1.7 Further Reading 9

2 Levelling 10
- 2.1 Principle 10
- 2.2 Instruments and Setting Up 11
- 2.3 Levelling Procedure 17
- 2.4 Errors and Accuracy 23
- 2.5 Level Grids 24
- 2.6 Longitudinal and Cross-sections 26
- 2.7 Contouring 28
- 2.8 Precise Levelling 28
- 2.9 Tests and Permanent Adjustments 33
- 2.10 Further Reading 35
- 2.11 Exercises 35

3 The Theodolite 37
- 3.1 Introduction 37
- 3.2 Construction 37
- 3.3 Setting Up 41
- 3.4 Angle Readings: Horizontal 45

3.5 Angle Readings: Vertical 50
3.6 Trigonometric Heighting 53
3.7 Other Uses of the Theodolite 55
3.8 Instrument Errors and Permanent Adjustments 55
3.9 Further Reading 61
3.10 Exercises 61

4 Linear Measurement 64
4.1 Taping: Description and Use 64
4.2 Taping: Corrections 66
4.3 EDM: Description and Operation 70
4.4 EDM: Problems, Corrections, Errors, Accuracy 74
4.5 EDM: Applications 79
4.6 EDM: Developments 80
4.7 Further Reading 84
4.8 Exercises 84

5 Control Surveys 86
5.1 Introduction 86
5.2 Coordinates 89
5.3 Traversing 95
5.4 Intersection 116
5.5 Resection 117
5.6 Networks 119
5.7 Geodetic Surveys 127
5.8 Height Control 135
5.9 Global Positioning System (GPS) 136
5.10 Further Reading 149
5.11 Exercises 149

6 Detail Surveys 153
6.1 Introduction 153
6.2 Linear Surveying 155
6.3 EDM Detailing 159
6.4 Optical Tacheometry 168
6.5 GPS Detailing 171
6.6 Contouring 172
6.7 Photogrammetry 177
6.8 Data Logging and Processing 178
6.9 Further Reading 183
6.10 Exercises 183

7 Errors and Adjustments 188
7.1 Types of Error 188
7.2 Analysis of Errors 190
7.3 Reliability of a Single Quantity 192
7.4 Weighting 195
7.5 Quantities Derived from Two or More Observed Quantities 197
7.6 Adjustment of Related Quantities 202
7.7 Variation of Coordinates 215
7.8 Matrix Solution 221
7.9 Error Ellipses 223
7.10 Tied Traverse Example 227
7.11 Control Survey Adjustment 233
7.12 Further Reading 234
7.13 Exercises 234

8 Area and Volume Measurement 238
8.1 Introduction 238
8.2 Regular Areas 239
8.3 Irregular Areas 240
8.4 Road Cross-sections 245
8.5 Volumes of Regular Solids 249
8.6 Volumes of Irregular Solids 249
8.7 Curved Irregular Solids 260
8.8 The Mass-Haul Diagram 262
8.9 Exercises 271

9 Setting Out – an Introduction 276
9.1 Preamble 276
9.2 The Method 276
9.3 Organization of Construction Works 279
9.4 The Site Engineer 280
9.5 Good Relations 281
9.6 Organization and Planning 283
9.7 Establishment and Maintenance of Equipment 283
9.8 Training Chainmen 285
9.9 Accuracy 286
9.10 Selection of Method 289
9.11 Types of Markers 289
9.12 Lasers 292
9.13 Control Setting Out 295
9.14 While in the Field 296

9.15 Sketch of Setting Out Information 296
9.16 Record Drawings 298
9.17 Further Reading 298

10 Roadworks I – Curve Calculations 299
10.1 Introduction 299
10.2 Horizontal Circular Curves: Nomenclature 299
10.3 Horizontal Circular Curves: Large Radius 301
10.4 Horizontal Curves: Small Radius 309
10.5 To Pass a Horizontal Curve through a Given Point 310
10.6 Horizontal Transition Curves 312
10.7 Composite Curve Example 324
10.8 Transition Origin Inaccessible 330
10.9 Compound Curves 331
10.10 Computer Design 331
10.11 Vertical Curves 331
10.12 To Pass a Vertical Curve through a Given Point 341
10.13 Coordinated Design 343
10.14 Further Reading 343
10.15 Exercises 343

11 Roadworks II – Setting Out 346
11.1 Initial Survey 346
11.2 Initial Setting Out 347
11.3 Earthworks 349
11.4 Formation and Sub-base 354
11.5 Road Base 358
11.6 Kerbs 359
11.7 Surfacing 360
11.8 Other Works 360

12 Drains and Pipelines 361
12.1 Introduction 361
12.2 Position in Plan 361
12.3 Levelling 363
12.4 Variations in Boning in 368
12.5 Manholes, Gullies, Air Valves, Wash-outs 371

13 Foundations, Temporary Works and Structures 373
13.1 Introduction 373
13.2 Piles – Load Bearing 373

13.3 Pile Caps, Column Bases, Footings 375
13.4 Deep Foundations 376
13.5 Temporary Works 377
13.6 Structures 379
13.7 Bridges 383
13.8 Precise Setting Out of Machinery – Autocollimation 384
13.9 Coordinate Measuring Systems 385

14 Underground and Marine Works 388
14.1 Underground Works – Introduction 388
14.2 Surface Survey 388
14.3 Transfer Underground 389
14.4 Underground Control 391
14.5 Control at the Face 396
14.6 Marine Works 399
14.7 Further Reading 401

Answers to Exercises 402

Index 409

Preface

Site Surveying was originally intended to be comprehensive in coverage, practical yet concise. Since the publication of the first edition a number of significant developments in surveying have occurred. These developments have speeded up and simplified many operations and it would be perverse (and unfair to students' pockets) to produce a revised text significantly longer than the original. To balance the inclusion of new material, the writing on processes nearly or totally obsolete has been greatly reduced or removed entirely. Relevance to current construction practice is still the principal criterion for selection of material.

The layout of the first edition has been retained. Recent developments in electromagnetic distance measurement (EDM) have been described and further applications of data logging are included. The measurement of distances and angles within a network is now a common way of carrying out a control survey and the book deals with the planning and execution of such work. Computer programs (using the principle of least squares) for the adjustment and analysis of control networks are widely available, and the underlying theory, the preparation of data for entry and the interpretation of results are comprehensively covered. The Global Positioning System (satellite surveying) is now sufficiently accurate and flexible to be an important tool for both surveying and setting out of construction works, and the system is covered in some detail. Those field operations seldom practised (triangulation and optical tacheometry for instance) are now dealt with briefly, while some calculation processes (log sine equal shifts adjustment, the osculating circle) have been removed completely.

I have tried to pay attention to readers' comments (which have been very helpful), adding further worked examples and providing extra clarification where necessary. Some traditional operations still merit inclusion. Use of the tape to fill in detail or to set out is described. Contouring is an apparently simple task, often poorly described, which presents problems in practice; I have tried to convey the spatial appreciation required in the field and the understanding of algorithms for automatic plotting to enable this operation to be carried out effectively.

The book is designed as a main text for undergraduate, and BTEC courses in construction subjects. It should also serve as a reference work on sites, where its theory can be complemented by essential practical experience. Students from other disciplines studying surveying, perhaps with a construction option, should also find the book of use.

Acknowledgements

In this second edition, I should like to acknowledge the support of Professor Clive Melbourne, my Head of School and the continual encouragement of Julia Burden, deputy publisher.

For the supply of, and permission to reproduce, photographs my thanks go to the following companies:

Geotronics Limited	(Figs 4.6, 4.7, 5.22(c), 6.17),
Hall and Watts Ltd	(Fig. 8.5(b)),
Laser Alignment	(Figs 9.7(b), 12.4(a)),
Leica UK Ltd	(Figs 2.4, 2.5, 2.6(c), 3.4(b), 4.4(a), 5.20, 5.22(a), 6.14(b), 6.16(b), 9.7(a), 13.8),
Sokkia Ltd	(Figs 2.3, 3.1(c), 3.4(a), 4.4(b), 5.22(d), 6.14(a), 6.16(a)),
Survey Supplies Ltd	(Figs 3.1(b), 4.8),
Trimble Navigation Europe Ltd	(Figs 5.22(b)).

My thanks also go to the Controller of Her Majesty's Stationery Office for permission to quote from TD9/93 *Highway Link Design Standard* published in 1993 by the Department of Transport, and to the County Surveyors' Society for permission to quote from *Highway Transition Curve Tables (metric)*.

I am grateful to my colleagues at Bolton who have made suggestions, particularly David Palin who has assisted my understanding and commented on parts of the text. Information provided by Trimble Navigation has been helpful, as have articles in *Civil Engineering Surveyor* on developments in equipment and methods.

Mildred Jones has typed the revisions to the text, and I acknowledge with much gratitude her help and advice on layout. Any errors are entirely my fault; comments on inconsistencies will be gratefully received.

My greatest thanks go to my family, who have again endured a period of neglect which they have borne, together with my grouchiness, with great patience.

John Muskett

Dedication

To Meryl and to Helen, Julia and Paul

1 Introduction

1.1 BACKGROUND

Surveying is the practice of taking measurements of features on, and occasionally above or below, the earth's surface to determine their relative positions. The practice may be more precisely described as land surveying to distinguish it from quantity surveying, building surveying and other forms of surveying.

Land surveys may be required for geographical, agricultural, geological, mineral, ecological, construction, land ownership or other purposes. Frequently the end product of a land survey is a drawn plan, although survey information can be stored in digital form.

The fieldwork usually takes the form of physical measurements on the land. Other methods include photography, measurements from orbiting satellites, measurement of radiation and inertial methods.

The Ordnance Survey is responsible for surveying and mapping Great Britain. Perhaps best known for its 1:50 000 maps, the Ordnance Survey produces a range of plans and maps suitable for a variety of purposes. In addition to conventional maps on paper, digital maps on computer disk are available, as are data on the coordinates of triangulation ('trig') points and bench marks. Such data must be obtained (from the OS) where surveys are to be tied in to the OS National Grid, or levelling is to be referred to OS (Newlyn) datum.

Of the divisions within land surveying, engineering surveying and topographic surveying (representation of land features) are relevant to site surveys, while aspects of geodetic surveying (accounting for the earth's curvature) must be considered in surveys over large areas and for connection of surveys to the National Grid. This book deals with those aspects pertinent to construction work, reference to deeper treatment or more specialized topics being made under the heading 'Further reading' in the appropriate chapters.

1.2 SITE SURVEYS

Site surveys are carried out for several reasons:

(1) To produce a plan of an area, often with contours.
(2) To produce sections of the ground.
(3) To determine land areas in plan.
(4) To determine volumes of earth or water.
(5) To set out construction works.
(6) To monitor ground and structural movement.

Survey measurements are of height differences, angles and distances. Chapters 2, 3 and 4 deal with the use of appropriate instruments for determining these quantities.

The principle to be adopted in surveying is that of 'working from the whole to the part'. Work should commence with a control survey to establish the positions of plan control stations and the levels of temporary bench marks throughout the site (Chapter 5). Measurements taken should be of adequate precision; the 'whole to part' method will reduce the likelihood of errors accumulating.

Within the control survey, detail surveying is then carried out (Chapter 6). Measurements are taken to relate points of detail, in plan, to the control stations and to determine their levels.

The surveyor's task will be to select suitable techniques and equipment, to use that equipment effectively, and periodically to check that the equipment is performing satisfactorily. Absolute accuracy is not possible and the surveyor will have to choose field processes and equipment which are acceptably precise. The surveyor should carry out independent checks to reveal the presence of mistakes in fieldwork or calculations. Once identified, mistakes can be rectified.

Systematic errors may occur from poor calibration of equipment, variation in operating conditions and other sources. The effects of a repeated systematic error are cumulative. The sources must be identified (by checking equipment, assessing the conditions) so that corrections can be applied to the measured quantities.

The remaining small (random) errors can sometimes be ignored; in other instances the measured quantities must be adjusted to produce mathematical consistency. Chapter 7 covers the analysis of errors and adjustment of quantities.

During the planning of construction works, the evaluation of plan areas and the determination of earthwork volumes will be required. Area and volume calculations will also be necessary as construction proceeds, for ordering and checking materials and to provide 'as constructed' measurements for payment purposes. Chapter 8 deals with the appropriate field and office work.

Surveying techniques must be used throughout a construction project to control the alignment of works in plan and in elevation ('setting out'). Chapter 9 gives general information on setting out, while Chapters 10 to 14 cover the setting out of particular types of construction.

1.3 ELEMENTS OF SITE SURVEYS

Figure 1.1 is a plan of some land which is to be developed. To allow design work to proceed, an accurate plan is required. The development will incorporate roadworks, construction of a school, provision of shops and houses and installation of the necessary drainage. A landscaped play area is also planned.

Control survey

A, B, C, D, E and F are control stations, typically pegs in the ground or road nails in asphalt, positioned by the surveyor. Stations A and E are of known coordinates, being tied in to existing roads. Lines AB, BC . . . EF, FA are the legs of a control traverse. Their horizontal lengths must be measured by taping or by electronic methods and the angle at each station between adjoining legs must be measured by theodolite. It may also be possible to use a GPS (Satellite) survey to fix the stations.

Calculations should be carried out to enable the stations to be plotted relative to east and north coordinate axes.

Detail survey

Some detail is shown located by radiating lines from station B. The distances can be found by taping or by using electronic or tacheometric methods. The bearings of the lines can be determined from theodolite measurement of the horizontal angles. Detail in the vicinity of leg DE can be located by taping perpendicular offset distances, or pairs of tie lines from the leg.

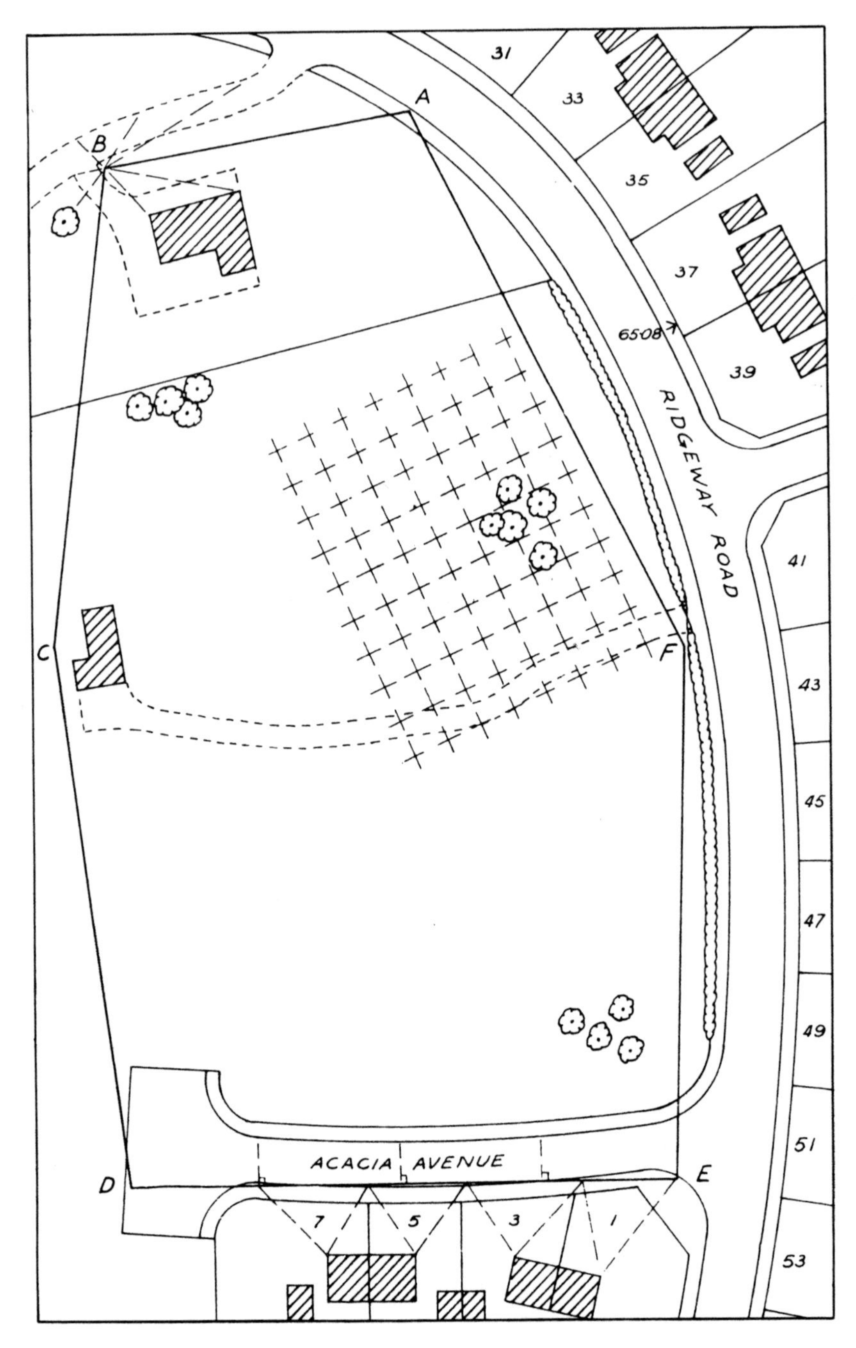

Fig. 1.1 Plan of development area.

Traditionally the detail survey, often requiring less precision than the control survey, was carried out after completion of the control survey. Nowadays, when electronic equipment is used, control and detail work may be undertaken concurrently.

Detail may also be surveyed by GPS (Satellite) methods.

Levelling

A control level survey around the stations should be carried out using an engineer's level and staff. Heights can be determined relative to the Ordnance Bench Mark on the garden wall of number 37, Ridgeway Road. Points on existing structures, road surface levels and manhole covers (to check drain levels) should also be levelled this way.

An engineer's level can also be used to produce a grid of ground levels (close to line FA) or to carry out contouring. Alternatively, ground levelling can be carried out by electronic or tacheometric methods.

Ground levels are required for the determination of a suitable vertical alignment for the road, for fixing the ground floor levels of the buildings and for the design of drainage to be efficient in use and economic in construction.

Plotting

Booking, calculating and plotting have, until recently, been performed manually. Modern electronic instruments have data output facilities so that a data logger can automatically record field observations. The logger can be connected to an office computer so that calculations and plotting can be performed automatically. Many surveys, however, are still processed manually, especially revisions and surveys of limited area.

Areas and volumes

Detail measurements enable the plan area of the land to be determined for costing and compensation purposes and to determine an acceptable number of properties to construct. Volumes will be required for earth movement calculations in connection with the roadworks and building foundations.

Setting out

At the start of construction works, control setting out stations must be fixed. It will not often be possible to set them to coincide with the surveying stations as they must be clear of the works and located to facilitate construction alignment.

Control setting out is followed by the setting out of markers for individual parts of the works to suit the construction programme.

1.4 RECONNAISSANCE

The first operation to be carried out when surveying or setting out starts on a new site is a reconnaissance:

(1) To determine suitable field processes and to select equipment and materials.
(2) To locate existing survey control points for plan and level.
(3) To plan control station positions.
(4) To plan the correlation between control and detail work.
(5) To assess potential problems, for example obstacles.
(6) To investigate land ownership and access.
(7) To ascertain the importance of man-made features such as roads, structures and services.

1.5 BOOKING

Observations should be booked systematically in a proper field book; for some processes there are specific booking routines; for others the surveyor may use his or her discretion. Bookings should be legible and unambiguous so that no confusion can arise during plotting. Faulty entries should not be erased but should be crossed out.

1.6 GOOD PRACTICE

Modern equipment allows very precise measurements to be made. Much of the labour of calculations has been eased by the introduction of computer programs. Field data can be automatically recorded and plotted. These developments cut down mistakes (human error) and reduce the effects of systematic errors. Nevertheless, unless used sensibly, modern equipment can produce results that are not within expected tolerances. There is no such thing as a perfect surveying measurement, and all errors may be magnified by unfavourable geometry. A knowledge of the physical characteristics of the quantities to be measured, of the equipment to be used, and of the conditions (e.g. atmospheric) under which the measurements are to be made is necessary. Frequently surveying methods that automatically apply compensation for possible errors can be used, and checks should be built into the measuring process.

In the handling of numbers also there is scope for misinterpretation, where the figures may suggest a more precise measurement of a quantity than has actually occurred.

Consider the measurement of the width of a room. Before measurement, a number of questions must be asked:

(1) Width at what height? (the walls may not be vertical).
(2) With or without skirting board?
(3) What sort of accuracy is required?
(4) For what purpose?

The answer to the final question may provide the answers to the others. If the purpose is to calculate the amount of paint for a ceiling, or to design central heating, a measurement within ±100 mm would suffice, and a single taped measurement reading within ±20 mm could be taken. If the purpose is to estimate for a fitted carpet, we need several taped measurement between skirting boards, and we need to check the squareness of the corners – if the room is slightly diamond shaped a greater width of carpet may be needed. We need to tape to ±10 mm, take the largest value and add (say) 50 mm for safety. If the purpose is to install fitted furniture across an end wall, we need taped measurements at different heights with readings to ±3 mm.

What is the condition of the tape – new, old, steel, fibre? Has it stretched with age or does it have kinks? At the measuring end where it must curve on contact with the wall, can one really read it to the nearest millimetre? Are the measurements perpendicular to the walls? (And are the walls parallel?)

It is possible with a steel tape to achieve a single measurement reading to the nearest millimetre. It would be possible to use other (electronic) methods to obtain a reading in even finer units. But would such measurements be valid? As representative of the width of the room, almost certainly not. It can be seen that until what is required is specified, a meaningful measurement is impossible. It is quite likely that 12 precise measurements will produce a range of 30 mm! A single quoted value should reflect this.

Consider a much longer measurement of (about) 300 m using an electromagnetic distancer. Such equipment may have a standard error of measurement of ±3 mm. However, if the instrument and reflecting prism are not precisely located at the ends of the line, further errors will occur. Also if the atmospheric temperature and pressure are not the same as those at which the instrument was calibrated, another error is likely. The cumulative effect of the errors could be as much as 10 mm, well in excess of the manufacturer's quoted error.

In the handling of figures the surveyor must be cautious. All measurements are of continuous, not discrete, quantities, and the accuracy of measurement is implied by the number of figures in the value. '5.1' suggests a value between 5.05 and 5.15, whereas 5.100 suggests millimetre accuracy. When figures are combined, the final amount should be expressed to comply with the least precise measurement.

Example 1.1

A baseline has been measured in several bays whose horizontal lengths were recorded as 73.165 m, 47.21 m, 55.387 and 91.4 m. How should the total length be quoted?

Solution

```
 73.165
 47.21
 55.387
 91.4
-------
267.162    which should be quoted as 267.2 m (one decimal place)
=======
```

Example 1.2

Calculate the area of a rectangular plot of land of
195.142 m length and 25.337 m width.

Solution

$195.142 \times 25.337 = 4944.312854$ which should be quoted as $4944.3\,m^2$ (5 significant figures).

Example 1.3

Find the plan length of a distance measured on the slope as 237.437 m with a zenith angle of 81°41′.

Solution

$237.437 \sin 81°41' = 234.9400545$ by calculator.

This is more difficult to round off; the angle has 4 significant figures (in effect); however at 81°, the sine changes far more slowly than the angle value. An error analysis (see Chapter 7) could be carried out, or the distance calculated for 81°40′30″ and 81°41′30″ (extremes of range). It can be shown that 234.94 m is a realistic value. In calculations, all the figures in the data may be used, but rounding off to the appropriate number of figures should be performed before the final quantity is quoted. A final '5' to be rounded should be rounded up if preceded by an odd number and down if preceded by an even number, e.g. $103.375 \rightarrow 103.38$ and $87.885 \rightarrow 87.88$.

Integer values ending in more than one zero can be expressed more precisely in exponents of 10, e.g. 123 000 (with 3 significant figures) = 123×10^3; 123 000 (with 5 significant figures) = $12\,300 \times 10$.

1.7 FURTHER READING

Department of Transport *Technical Memorandum H5/78* (1978) *Model Contract Document for Topographical Survey Contracts*.

Worthington, B.D.R. and Gant, R. (1983) *Ordnance Survey Mapwork*. London: Macmillan.

2 Levelling

2.1 PRINCIPLE

Levelling is the determination of variations in altitude (or level) of points. It is usually performed with an 'engineer's level' and a levelling staff; this is known as 'spirit levelling'.

A horizontal line (or plane) of sight (line or plane of collimation) is established with a telescope (fitted with cross-hairs) which can be turned about a vertical axis. The difference in consecutive readings taken on a vertical staff (at A and B) gives the level difference (2.635 − 1.470 = 1.165) between the two points (Fig. 2.1(a)). Over a limited distance, the horizontal plane may be taken as level; correctly, a level surface is spheroidal, following the earth's shape (Fig. 2.1(b)). The effects of curvature are covered in Section 2.8.

In the United Kingdom, level values are often referred to Ordnance (Survey) Datum, which is mean sea level at Newlyn (Cornwall, England) as determined between 1915 and 1921. From this datum the Ordnance Survey have established Ordnance Bench Marks (OBMs) at regular intervals over the country. They are generally lines cut in the vertical faces of stone or brick permanent structures, up-to-date values being obtainable from the Ordnance Survey (Fig. 2.2). Bolt and pivot bench marks are also found. A chiselled arrow points to the mark.

Where no OBM exists close to a construction site, a temporary bench mark (TBM) must be established. A horizontal mark, such as a pencil line, over paint or wax crayon on a vertical surface (wall or column) is satisfactory, as is a bolt or steel rod driven into the ground and surrounded with concrete, provided that the ground is stable. Levelling should be carried forward from the nearest OBM to establish the TBM value.

Where the heights of points above a datum have been determined, the values are known as 'reduced levels'. If levelling is tied in to an

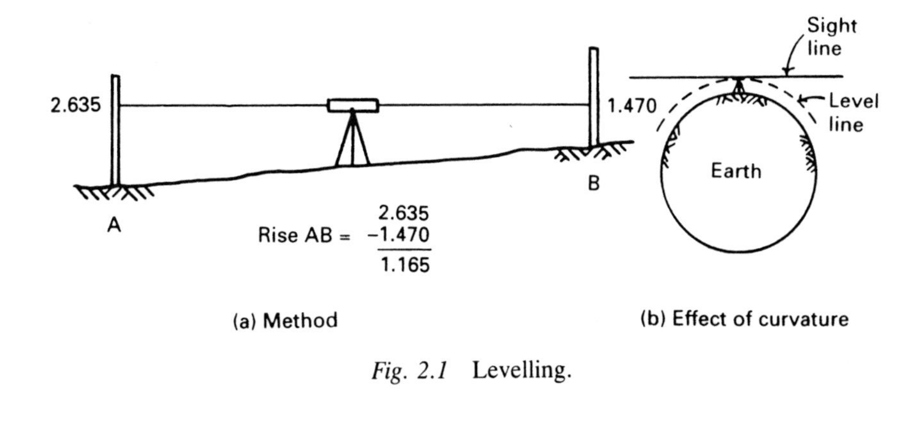

Fig. 2.1 Levelling.

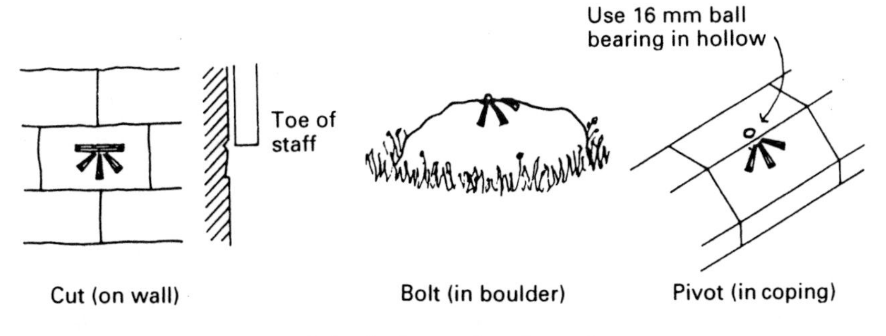

Fig. 2.2 Ordnance bench marks.

OBM, the reduced levels are quoted as 'above Ordnance Datum' and AOD may be written after the value.

On some construction sites a site datum is used, often so that 100.000 m represents ground floor finished level.

GPS (satellite surveying) can be used for levelling. Levels so obtained are related to a reference ellipsoid and not to a sea level (or other) datum. Within a limited area, correspondence between the two datums can be achieved by interpolation between points levelled by both methods (see Chapter 5, Sections 5.7 to 5.9).

2.2 INSTRUMENTS AND SETTING UP

Manufacturers categorize instruments as 'builders'', 'surveyors'', 'engineers'', and 'precise' levels, the terms being a comparative indication of the precision of levelling possible with them. For construction surveying, an instrument with a manufacturer's standard error of no more than 2.5 mm per km length is satisfactory. In their method of construction and operation, there are three types of level currently in use.

The tilting level

The tilting level is the oldest established. An initial coarse levelling is performed when the instrument is first set up, and the telescope is finely levelled before each sighting of the staff. Figure 2.3 shows a tilting level.

The telescope, with internal focusing, is designed to produce a real image in the plane of the diaphragm. The diaphragm is made of thin glass with cross-hairs etched on it. The image and cross-hairs are viewed together through a twin lens Ramsden eyepiece. Older instruments produce an inverted image, whereas modern instruments have an extra lens or prism to erect the image. Magnification of 25× is provided.

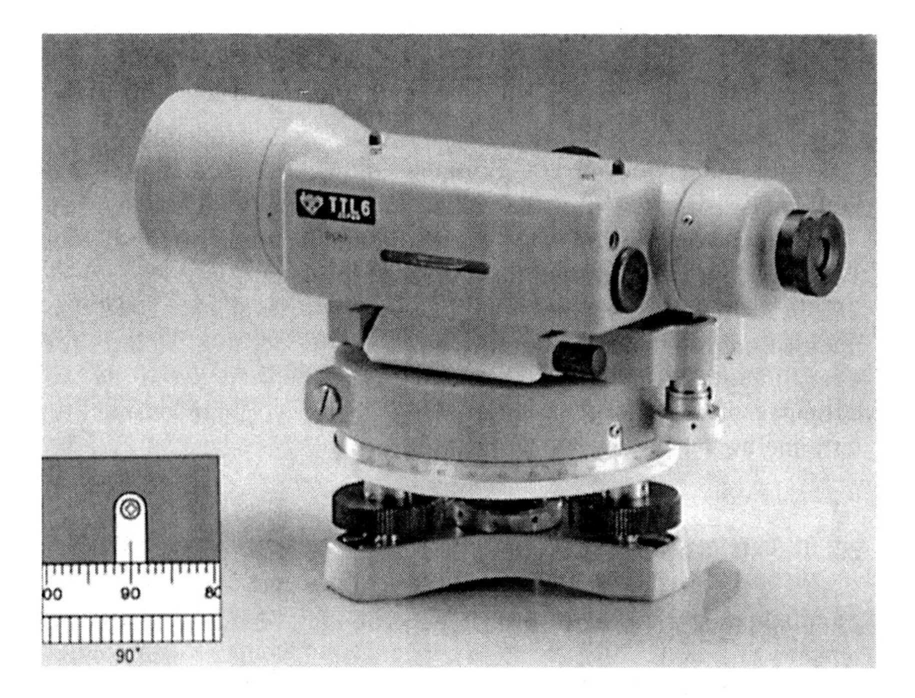

Fig. 2.3 Sokkia TTL 6 tilting level.

To the side of the telescope is fitted a levelling bubble. On older levels this can be seen from the eyepiece position via a mirror; most instruments now allow a split image of both ends of the bubble to be viewed through an auxiliary eyepiece alongside the telescope eyepiece, or directly through the main eyepiece. The telescope is mounted on a stage by means of a horizontal pivot; a return spring keeps the eyepiece end of the telescope in contact with a tilting screw which provides the fine levelling facility. The stage can rotate about a vertical axis above the tribrach, movement being controlled by a clamp and a fine motion (tangent) screw. Some manufacturers eliminate the clamp and provide a friction bearing.

In addition to housing the clamp, tangent and tilting screws, the stage is equipped with a circular 'bullseye' bubble. The coarse levelling is achieved by centring this bubble using the three footscrews of the tribrach. Under the footscrews, the baseplate of the level is secured to a tripod by a bolt passing through the flat top of the tripod. On instruments with a 'quicksetting' base, a ball and socket arrangement replaces the footscrews. Release of a clamp or of the securing bolt allows rotation for levelling.

Setting up the tilting level

(1) Open and extend to a suitable length the tripod legs. Place the tripod in a convenient position with its head approximately level, the legs spread out and made firm. On smooth surfaces care must be taken to prevent the legs from sliding outwards; on soft earth the surveyor must tread the legs into the ground.
(2) Bolt the instrument centrally on the tripod. Turn two footscrews in opposite senses until the bullseye bubble is centred along the line of the two screws. Centralization of the bubble along a line perpendicular to this line is achieved by turning the third screw. Further small adjustments are likely to be necessary until the bubble is truly central. The direction of bubble movement is the same as that of the operator's left thumb or right forefinger. The coarse levelling has now been carried out and should not need repeating until the instrument is moved or disturbed.
(3) Focus the cross-hairs by rotating the eyepiece when sighting a plain background through the telescope. Release the clamp and turn the telescope to sight the staff, using the 'gunsights' on top of the telescope to achieve an approximate pointing. With the clamp now locked, adjust the tangent screw to obtain coincident images

of the staff and vertical cross-hair when looking through the telescope. The telescope should be focused to eliminate parallax. Parallax occurs when staff and cross-hairs are not focused together; it can be detected by the surveyor moving his head while looking through the telescope – differential movement of staff and cross-hairs will be observed. Non-elimination may cause errors in staff readings.

(4) Turn the tilting screw to centralize the telescope bubble, and take a reading of the staff and book it. It is good practice to repeat this process to check the reading. The bubble must be checked (and recentred if necessary) immediately before every reading. If the telescope is turned to sight a fresh staff position, the bubble will almost certainly require resetting.

The automatic level

This type of instrument is rapidly superseding the tilting level. In appearance it is a simpler version of the older type of instrument, but there is no tilting facility, the telescope being cast integrally with the stage. A combination of fixed and suspended prisms in the optical train of the telescope ensures that, provided the telescope is approximately level, as shown by the bullseye bubble, a horizontal collimation line will pass through the cross-hairs. As with the tilting level, the tribrach may be equipped with footscrews or with a ball and socket quicksetting base. Figure 2.4 shows an automatic level.

The automatic level is quicker in use than the tilting level and free from operator errors in setting the telescope bubble. However, in conditions of wind or vibration the staff image may oscillate, negating the instrument's advantages.

Setting up the automatic level

(1) Open and extend to a suitable length the tripod legs. Place the tripod in a convenient position with its head approximately level, the legs spread out and made firm. On smooth surfaces care must be taken to prevent the legs from sliding outwards; on soft earth the surveyor must tread the legs into the ground.

(2) Bolt the instrument centrally on the tripod. Turn two footscrews in opposite senses until the bullseye bubble is centred along the line of the two screws. Centralization of the bubble along a line

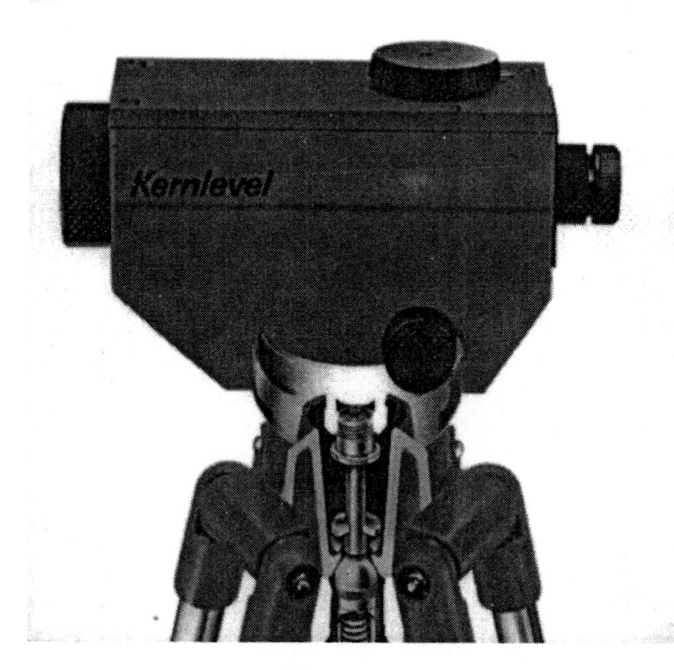

Fig. 2.4 Leica automatic Kernlevel.

perpendicular to this line is achieved by turning the third screw. Further small adjustments are likely to be necessary until the bubble is truly central. The direction of bubble movement is the same as that of the operator's left thumb or right forefinger. The coarse levelling has now been carried out and should not need repeating until the instrument is moved or disturbed.

(3) Focus the cross-hairs by rotating the eyepiece when sighting a plain background through the telescope. Release the clamp and turn the telescope to sight the staff, using the 'gunsights' on top of the telescope to achieve an approximate pointing. With the clamp now locked, adjust the tangent screw to obtain coincident images of the staff and cross-hairs when looking through the telescope. The telescope should be focused to eliminate parallax. Parallax occurs when staff and cross-hairs are not focused together; it can be detected by the surveyor moving his head while looking through the telescope – differential movement of staff and cross-hairs will be observed. Non-elimination may cause errors in staff readings.

(4) The staff reading can now be taken directly and booked. To check

that the automatic compensator is functioning correctly, press the checking button if fitted, or given half a turn to a footscrew under the telescope. The cross-hairs should appear to move off the staff reading and then slowly return to it.

The digital level

A third type of level is the more recently developed digital level introduced by Leica. Mechanically it is similar to an automatic level. Instead of a conventionally graduated staff, one with 'bar code' markings is used. The telescope is pointed at the staff which is viewed through the telescope normally and a portion above and below the cross-hairs is scanned by an image processing system in the instrument. The cross-hair value appears on a liquid crystal display, as does the horizontal distance; they are also stored by the instrument. Reduced levels are automatically calculated and displayed. The digital level can accept a memory card (see Section 6.8 Data logging and processing) so that more permanent storage of observations can be effected. The card can subsequently be coupled with an office computer and the levels can either be linked with other survey work, or an adjustment process can be carried out. Leica produce two models: the NA2002 and the precise counterpart, the NA3000, see Fig. 2.5.

Setting up the digital level

Steps 1 to 3 for the automatic level should be followed. After a key is pressed, the reading and horizontal distance are displayed in turn. A keyboard allows various options of calculation and storage to be carried out. The reverse of the staff is conventionally figured, so that an optical check is possible.

The range is between 1.8 m (to give sufficient length of the staff visible in the telescope for scanning) and 100 m. The manufacturer's standard deviation is ±1.5 mm for 1 km of double run levelling carried out automatically with the NA2002. Optical levelling is less precise; use of the NA3000 with an invar bar-coded staff is more precise. Distance measurement accuracy is between 3 mm and 5 mm per 10 m length.

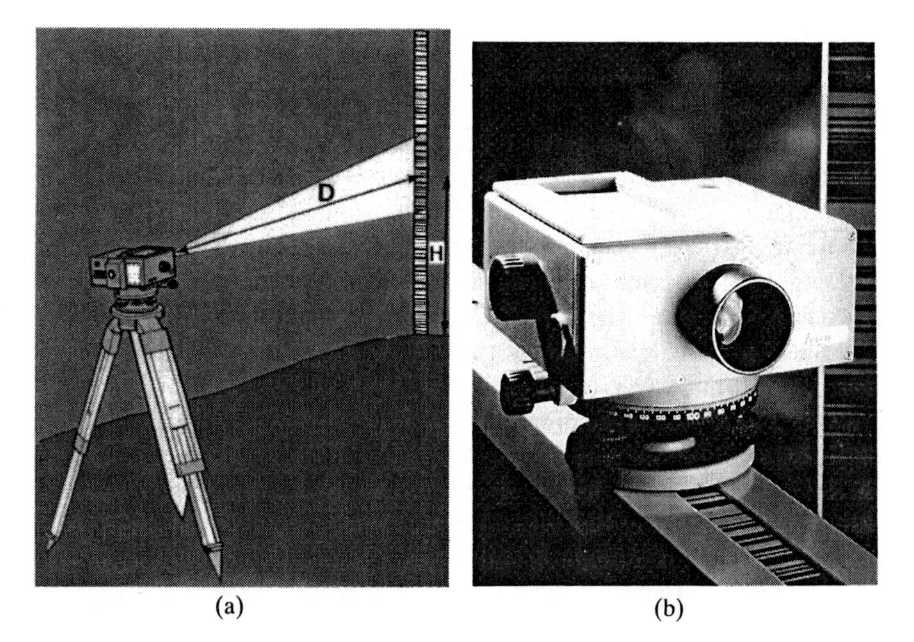

(a) (b)

Fig. 2.5 (a) Leica NA2002; (b) Leica NA3000.

Adjustments

The steps in setting up a level are known as the temporary adjustments. Periodically the bubble and/or compensator accuracy must be checked. The checks can be carried out in the field and are known as the permanent adjustments (see Section 2.9).

2.3 LEVELLING PROCEDURE

The instrument should be set up on stable ground in a position from which as many points to be levelled (staff stations) as possible can be sighted. Work should commence with a sight to a bench mark and should finish with a sight to a bench mark or other point of known altitude. It is rare that all levelling can be accomplished from one instrument station. Obstructions, rise or fall of ground, length of sight, will require a fresh instrument station to be selected. Before the instrument is moved, a staff station should be chosen which is visible from old and new instrument stations. Such a point is called a *change point* and it should be especially stable and visible. A foresight is taken from the old station, the instrument is set up at the new station and a backsight is taken. Both sights are booked on the same line of the field book.

Sights should not exceed 60 m length, though for communication on site between surveyor and staff holder, 40 m is a better limit. Ideally, to minimize collimation errors, all sights should be equal. This will rarely be possible, but cumulative errors can be reduced by equalizing backsight and foresight distances at each instrument station and taking particular care over such readings. Lines of sight should be at least 1 m above ground level. Lower 'grazing' rays may be subject to refraction by warm air layers at ground level.

Some points above the instrument can be levelled by inverting the staff, for example the undersides of beams and floors. Levels can also be set out above the height of the instrument by such a method: datum levels near the tops of columns can be marked out. Whenever the inverted staff technique is being used, readings should be booked as negative and treated as such when the arithmetic is carried out.

Reading the staff

Staves are usually 4 m high in three telescopic sections. Folding models and longer and shorter ones are also manufactured. Metric staves are marked in divisions of 10 mm or, less frequently, divisions of 5 mm. The standard metric 10 mm staff has 10 mm bars marked 10 mm apart on the white face of the staff in red (odd metres) and black (even metres). Values are marked every 100 mm, being in metres and decimals (e.g. 0.7, 2.3). The underside of the number is the point to which it refers. A vertical bar joins the three bars making up the lower 50 mm between each marked value, giving the effect of a capital 'E' (Fig. 2.6(a) and (b)).

Figure 2.6(c) shows the bar-coded staff for use with a Leica digital level.

Booking and reducing

A proper surveyor's level book should be used for recording all the readings. Columns are laid out in one of the two patterns shown in Fig. 2.7.

Each line in a book refers to a staff station.

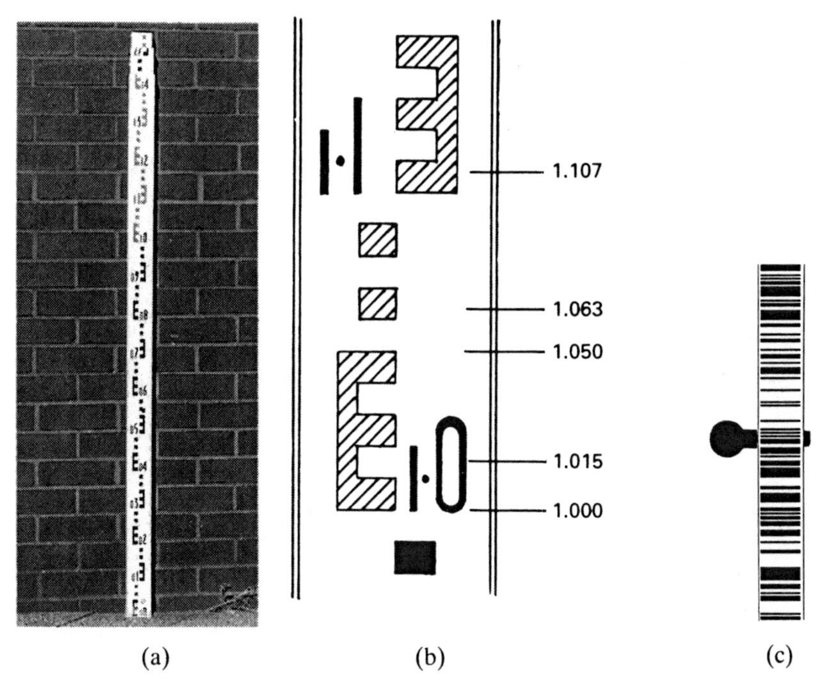

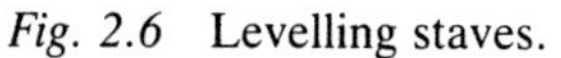
Fig. 2.6 Levelling staves.

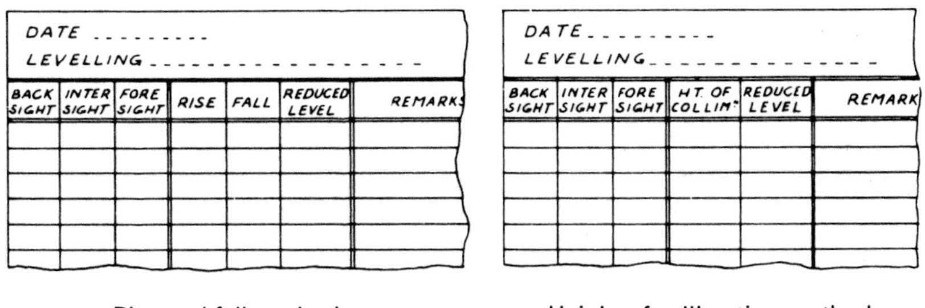
DATE ....
LEVELLING ....

BACK SIGHT	INTER SIGHT	FORE SIGHT	RISE	FALL	REDUCED LEVEL	REMARKS

DATE ....
LEVELLING ....

BACK SIGHT	INTER SIGHT	FORE SIGHT	HT. OF COLLIMⁿ	REDUCED LEVEL	REMARK

Rise and fall method

Height of collimation method

Fig. 2.7 Level book pages.

- first sight from an instrument station is a *backsight*
- further sights are *intermediate sights* (‘*intersights*’), *except* that final sight from an instrument station is a *foresight*
- book to three decimal places, estimating to 1, 2 or 5 mm according to length of sight and accuracy required.

Rise and fall method of reducing

The reduced level of each staff station is calculated from the level of the previous station by adding the *rise*, or subtracting the *fall*. The rise (or fall) is the difference between staff readings of points levelled consecutively, and may be defined as:

Rise = foresight or intersight booked on one line
subtracted from
intersight or backsight on line above

(If the rise is negative it is a fall and is booked in the appropriate column.)

Example 2.1

Figure 2.8 shows an elevation of levelling and bookings.

$$
\begin{aligned}
\text{Rise of B from A} &= 2.100 - 1.650 = 0.450 \\
\therefore \text{reduced level of B} &= 85.160 + 0.450 = 85.610 \\
\text{Rise of C from B} &= 1.650 - 3.230 = -\,1.580 \\
&= 1.580 \text{ (fall)} \\
\therefore \text{reduced level of C} &= 85.610 - 1.580 = 84.030 \\
\text{Rise of D from C} &= 1.210 - 1.370 = -\,0.160 \\
&= 0.160 \text{ (fall)} \\
\therefore \text{reduced level of D} &= 84.030 - 0.160 = 83.870
\end{aligned}
$$

Inverted staff readings are handled the same way, the negative sign being taken into account.

Checks on the arithmetic should be carried out. From an inspection of Fig. 2.8, it can be seen that:

Sum of the backsights minus sum of the foresights
should equal
sum of the rises minus sum of the falls
should equal
reduced level of final station minus reduced level of first station

These checks can be seen at the bottom of the table of bookings.

Where levelling is carried over the page of a level book, two approaches are possible:

(1) Continue directly on to the next page.

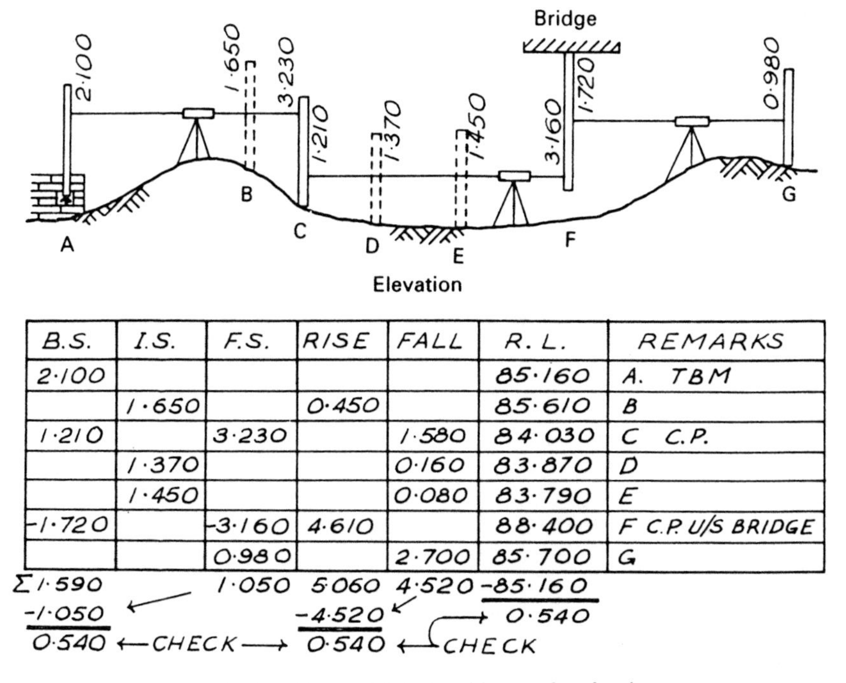

B.S.	I.S.	F.S.	RISE	FALL	R. L.	REMARKS
2·100					85·160	A. TBM
	1·650		0·450		85·610	B
1·210		3·230		1·580	84·030	C C.P.
	1·370			0·160	83·870	D
	1·450			0·080	83·790	E
-1·720		-3·160	4·610		88·400	F C.P. U/S BRIDGE
		0·980		2·700	85·700	G
Σ 1·590		1·050	5·060	4·520	-85·160	
-1·050			-4·520		0·540	
0·540 ← CHECK →			0·540 ← CHECK			

Fig. 2.8 'Rise and fall' booking and reduction.

(2) Write the last entry as a foresight, repeating it as a backsight on the next page; each page can then have the arithmetic checks performed at the bottom.

It is desirable for all checks to be carried out in the field. The widespread adoption of the pocket calculator appears to have misled a generation into thinking that arithmetic cannot be performed mentally. Unless a new breed of high-speed chainmen has evolved, the surveyor should have time to reduce the levels as work proceeds.

The checks on arithmetic should tally exactly, otherwise a mistake in addition or subtraction will have occurred.

The validity and accuracy of the fieldwork can only be assessed by taking the final sight on to a point of known level (closing the levelling), preferably the point from which the work originated.

Collimation (HPC) method

In this method, the reduced level of the instrument line of sight (or height of plane of collimation) is calculated, and the staff readings are subtracted from this value to determine the reduced levels. Thus:

Known reduced level of starting bench mark, or deduced value of a change point } + backsight = collimation level

Collimation level minus intersight or backsight = reduced level

Example 2.2

As for rise and fall – see Fig. 2.9.

Sample calculations:

85.160 + 2.100 = 87.260 height of collimation at first instrument station
87.260 − 1.650 = 85.610 reduced level of B
87.260 − 3.230 = 84.030 reduced level at C

Again, arithmetic checks should be carried out.

Sum of the backsights minus sum of the foresights
should equal
reduced level of final station minus reduced level of first station

As with the rise and fall method, calculations should be performed in the field, and the work should be 'closed'.

B.S.	I.S.	F.S.	H.P.C.	R.L.	REMARKS
2·100			87·260	85·160	A T.B.M.
	1·650			85·610	B
1·210		3·230	85·240	84·030	C C.P.
	1·370			83·870	D
	1·450			83·790	E
1·720		3·160	86·680	88·400	F C.P. U/S BRIDGE
		0·980		85·700	G

Σ 1·590   1·050   −85·160
−1·050   0·540
0·540 ←→ CHECK ←→

Fig. 2.9 Collimation booking and reduction.

Comparison of methods

(1) A check on intersight arithmetic is provided by the rise and fall method, but not by the collimation method.
(2) In setting out specific levels (pegs in the ground or marks on structures), the required staff readings are more easily calculated by the collimation method.

Thus for control surveying and checking of TBMs in setting out, the rise and fall method is preferable, while for site setting out the collimation method will be simpler.

2.4 ERRORS AND ACCURACY

Errors in levelling

Source	Prevention or reduction
Staff not vertical	Take lowest reading with the staff being swayed gently backwards and forwards through the vertical position. (Not for readings less than 1.000 m.) OR Use a staff with a plumbing bubble.
Parallax	Ensure images of staff and cross-hairs are in same plane.
Reading estimation errors	Limit distances to 60 m. For precise work, use a parallel plate micrometer.
Bubble not levelled	Check and re-level.
Change point errors	Ensure that staff remains in position during foresight and backsight readings.
Staff not properly extended	Check locking of extensions.
Ground instability	Set up level on a firm, stable surface. Use a change plate (or flat stone) under the staff.
Bubble error (tilting level)	Carry out 'two peg test'.
Compensator error (automatic level)	Carry out 'two peg test'.
Refraction	Avoid grazing rays.

Accuracy in levelling

Within a limited area, the allowable closing error is taken to be proportional to the square root of the number of instrument stations; for long levelling runs, proportional to the square root of the distance levelled.

Recommendations are:

$\pm 5(n)^{1/2}$ mm
$\pm 12(K)^{1/2}$ mm in flat or undulating country
$\pm 24(K)^{1/2}$ mm in mountainous areas

Where n is the number of instrument stations, K is the horizontal distance levelled in kilometres.

Distribution of closing error

In some instances a levelling circuit with an acceptable closing error can be left; at other times adjustment is required to produce consistent results. Level runs can be adjusted in accordance with the distances covered if these have been measured. Alternatively the closing error can be assumed to have been generated equally at each instrument station, and an incremental adjustment made according to instrument position. Thus if a +10 mm error has occurred over five instrument stations, the levels of points sighted from the first one (by intersights and by the foresight) should be adjusted by −2 mm, points sighted from the second one by −4 mm and so on. For a −12 mm error over eight stations, +2, +3, +4, +6, +8, +9, +10 and +12 mm adjustments would be satisfactory. A control network with cross runs may warrant a least squares adjustment (Chapter 7).

Example 2.3

The levelling circuit in Table 2.1 has a closing error of 10 mm. The total distance covered was 1.6 m. The levels require adjusting.

2.5 LEVEL GRIDS

Frequently a rectangular grid of levels is required over an area of land planned for construction. The levels are to enable an earthwork volume (of cut or fill) to be calculated. The main problem in carrying

Table 2.1

Backsight	Intersight	Foresight	Rise	Fall	Reduced level	Adjustment	Adjusted reduced level	Remarks
1.385					126.980			TBM'X'
	0.981		0.404		127.384	−0.002	127.382	
	1.234			0.253	127.131	−0.002	127.129	
2.992		2.487		1.253	125.878	−0.002	127.876	
	1.933		1.059		126.937	−0.003	126.934	
	2.015			0.082	126.855	−0.003	126.852	
	2.505			0.490	126.365	−0.003	126.362	
2.014		2.969		0.464	125.901	−0.003	125.898	
	3.131			1.117	124.784	−0.005	124.779	
2.911		1.996	1.135		125.919	−0.005	125.914	
1.895		2.276	0.635		126.554	−0.007	126.547	
	2.408			0.513	126.041	−0.008	126.033	
	2.880			0.472	125.569	−0.008	125.561	
	3.180			0.300	125.269	−0.008	125.261	
2.200		2.845	0.335		125.604	−0.008	125.596	
	1.749		0.451		126.055	−0.008	126.047	
		0.814	0.935		126.990	−0.010	126.980	TBM'X'
13.397		13.387	4.954	4.944	−126.980			
−13.387			−4.944		0.010			
0.010			0.010					

Closing error = +10 mm $5(n)^{1/2}$ = 12.2 mm. $12(K)^{1/2}$ mm. The closing error is therefore acceptable. For six stations, increment = −10/6 mm; adjust according to instrument station by −2, −3, −5, −7, −8 and −10 mm.

out the fieldwork is that of marking the grid intersection points. Rarely will there be a sufficient number of ranging rods available, nor will there always be enough time to peg out each point. The following methods are suggested.

(1) Points on two grid lines east–west and two grid lines north–south are marked with ranging rods. The chainman, with a staff, lines himself in. Ground levels at the grid points are taken (Fig. 2.10(a)).
(2) Ranging rods are positioned at grid points along the extreme lines on opposite sides of the area. The chainman paces between opposite rods to locate intermediate points. If there is an insufficient number of rods, alternate rods can be omitted, the chainman having to judge the position of the missing rods (Fig. 2.10(b)).

A plan is drawn to a suitable scale, each ground level being written beside the appropriate grid intersection point.

Calculations of the volumes involved are covered in Chapter 8.

2.6 LONGITUDINAL AND CROSS-SECTIONS

For the planning of construction works, particularly earthworks and

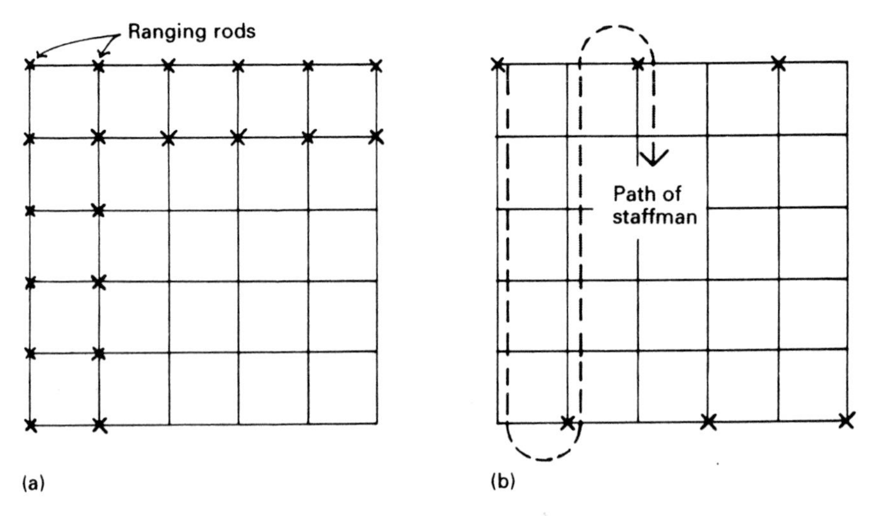

Fig. 2.10 Grid levelling.

vertical alignment, sections longitudinally and crosswise are often required. Ground levels are taken at regular intervals along the centrelines of proposed roads, railways and pipelines and at intervals at cross-sections. Sections are also taken along grid lines of buildings and industrial structures.

The ground profile is plotted, and construction details can be superposed. Levels, distances/chainages (increasing from left to right) are tabulated directly below the drawn profile. On longitudinal sections, an exaggerated vertical scale is often used (Fig. 2.11).

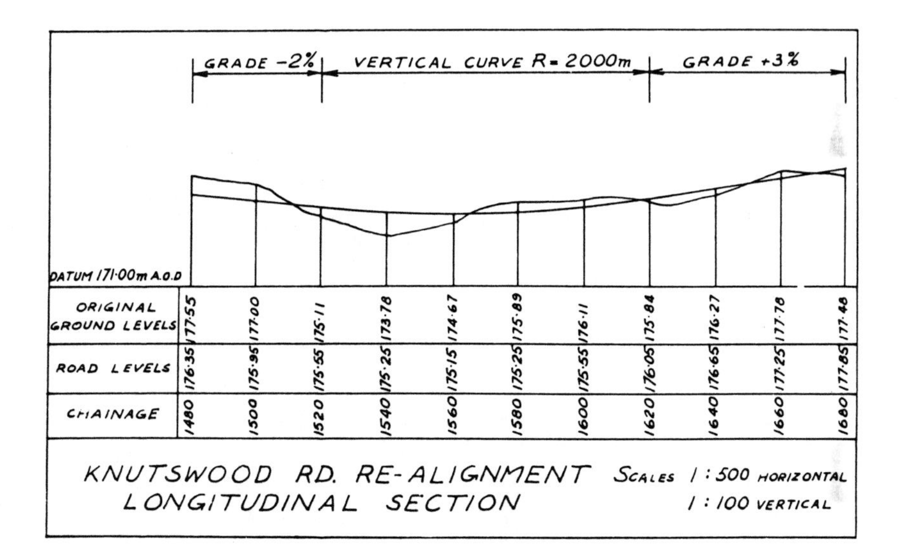

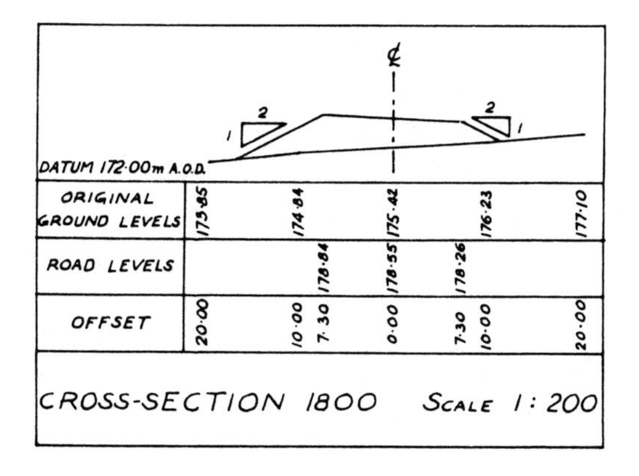

Fig. 2.11 Longitudinal and cross-sections.

2.7 CONTOURING

If the ground is fairly level and if the instrument is equipped with a horizontal circle, an engineer's level can be used for contouring. A theodolite is usually preferable, allowing inclined sights to be taken over undulating ground, and a better determination of the bearing to be made with the more precise horizontal circle. Contouring is explained in Chapter 6.

2.8 PRECISE LEVELLING

Control levelling must be carried out to establish reliable bench marks on site. The principles of levelling should be followed, and bench marks established at regular intervals on firm points; parts of existing structures or steel rods driven well into the ground and concreted in are suitable. The accuracy of levelling can be improved by the use of precise levels and staves, and account should be taken of the earth's radius and the effect of atmospheric refraction.

Precise levels

In comparison with an engineer's level, a precise level has greater magnification, a more sensitive bubble (tilting level) or compensator (automatic level) and a parallel plate micrometer. This is a piece of plane glass mounted in front of the objective lens on a horizontal axis about which it can pivot. If the glass plate is rotated from the vertical position, the line of sight is refracted and displaced upwards or downwards. By adjusting the micrometer, the surveyor can obtain coincidence of a staff division and the cross-hair, the control knob being graduated to indicate the amount of vertical displacement, as in Fig. 2.12.

The surveyor thus turns the control knob to move the plate from the neutral (vertical) position so that the line of sight is displaced downwards. The micrometer reading is then added to the staff reading. Micrometers can be fixed or demountable according to the model of level. Before using a particular model for the first time, the surveyor should check the graduation interval by seeing how many divisions correspond to a full staff division. A typical interval is 0.5 mm.

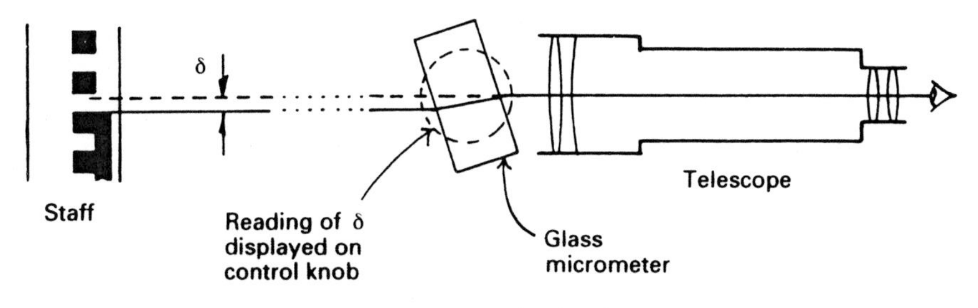

Fig. 2.12 Precise level.

Precise staves

A precise staff has the divisions scribed on an invar strip, making errors from thermal expansion or contraction negligible. The strip is mounted so that it can slide in the main wood or metal part of the staff on which the figures are marked. To help prevent reading errors, there is sometimes provided a second set of readings which does not have zero at the bottom of the staff. Both sets of readings should be observed and the levelling computed twice; the results should be identical.

A levelling bubble is always provided on a precise staff, and supporting stays are fitted by some manufacturers.

Curvature

A level line or surface is one with a constant radius about the earth's centre. A surveyor's level, when correctly set up, provides at best a tangential line or plane. A negative correction, c, should be applied to all staff readings, although over short distances this is negligible.

Referring to Fig. 2.13(a), where D is the sight distance, h is the height of instrument above mean sea level, R is the radius of the earth and c is the curvature correction, and applying Pythagoras' theorem:

$$(R + h)^2 + D^2 = (R + h + c)^2$$

$$R^2 + 2Rh + h^2 + D^2 = R^2 + 2Rh + 2Rc + 2hc + h^2 + c^2$$

$2hc$ and c^2 terms are insignificant, thus:

$$D^2 = 2Rc$$

$$c = \frac{D^2}{2R}$$

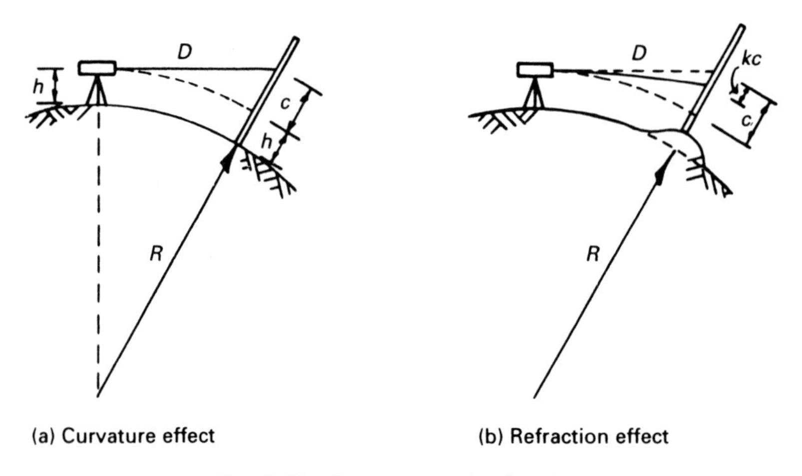

Fig. 2.13 Curvature and refraction.

Table 2.2

Distance (m)	Curvature correction (mm) to staff readings
113	−1
160	−2
195	−3
252	−5

Table 2.2 shows typical curvature corrections.

Refraction

It is actually rare that a line of sight of sufficient length to make curvature corrections necessary is a true tangent. As it passes through air with layers of differing temperatures and pressures, the line will be refracted. Usually it will curve downwards (see Fig. 2.13(b)) and is taken to have a constant radius.

$$\frac{\text{Earth radius}}{\text{Refraction radius}} = k \text{ (coefficient of refraction)}$$

The effect is to reduce the curvature correction by kc so that:

$$\text{Combined correction} = \frac{D^2(1 - k)}{2R}$$

It is believed that a typical value of k is 0.14. In some surveying literature k is defined to be half the amount given here: the corresponding typical value is 0.07. Care must be taken to establish which coefficient applies. In practice the refraction effect often varies, even over the duration of a levelling exercise, and it may be unwise to assume a coefficient. Lines can even be refracted upwards so that the error due to curvature is increased. If sight distances are kept short, or are made equal, no corrections are needed. Where unequal long sights occur, curvature and refraction corrections can be eliminated by reciprocal levelling.

Reciprocal levelling

Where there is no possibility of equalizing sight distances, for example levelling across a wide river or ravine, reciprocal levelling may be employed. The method makes use of two instruments each set up close to a station. The rise (or fall) is determined according to each instrument and a mean is taken, the corrections being equal and opposite and cancelling out. From the corrected levelling, the refraction coefficient can be determined and used in subsequent single levelling over a limited period.

Referring to Fig. 2.14:

Rise AB $= s_1 - (s_2 - c + kc)$
Rise BA $= s_3 - (s_4 - c + kc)$
i.e. Rise AB $= -s_3 + (s_4 - c + kc)$
Mean rise AB therefore $= 0.5(s_1 - s_2 - s_3 + s_4)$.

If only one instrument is available, it must be transferred quickly to the second position before the refraction effect can change. With a

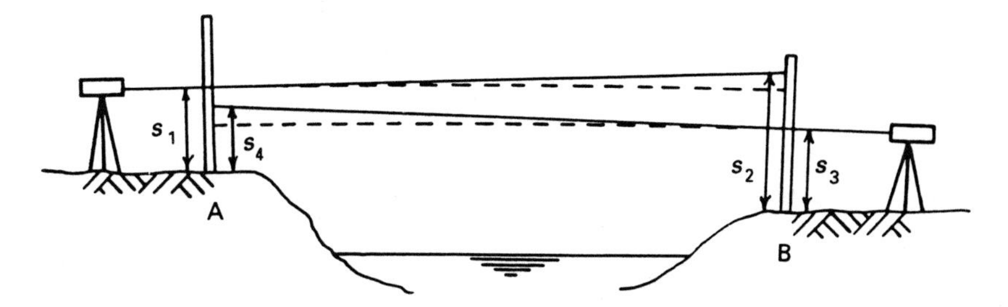

Fig. 2.14 Reciprocal levelling.

single instrument, the effects of a collimation error are also compensating, although the method cannot then be used to determine the refraction coefficient.

If a collimation error is suspected when two instruments are in use, they must be exchanged with each other, the process repeated and a mean taken of four rises.

Example 2.4

Simultaneous reciprocal observations across a river were made with two engineers' levels in good adjustment. An instrument was set up a short distance to the side of each station, the plan distance between the stations being 450 m. The reduced level of station A was known to be 94.360 m AOD. From the readings given below, calculate the reduced level of B and deduce the coefficient of refraction. The earth's radius may be taken as 6370 km.

Staff at	Instrument close to A	Instrument close to B
A	1.135	0.638
B	3.405	2.880

Solution

Apparent rise AB = 1.135 − 3.405 = −2.270
Apparent rise BA = 2.880 − 0.638 = 2.242
Mean rise AB = 0.5(−2.270 − 2.242) = −2.256
RL of B = 94.360 − 2.256 = 92.104 m AOD

For levelling from A:

Rise AB = 1.135 − (3.405 − combined correction) = −2.256

$$\text{Combined correction} = 0.014 = \frac{(450)^2(1 - k)}{2 \times 6370 \times 1000}$$

$$\text{Refraction coefficient} = 1 - \frac{0.014 \times 2 \times 6370 \times 1000}{(450)^2}$$

$$= 0.12$$

2.9 TESTS AND PERMANENT ADJUSTMENTS

For a level to give reliable results, the bubbles (and compensator of an automatic level) must be correctly adjusted. Tests to investigate these (permanent) adjustments should be carried out regularly – once a month on a busy site.

Bullseye bubble

Although a small error in the bullseye bubble can be tolerated, with the tilting level repeated fine levelling with the tilting screw will be needed; with the automatic level there is the risk of the telescope being tilted outside the range of the compensator.

Test

Plant tripod, fix instrument, level it with the footscrews according to the bullseye bubble. Turn the telescope horizontally through 180°. The bubble should stay central.

Adjustment

If the bubble does not stay central, bring the bubble halfway back to the central position with the footscrews. Turn the bubble-adjusting screws until the bubble case is set with the bubble central. Turn the telescope through 180° again, check and re-adjust if necessary. Repeat the procedure until the bubble remains central. Note that bullseye bubbles are usually secured by three small screws which can easily be broken if overtightened. All three should be loosened initially and then gently tightened until the bubble is set correctly.

Two peg (collimation) test

Collimation errors will almost certainly produce levelling of unacceptable accuracy.

Test

Set the instrument up on level ground midway between two clear,

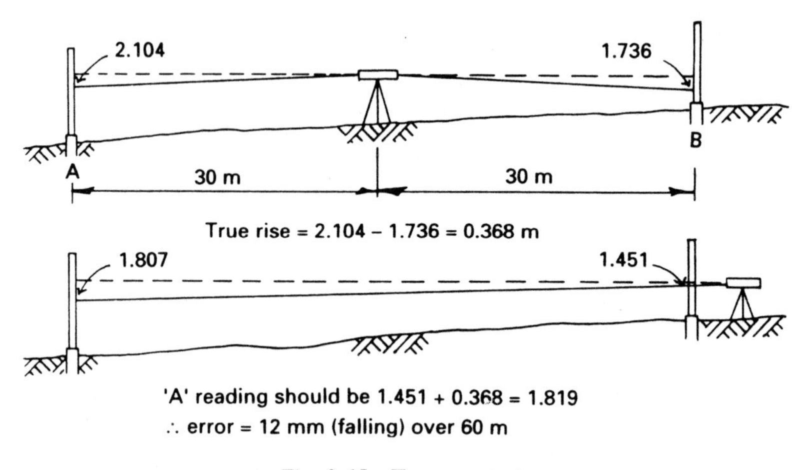

Fig. 2.15 Two peg test.

stable points (pegs) 60 to 70 m apart, as in Fig. 2.15 (distances may be paced). Take a staff reading at each point: the difference will be the *true* level difference, equal collimation errors cancelling each other out.

Move the instrument close to one point (about 3 m) and beyond it on the line of the points. Take a staff reading at the near point.

Using the true level difference, calculate the required staff reading at the far point. Take the reading and compare the two; if they correspond, the instrument is correctly adjusted.

Collimation error is revealed if actual and theoretical readings do not agree, and may be quoted in millimetres for the distance (far sight distance minus near sight distance) up or down. If it is not possible to adjust the instrument, a correction, according to each sight distance should be applied to subsequent levelling.

Adjustment

If the actual reading does not agree with the theoretical one (a discrepancy of 1 or 2 mm may be tolerated), an adjustment will be required.

Tilting level. Tilt telescope with tilting screw until theoretical staff reading is observed. Adjust the telescope bubble tube to make the bubble central. The adjustment is carried out by turning screws, locknuts or threaded collets using a screwdriver, spanner or tommy-bar.

Automatic and digital levels. Unscrew the diaphragm screw cover by the telescope eyepiece. Adjust the diaphragm screws until the required staff reading is observed. Do not tighten a screw until the opposing one has been loosened. Take care also not to rotate the diaphragm when adjusting it; the horizontality of the cross-hair can be checked by sighting the staff and rotating the telescope with the tangent screw.

Sight the staff at the near station again and, if the reading has changed, recalculate the required reading at the far point. Again read the far staff; if the readings still do not coincide (they should be much closer), repeat the adjustment and check again. Keep repeating the procedure until the error is removed (within 1 or 2 mm). Usually the adjustment will have to be carried out two or three times. Replace the diaphragm cover on an automatic level.

2.10 FURTHER READING

Cooper, M.A.R. (1982) *Modern Theodolites and Levels*. 2nd edn. Oxford: Blackwell Science.

2.11 EXERCISES

Exercise 2.1

Figure 2.16 is a sketch plan showing the results of a levelling exercise, the surveyor having forgotten the accepted methods of booking readings. Work commenced with a sight to the temporary bench mark from instrument station 1, and continued via instrument stations 2, 3 and 4, concluding with a sight to the same temporary bench mark.

(a) Draw a booking page, enter the staff readings and reduce the levels by an accepted method. Note any closing error.
(b) Briefly explain the possible sources of error in levelling, suggesting reasons for the occurrence of any error in part (a).

Exercise 2.2

(a) A modern but neglected tilting level has been discovered in a

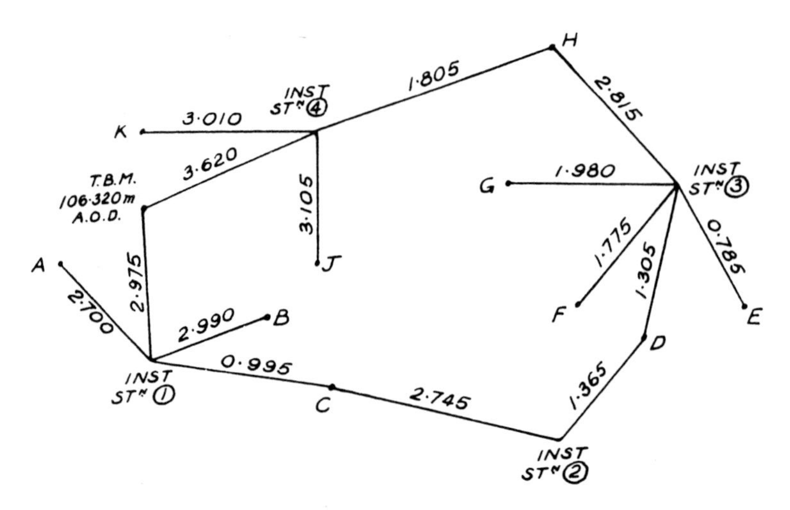

Fig. 2.16 Plan of levelling.

dusty corner of a site cabin. As site engineer, you have been asked to report on its condition. Explain which aspects of its construction should be examined, and briefly describe the tests to determine its accuracy if the instrument is to be used on site.

(b) List the steps that should be taken when levelling to prevent or minimize:
  (i) Collimation errors.
  (ii) Errors from a non-vertical staff.
  (iii) Booking errors.
  (iv) Undiscovered accumulative errors.

(c) During site levelling, the opportunity was taken to check the instrument by sighting two bench marks of known value. From the figures shown below, determine:
  (i) Whether the instrument is in adjustment.
  (ii) Without adjusting the bubble, what staff reading should be observed to set a peg which is 45 m from the instrument to a reduced level of 50.795 m AOD.

Bench mark	Value (m AOD)	Distance from instrument	Staff reading (m)
X	49.315	1 m	3.175
Y	51.055	61 m	1.455

3 The Theodolite

3.1 INTRODUCTION

The theodolite is designed to measure angles. A horizontal angle is the angle in plan subtended by two distant points at the station where the theodolite is set up. A vertical angle is one between a distant point and a vertical axis (zenith angle) or a horizontal axis (angle of elevation or depression) at the geometric centre of the theodolite. (The geometric centre is the intersection point of the vertical and horizontal axes of rotation and the line of collimation of the telescope; see Fig. 3.1(a).)

Angle measurement is carried out as part of control and detail surveys. In addition, on construction sites, the theodolite is used to set out angles, to extend straight lines and to check verticality ('plumbing'). The theodolite is frequently used in combination with electromagnetic distance measuring equipment.

3.2 CONSTRUCTION

The principal component of a theodolite is a telescope equipped with cross-hairs to enable precise sighting of distant points. The telescope can rotate about vertical and horizontal axes, the angles of rotation being measurable on graduated circles. The theodolite is mounted on a tripod so that it can be set with its axes correctly aligned and its geometric centre over a station.

The precision of a theodolite depends (apart from on the quality of construction) on the magnification of the telescope, the lowest direct reading which it can make (the least count) and the plate bubble sensitivity. The sensitivity is quoted as the number of seconds corresponding to one division (2 mm) on the bubble tube of dislevelment.

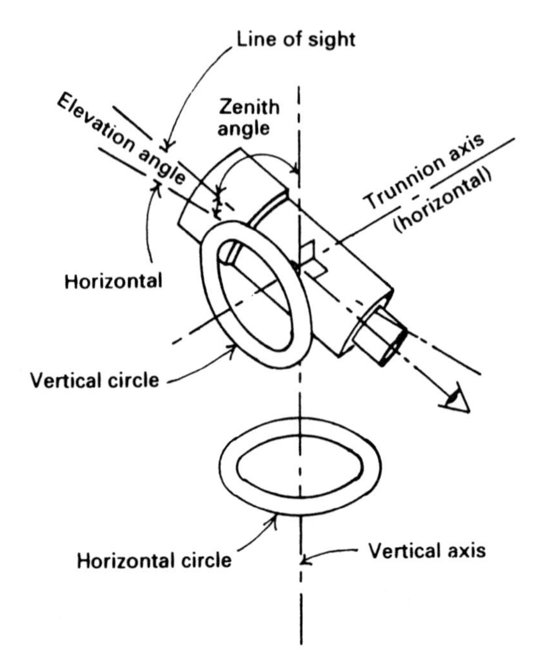

Fig. 3.1 Theodolites. (a) Geometry.

Figures 3.1(b) and (c) show modern theodolites which have the following features.

Telescope

A telescope provides about 30× magnification with internal focusing producing a real erect image which can be made coincident with the cross-hairs. A gunsight allows an initial approximate pointing to be made. There are horizontal and vertical clamps to lock the telescope; with them locked, tangent screws can be turned to give slow limited movement for exact sighting of a point. The circle viewing eyepiece is usually located next to the telescope eyepiece. The barrels of both can be rotated to provide clear images of the circle graduations and of the cross-hairs.

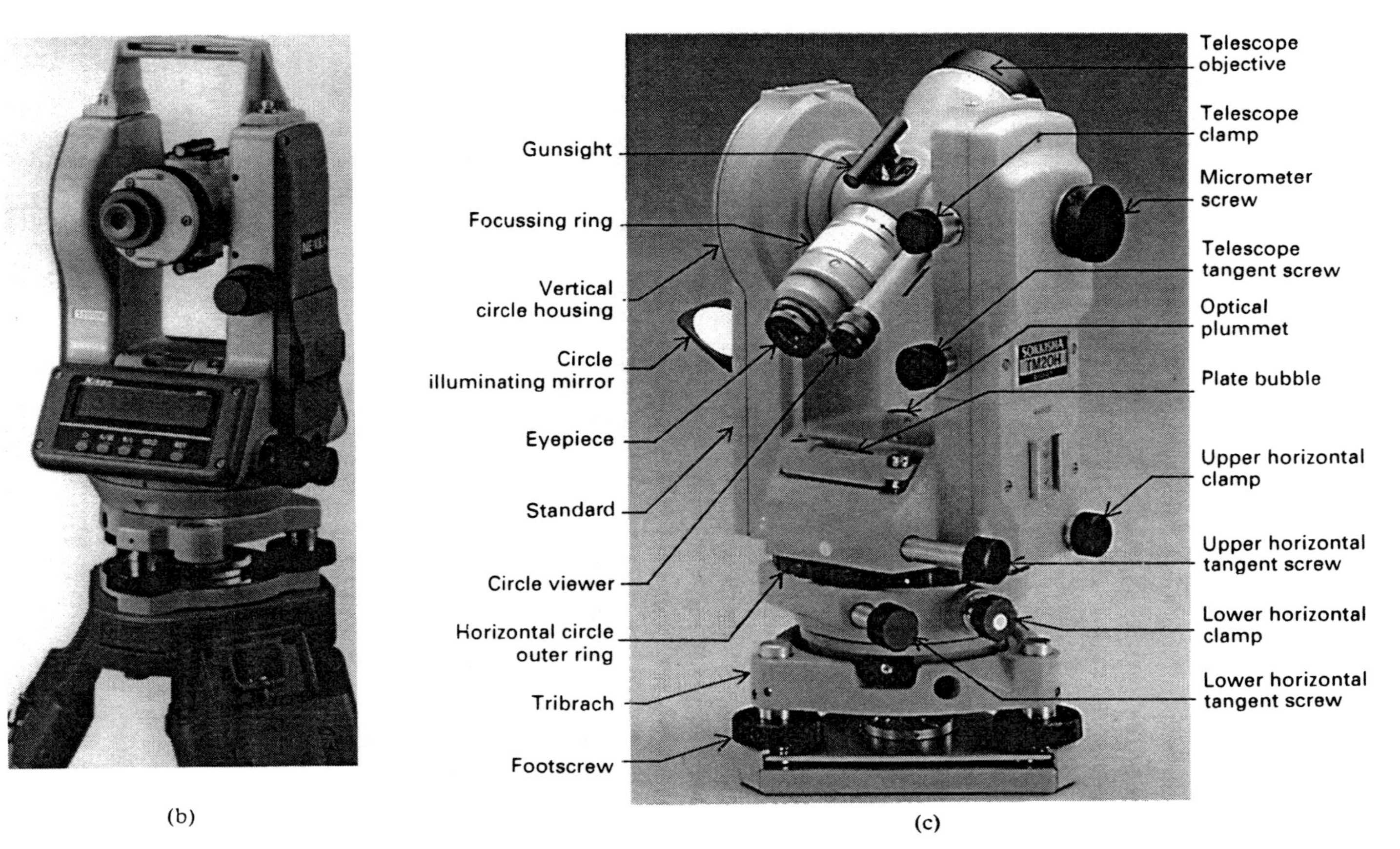

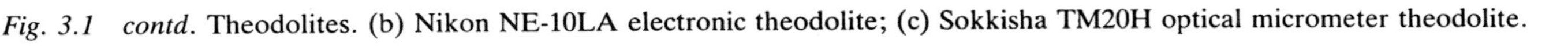

Fig. 3.1 contd. Theodolites. (b) Nikon NE-10LA electronic theodolite; (c) Sokkisha TM20H optical micrometer theodolite.

Circles

Optical theodolite

Graduated glass circles are enclosed; they are viewed through the auxiliary eyepiece. A small mirror must be adjusted to provide adequate circle illumination by natural light. Many instruments have an optical micrometer controlled by a screw to enable angles to be read with sufficient precision.

Electronic theodolite

Glass circles are used for stability; electronic codings are scanned and a liquid crystal display of the angles is provided on the standard or plate housing.

Standards

The telescope is supported by two standards and can rotate about a horizontal trunnion axis running between them. The clamp and tangent screws for controlling this rotation are located on one of the standards, as are the circle illumination mirror and the micrometer control screw. The vertical circle is inside one of the standards and on older instruments an altitude bubble for indexing the vertical circle was mounted at the top of this standard. A clipping screw was fitted to the same standard to allow centring of this bubble. Modern instruments achieve the indexing automatically using a pendulum and damper inside the standard.

Plate assembly

The standards project up from the plate assembly. This contains the horizontal clamp and tangent screw, the horizontal circle with a means of setting it (often a second clamp and tangent screw), an optical plummet and the plate bubble. The plate assembly above the circle, the standards and telescope are known as the *upper motion* or the *alidade*.

Tribrach

The tribrach supports the plate assembly which can rotate horizontally above it. Three footscrews allow the instrument to be levelled so that the vertical axis is truly vertical. The plate bubble is used to check this levelling; a small bullseye bubble is often provided to enable a quick approximate levelling to be made. The tribrach is the lowest part of the theodolite and is bolted to the top of a tripod before use. Many theodolites can be detached from their tribrachs, facilitating the interchange of instruments, targets and prisms in the field.

Tripod

Tripods have pointed feet and telescopic legs to facilitate the setting of them firmly over stations so that a theodolite can easily and precisely be levelled and centred. The flat head and securing bolt allow the theodolite to be moved a limited amount sideways to achieve precise centring over the station as checked through the optical plummet in the plate assembly. There is also provision for hanging a plumb-bob either from the tripod or the theodolite. Some older tripods have a threaded head on to which the theodolite is screwed; release of a clamp allows centring above the footscrews.

Adjustments

The steps in setting up the instrument are known as the temporary adjustments. Periodically the geometry of the instrument must be checked; the checks can be carried out in the field and are known as the permanent adjustments (see Section 3.8).

3.3 SETTING UP

For readings of angles to be valid, the theodolite must be set up with the vertical axis truly vertical (as shown by the plate bubble) and with this axis over the station (as shown by the optical plummet).

There are two methods of setting up; the one first described does not need a plumb-bob, but works best when the instrument is set up over a point at the same level as the tripod feet.

(1) Open the legs of the tripod and adjust the length of them so that the instrument, when fitted, will be at a convenient height. Clamp the legs and by estimation set the tripod over the station.
(2) Remove the theodolite from its box, noting how it was fitted, bolt it centrally on the tripod head and set the footscrews to the middle of their travel.

Method 1

(3) Tread one tripod foot into the ground and, looking through the optical plummet, position the other two legs so that the station appears central. Tread in the other two feet. On a hard smooth surface, the tripod feet must be prevented from sliding outwards; bricks or concrete blocks can be placed against them.
(4) If the station appears no longer central in the plummet, turn the footscrews until it reappears central.
(5) Make the instrument level by shortening or lengthening the tripod legs. Check with the bullseye bubble, if fitted. Otherwise set the plate bubble parallel to the line between two legs, shortening or lengthening them to centre the bubble; turn the alidade through 90° and recentre the bubble by adjusting the third leg. Repeat this process until the bubble is within two divisions of central. Sight through the optical plummet; if the instrument is not still approximately centred, recentre it by slackening the securing bolt (or centring clamp), sliding the theodolite horizontally and tightening the bolt/clamp.
(6) Finely level the theodolite using the footscrews. Set the plate bubble parallel to the line through two footscrews and turn these screws in opposite senses to bring the bubble central. Turn the alidade through 90° and recentre the bubble using the third footscrew. Return the bubble to its original position and repeat the procedure until the bubble remains central in both positions. The bubble follows the movement of the left thumb.

    If the bubble is correctly set in its mounting, it will remain central whatever its disposition. This may be checked by turning the alidade through 180° from the first position. If the bubble moves off centre, it should be brought halfway back with the footscrews and further adjustments made (with all footscrews) so that the bubble remains in this (off-centre) position. Figure 3.2 shows the plate with the bubble in its various positions.
(7) Precisely centre the theodolite. Loosen the securing bolt (or

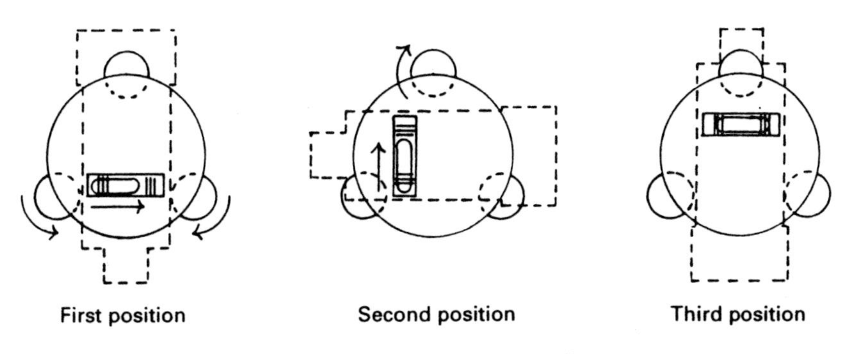

Fig. 3.2 Levelling the theodolite.

centring clamp) and carefully slide the instrument sideways till the station appears central in the optical plummet (the plummet should be correctly focused to eliminate parallax). Where centring is below the footscrews, take care not to rotate the instrument or the levelling will be disturbed. Tighten the bolt/clamp.

(8) Repeat steps (6) and (7) until the theodolite remains levelled and centred.

(9) Sighting a plain background, rotate the eyepiece to focus the cross-hairs.

Method 2

(3) Hang the plumb-bob under the theodolite. Move the tripod legs to bring the bob over the station and to level the tripod head. (The legs should be moved circumferentially for levelling and radially for centring.) Tread the legs into the ground. If this moves the plumb-bob off centre, shorten or lengthen the tripod legs to bring it back. On a hard, smooth surface, the tripod feet must be prevented from sliding outwards; bricks or concrete blocks can be placed against them.

(4) Finely level the theodolite using the footscrews. Set the plate bubble parallel to the line through two footscrews and turn these screws in opposite senses to bring the bubble central. Turn the alidade through 90° and recentre the bubble using the third footscrew. Return the bubble to its original position and repeat the procedure until the bubble remains central in both positions. The bubble follows the movement of the left thumb.

If the bubble is correctly set in its mounting, it will remain central whatever its disposition. This may be checked by turning the alidade through 180° from the first position. If the bubble moves off centre it should be brought halfway back with the footscrews and further adjustments made (with all footscrews) so that the bubble remains in this (off-centre) position. Figure 3.2 shows the plate with the bubble in its various positions.

(5) Precisely centre the theodolite. Loosen the securing bolt (or centring clamp) and carefully slide the instrument sideways till the station appears central in the optical plummet (the plummet should be correctly focused to eliminate parallax). Where centring is below the footscrews, take care not to rotate the instrument or the levelling will be disturbed. Tighten the bolt/clamp.

(6) Repeat steps (4) and (5) until the theodolite remains levelled and centred.

(7) Sighting a plain background, rotate the eyepiece to focus the cross-hairs.

Although there are fewer steps in method 2, method 1 is usually quicker and more reliable in establishing the initial, approximate position of the tripod. It is the same principle as that used in setting up a Kern theodolite equipped with a 'plumbing tripod'.

Packing the instrument

Release all clamps, bring each footscrew and tangent screw to the middle of its run, line up any alignment marks, place carefully in the case, pack any accessories and gently close the lid. Only fasten the lid if it closes easily. A wet instrument should be left to dry before being packed.

Telescope positions

The telescope can be in two positions for observations: with the vertical circle to the left of the telescope, as seen by the surveyor (face left) or to the right (face right). It is usual to commence with face left readings, when the controls are close to hand, and repeat observations on face right. The face may not be apparent on electronic instruments – markers (e.g. I and II) may be provided.

3.4 ANGLE READINGS: HORIZONTAL

The horizontal angle is that subtended by two distant stations. The horizontal circles of theodolites are graduated in a clockwise sense (in plan) and measured angles are conventionally quoted clockwise ($A\hat{B}C$ is the clockwise angle at B from A to C). It is usually convenient to swing the alidade clockwise from first to second station for the first measurement of an angle. The value of the clockwise angle is then the first reading subtracted from the second. Should the second reading be smaller than the first, 360° must be added to it, for it means that the index will have passed the 360°/0° division on the circle. If the circle is zeroed at the first station, the reading at the second station will be a measure of the angle (Fig. 3.3).

The first station sighted is taken as a reference object (RO). Ideally it is the most permanent, most distant and most sharply defined station. An electronic instrument can easily be set to zero when pointed to the RO; with an optical theodolite it is better practice to set the circle to an arbitrary value a small amount above zero.

Note that the circle is, in fact, a 360° protractor. Once it has been set, any subsequent pointing will yield the same reading after swinging left (anticlockwise) as after swinging right (clockwise).

Procedure

(1) Ensure horizontal circle is clamped to tribrach.
(2) Release the horizontal and vertical clamps and swing the telescope to sight the target. (This initial sighting can be more easily achieved by moving a short distance back from the eyepiece and using the 'gunsights' on the telescope.)
(3) Look through the telescope and focus it on the target. Lock

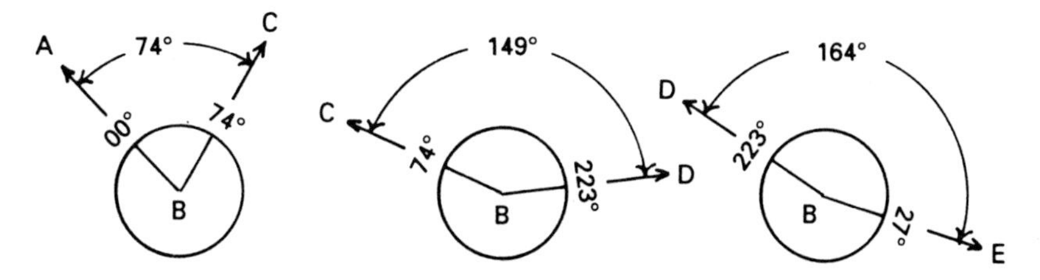

Fig. 3.3 Horizontal angles.

the clamps and use the tangent screws to bring the target into coincidence with the vertical cross-hair. Check that parallax has been eliminated.

(4) Read the horizontal circle. Book the value.
(5) Release the clamps (be sure to keep the circle clamped to the tribrach), and turn the alidade and move the telescope in elevation to sight the second target. Repeat the sighting, reading and booking procedure.

Target sighting

It is important for the target to be sighted as precisely as possible. Focus the telescope and cross-hairs so that there is no parallax between target and cross-hair images. A pencil held vertically with the point on the target will assist sighting, as will a white background (open page of field book) behind the target. Lines of sight should be at least 1 m above ground level. Lower 'grazing' rays may be refracted laterally by warm air layers, and angular errors may result.

If the target cannot be sighted directly:

(1) Get the chainman to hold a spirit-level over it, indicating which side is to be sighted. Or:
(2) Get the chainman to hold a plumb-bob over the station, using ranging rods or pieces of timber to make a temporary tripod or support. Or:
(3) Set a tripod and plumb-bob over it. Or:
(4) Set a tripod and surveying target over it (for precise work).

Angle reading

The smallest figured division on the circle of a theodolite is its 'least count', 20 seconds, 10 seconds, 6 seconds or 1 second in increasing order of precision. There are three systems of circle reading:

(1) *Direct*. The circle is marked in degrees and the minute and 20 second intervals are read off the index scale against the degree division. It is not generally possible to estimate more finely than to the least count.
(2) *Electronic*. Readings to the least count are automatically displayed.
(3) *Optical micrometer*. The glass micrometer is a device to facilitate the reading of intervals smaller than can be scribed on the circle

or index. The ray of light from the circle divisions can be moved sideways by passing it through a plane sheet of glass which can be rotated. The surveyor turns the micrometer screw until a division on the circle coincides with the fixed index mark. The amount of rotation is proportional to the sideways movement (in minutes and seconds) and is displayed on a micrometer scale viewed through the circle telescope. The micrometer divisions can often be further divided by estimation, giving smaller increments than the least count.

The majority of optical (glass circle) theodolites are of this type, and the young surveyor/engineer should master them as soon as possible. Confusion is sometimes experienced by the beginner when the micrometer knob is turned; although there is apparent movement of the circle, neither instrument nor circle is rotating. On one second instruments, diametrically opposite sections of the circle are viewed simultaneously; concidence of the divisions is obtained with the micrometer screw and checked in a small 'window' close to the main circle display.

Micrometer operation

(1) Refer to Fig. 3.4(a). Turn the micrometer screw until the micrometer reads zero. (Not strictly necessary, but recommended for beginners – an initial estimate of the minutes can be made on the circle scale giving a rough check on the subsequent minutes and seconds, and the movement of the scales during rotation of the micrometer screw can be observed.)
(2) Turn the screw until the circle scale shows coincidence of a division and the index marker; on a one second instrument, obtain coincidence of diametrically opposite divisions (Fig. 3.4(b)). (If the screw is turned the wrong way, movement will cease before coincidence is achieved.) Read circle and micrometer scales, adding readings to get the angle values.

Note: The horizontal and vertical scales usually appear close to each other, therefore:

- take care to read the correct circle
- where the micrometer serves both circles, reset it to read a vertical angle after a horizontal one.

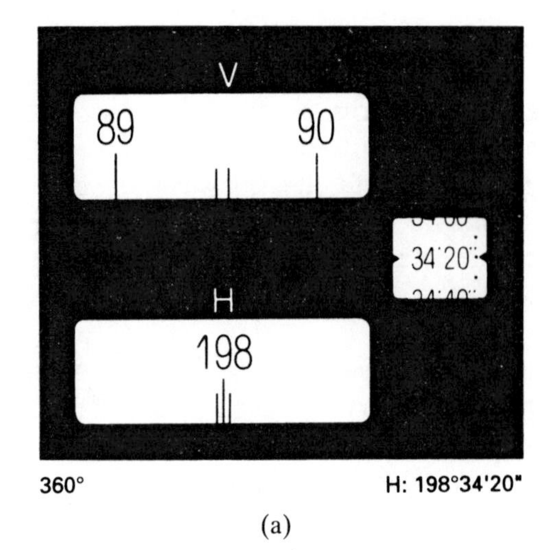

(a)

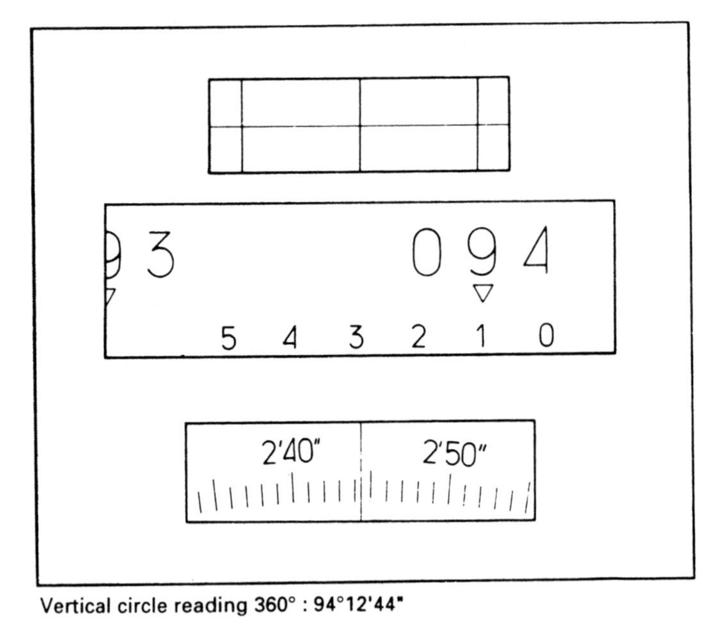

(b)

Fig. 3.4 Theodolite reading systems. (a) Sokkisha TM20H; (b) Wild T2.

Circle setting

When the circle has to be set to a specific initial value, turn the micrometer screw to give the desired minutes and seconds and then bring the circle to the required (coarse) reading using clamp and tangent screws or the circle setting screw. For an instrument with separate upper and lower clamps and tangent screws, set the reading before the theodolite is pointed to the initial station, lock the upper clamp and use the lower clamp and tangent screw for target sighting. Use a similar process on a theodolite with a repetition clamp, taking care to clamp the circle to the alidade before sighting the target.

Setting out

To set out an angle, set the circle (or take a reading) with the telescope pointing to the reference object. With the (upper) horizontal clamp released, turn the alidade to 'lay off' the required direction. Set the micrometer first, then use clamp and tangent screws to obtain the required coarse (main circle) reading.

Rounds of angles – procedure and booking

Angles should be observed more than once so that:

(1) Inconsistent values can be identified and rejected.
(2) A mean of several consistent values can be taken, giving a more reliable measure than a single one.
(3) For optical reading instruments both faces and different parts of the circle can be used, providing compensation for some instrumental errors.

Optical reading theodolites

Take a round of angles by commencing sighting the RO on face left (with the circle set to give an arbitrary reading a small amount above zero), swinging right (clockwise in plan) and sighting the other stations in turn. At the final station, transit the telescope to the face right position without disturbing the circle; observe the stations in reverse order swinging left, finally sighting the RO.

Each subsequent round should be commenced at 180°/n further (clockwise) around the circle for a total of n rounds. The micrometer

Table 3.1

Readings				Calculations	
		Face			
Inst at	Sighting	Left	Right	Mean direction	Reduced angle
I	A	00°03′42″	180°03′44″	00°03′43″	00°00′00″
Round 1	B	67°24′37″	247°24′33″	67°24′35″	67°20′52″
	C	116°21′03″	296°20′55″	116°20′59″	116°17′16″
I	A	90°41′18″	270°41′21″	90°41′20″	00°00′00″
Round 2	B	158°02′12″	338°02′14″	158°02′13″	67°20′53″
	C	206°58′30″	26°58′35″	206°58′32″	116°17′12″

on such instruments should be initially re-set so that readings throughout its range can be utilized. Table 3.1 shows booking and reduction.

Corresponding face left and right readings should differ by 180° within about twice the least count of the instrument. If this is the case, subtract 180° from (or for values less than 180°, add 180° to) the face right reading and take the mean. This gives a mean direction for each pointing. The reduced angle is found by subtracting the mean RO direction. The reduced angles from several rounds can be further meaned.

Electronic theodolites

Some instruments require the alidade to be turned through 360° before angles are displayed.

The initial reading can always be set (by pressing a key) to zero. Many instruments have axis compensation to eliminate the need for face right readings. A periodic calibration can establish any lack of adjustment and automatically apply a correction. Full circle scanning and meaning can also eliminate the need for additional rounds to provide compensation for circle eccentricity. Extra sightings do, however, also provide checks and compensation for pointing errors and should increase the surveyor's confidence, particularly for manual booking.

3.5 ANGLE READINGS: VERTICAL

Modern instruments have 360° vertical circles with zero representing

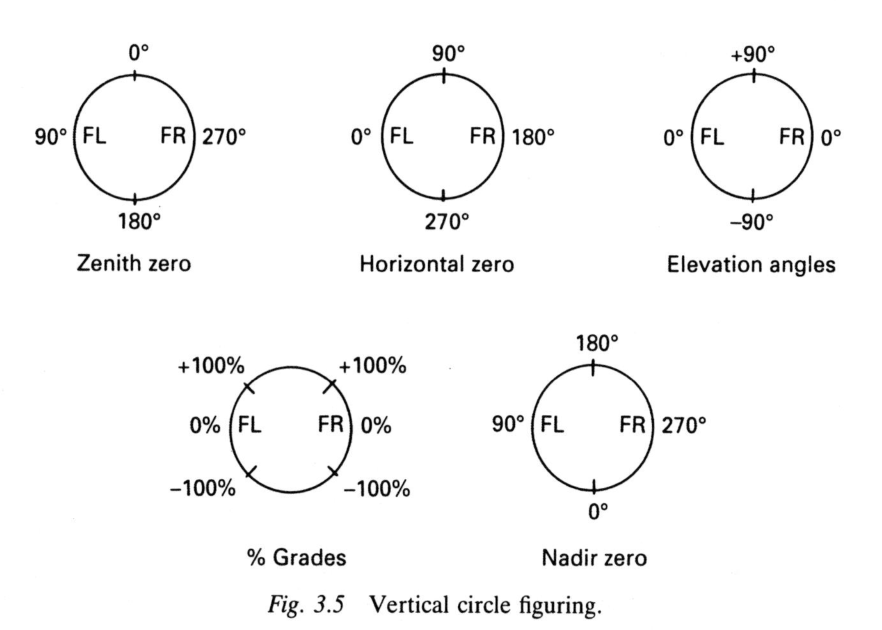

Fig. 3.5 Vertical circle figuring.

(on face left) a vertically upward, a horizontal or a vertically downward sight, according to manufacture. On modern instruments the first type (zenith zero) is the most common. Electronic instruments can often be switched between zenith angles, horizontal angles and percentage grades. Figure 3.5 shows the most common figuring of vertical circles.

Indexing

Older instruments are fitted with an altitude bubble connected to the vertical circle. It is located on one of the standards and is levelled with a clipping screw which is distinguished from other screws by being fluted rather than knurled. Before vertical angles are measured, circle indexing must be carried out by centring the altitude bubble with the clipping screw, giving greater precision than obtainable with the plate bubble.

Modern optical instruments have automatic vertical indexing provided by a damped pendulum connected to the circle.

Some older electronic theodolites are fitted with a bubble on the telescope which must be set level before an indexing key is pressed. Most electronic instruments have automatic indexing; a warning is displayed if the instrument is disturbed so as to be beyond the range of the compensator. On some instruments the telescope must be transitted (over the top) before readings are displayed.

Angle measurement

(1) Note circle zero position and figuring.
(2) Index the vertical circle (if not automatic).
(3) Accurately sight target, read and book angle.

Notes

(1) The micrometer is used as for horizontal circle readings.
(2) One second instruments give a mean of diametrically opposite readings.
(3) Take a mean of readings on both faces. If this is not possible (when a telescope mounted EDM is attached, for example), take extra care. The improved accuracy from meaned readings is frequently not needed for vertical angle measurement. Exceptions are:

(a) where steep sights are taken:

Zenith angle	Error in angle	Proportional error in slope corrected distance
85°	±1′	1 in 39 000
80°	±1′	1 in 19 000
70°	±1′	1 in 9 400

(b) where the angle is to be used for trigonometric heighting:

Zenith angle	Error	Error in levelling relative to distance
90° to 50°	± 1′	1 in 3 400– 2 000
	±10″	1 in 10 300– 6 100
	±20″	1 in 20 600–12 100

3.6 TRIGONOMETRIC HEIGHTING

Trigonometric heighting (or levelling) can be carried out using a theodolite. Over long distances there will be errors from the effects of curvature and refraction. The method is convenient if an EDM is attached to the theodolite for distance measurement; for control work precise determination of the vertical angle is necessary.

Corrections can be evaluated as linear amounts, as in precise levelling (Chapter 2), or as angular amounts, to be applied to the vertical angles.

Figure 3.6 shows trigonometric heighting being carried out between station A (with an instrument height of h_I) and station B (with a target height of h_T). The zenith angle has been measured as α and the slope distance has been determined as D. Angular corrections of $\hat{c}$ (curvature) and $k\hat{c}$ (refraction, taken as opposite in sign to $\hat{c}$) are to be applied with k, the coefficient of refraction, as defined for precise levelling (earth radius divided by refraction radius). The radius of the earth is R.

$$\begin{aligned}\text{Rise AB} &= h_I + D\sin[90 - \alpha + \hat{c}(1 - k)] - h_T \\ &= h_I + D\cos[\alpha - \hat{c}(1 - k)] - h_T\end{aligned}$$

The curvature correction will be half the angle subtended at the earth's centre by the distance AB. The slope distance D may be approximated to an arc, hence:

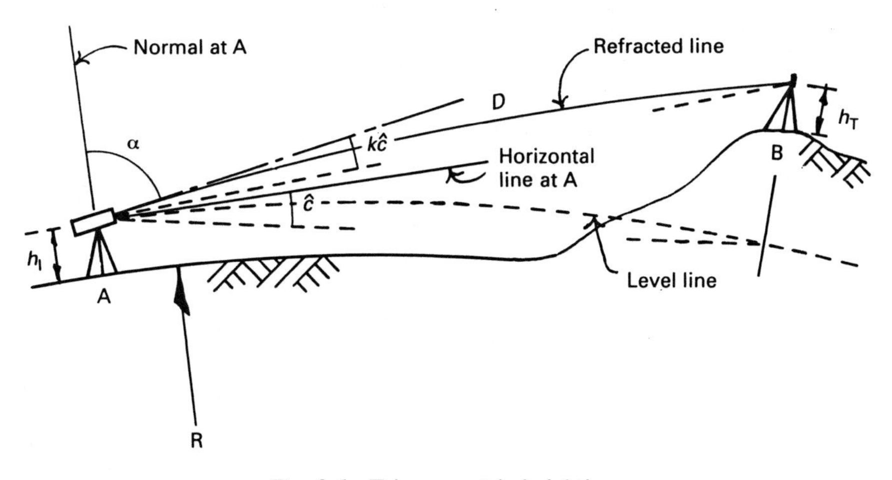

Fig. 3.6 Trigonometric heighting.

$$\hat{c} \text{ (radians)} = \frac{D}{2R}, \text{ or } \hat{c}'' = \frac{206\,265\,D}{2R}$$

$$\text{Combined correction (seconds)} = \frac{206\,265 \times D(1-k)}{2R}$$

A typical value of k is 0.14. (In some surveying literature k is defined to be half the amount given here; a corresponding typical value of 0.07 is quoted. Care must be taken to establish which coefficient applies.) The value of k can vary, even during measurement, and it may be better to carry out reciprocal (or double-ended) trigonometric heighting whereby curvature and refraction corrections are eliminated.

Reciprocal trigonometric heighting

This process involves simultaneous readings from both stations. If the zenith angle at B is β:

$$\text{Rise BA} = h_{I_B} - h_{T_A} + D\cos[\beta - \hat{c}(1-k)]$$

Hence:

$$\text{Rise AB} = -h_{I_B} + h_{T_A} - D\cos[\beta - \hat{c}(1-k)]$$

From A:

$$\text{Rise AB} = h_{I_A} - h_{T_B} + D\cos[\alpha - \hat{c}(1-k)]$$

Mean rise AB therefore:

$$= 0.5(h_{I_A} - h_{I_B} + h_{T_A} - h_{T_B}) + 0.5D\{\cos[\alpha - \hat{c}(1-k)] - \cos[\beta - \hat{c}(1-k)]\}$$

$$= 0.5(h_{I_A} - h_{I_B} + h_{T_A} - h_{T_B}) + D\sin\frac{\alpha + \beta - 2\hat{c}(1-k)}{2}\sin\frac{\beta - \alpha}{2}$$

Usually $\frac{\alpha + \beta}{2} \simeq 90°$ and $\hat{c}(1-k)$ is small (a few seconds), hence the first sine term can be taken as one. Then:

$$\text{Rise AB} = 0.5(h_{I_A} - h_{I_B} + h_{T_A} - h_{T_B}) + D\sin\frac{\beta - \alpha}{2}$$

As with reciprocal levelling, the coefficient of refraction can be deduced from reciprocal trigonometric heighting and used for single-ended observations over a limited period.

3.7 OTHER USES OF THE THEODOLITE

The applications of the theodolite in control surveying, detail surveying and setting out are covered in Chapter 5, Chapter 6 and Chapters 9 to 14.

3.8 INSTRUMENT ERRORS AND PERMANENT ADJUSTMENTS

For a theodolite to provide acceptable results, the axes must bear the correct relationships to each other, the bubbles must be correctly set, the optical plummet must give reliable centring, vertical indexing must be satisfactory and there should be no eccentricity of the circles. There are well-established procedures to check these requirements, and adjustments for most of them are possible on site.

The following tests should be carried out in the order shown.

(1) Plate bubble test

Requirement: The vertical axis should be truly vertical when the plate bubble is central.

Test: Set the instrument up on a firmly positioned tripod and level it with the footscrews and plate bubble, setting the bubble first parallel to and then at right angles to the line between two footscrews. Turn the alidade through 180° from the first position; the bubble should remain central (Fig. 3.2 – third position). If it does not, an adjustment is needed.

Adjustment: Bring the bubble halfway back to the central position using the footscrews. Remove the remaining error using the adjusting screws/nuts. Repeat the complete levelling and testing procedure until the plate bubble remains central with the alidade in all positions.

(2) Horizontal collimation test

Requirement: The line of collimation should be perpendicular to the trunnion axis.

Test: With the instrument set up and levelled, sight a distant target at the same level as the theodolite. Read the horizontal circle and transit the telescope (turn it over the top) and rotate it to sight the same target on the opposite face. The horizontal circle reading should differ by 180° from the first one; if not, an adjustment is needed.

Adjustment: Find the correction by subtracting the difference of the two readings from 180° and halving the result. Apply it (with the appropriate sign) to the second reading and turn the (upper) horizontal tangent screw until this reading is observed.

Example 3.1

$$
\begin{aligned}
\text{Face left reading} &= 23^\circ 12' 40'' \\
\text{Face right reading} &= 203^\circ 13' 20'' \\
\text{Difference} &= 180^\circ 00' 40'' \\
\text{Correction} &= \frac{180^\circ 00' 00'' - 180^\circ 00' 40''}{2} = -20'' \\
\text{Corrected reading} &= 203^\circ 13' 00''
\end{aligned}
$$

When the instrument has been set to the corrected reading, the target and cross-hairs will no longer coincide. Unscrew the diaphragm cover to give access to the adjusting screws. Adjust the screws with a screwdriver or spanner as appropriate. Before tightening a screw, slacken the opposing one. Avoid rotating the diaphragm; check verticality of the cross-hair by sighting a fine point and adjusting the telescope elevation with the tangent screw. The vertical hair should remain coincident with the point; if it does not, rotate the diaphragm into correct alignment. Repeat the test and adjustment until all error has been removed. Replace the diaphragm cover.

(3) Trunnion axis (spire) test

Requirement: The trunnion (or transit) axis should be perpendicular to the vertical axis.

Test: Set up and level the theodolite so that a clearly defined elevated target (a church spire, for example) at a distance of about 100 m can be viewed. The angle of elevation should be 30° to 40°. Accurately sight the target and take the horizontal circle reading. Change the face of the instrument and repeat the operation. If the two readings differ by 180°, the adjustment is satisfactory.

Adjustment: If the readings do not differ by 180°, find the correction by halving the difference subtracted from 180°. Apply this correction (with the appropriate sign) to the angle read on the second sighting by turning the (upper) horizontal tangent screw. The vertical cross-hair will now have moved out of coincidence wiith the target. Re-establish coincidence by raising or lowering one end of the trunnion axis. On many modern theodolites, this is not possible in the field; return the instrument to the supplier or manufacturer. Repeat the test and adjustment until all error has been removed.

(4) Horizontal circle eccentricity test

Requirement: The circle and alidade centres should coincide.
Test: Set up and level the theodolite. Sight a distant station and read the horizontal circle on face left. Transit and rotate the telescope and repeat the reading on face right. Shift the horizontal circle by about 45° and repeat the process three times, so that all parts of the circle have been used. If there is eccentricity, some face left and face right readings will correspond, while others will not. A pattern of want of correspondence will emerge, the greatest inconsistencies occurring at values 90° away from those with good correspondence.

Adjustment: Return the instrument to the suppliers or manufacturers.

Note: Test 4 is linked with tests 2 and 3, so that circle eccentricity will affect values observed in the collimation and spire tests. If an instrument is sent away for the removal of eccentricity, tests 2 and 3 must be repeated.

(5) Vertical circle indexing

Requirement: The vertical circle should be correctly indexed.
Test: Set up and level the theodolite. Where an altitude bubble is

fitted, centralize it with the clipping screw; for an electronic instrument with a telescope bubble, tilt the telescope to centralize the bubble and press the indexing key.

Sight a target at a distance of about 100 m and read the vertical angle. Change the face, resight the target and reread the vertical angle. It the angles sum to 360° (180° or 540° for horizontally zeroed circles), the adjustment is satisfactory.

Adjustment: If the sum does not equal 360° (or 180° or 540°), find the correction by subtracting the sum of the readings from 360° (or 180° or 540°) and halving the result. The method of adjustment depends on the type of instrument.

- *Optical type with altitude bubble*: With the telescope still sighted on to the target, turn the clipping screw until the corrected reading is observed. The altitude bubble will have moved off centre; bring it back with the bubble adjusting screws. Repeat the test and adjustment until all error has been removed.
- *Optical type with automatic indexing*: Turn the telescope tangent screw until the corrected angle reading is observed. Adjust the cross-hairs to obtain coincidence with the target, taking care not to disturb the horizontal collimation. Repeat the test and adjustment until all error has been removed.
- *Electronic type with telescope bubble*: Set the telescope horizontal by adjusting until the vertical scale reads 90° (or 0°) plus the correction (with the appropriate sign). Use the adjusting screws to re-centre the telescope bubble. Repeat the test and adjustment until all error has been removed.
- *Electronic type with automatic indexing*: Follow the manufacturer's instructions.

(6) Optical plummet test

Requirement: The optical plummet collimation and vertical axis should coincide.

The majority of manufacturers locate the plummet in the plate assembly; some instruments, notably those made by Leica, have the plummet in the tribrach.

Test (plummet in plate): Set the theodolite on the tripod and mark the ground point defined by the optical plummet. Turn the instrument horizontally through 180° and make a fresh mark; if the marks coincide, the adjustment is satisfactory.

Test (plummet in tribrach): There is a difficulty in that the theodolite must be kept fixed while the tribrach is rotated. One way of achieving this is to lay the theodolite on its side and to clamp it close to the edge of a stable bench so that the tribrach can be rotated. Take care not to damage the standards or any knobs. 'Plumb' a mark on a nearby wall, turn the tribrach through 180° and make a further plumbing; the marks should coincide.

Adjustment: Make a mark midway between the points defined by the plummet before and after rotation of 180°. Turn the capstan screws to bring the plummet index on to this mid-point. Repeat the test and adjustment until all error has been removed.

Note: Kern instruments are often fitted with centring tripods. A bullseye bubble on a plumbing pole is used to determine verticality. Check the bubble by rotating the plumbing pole through 180° after centring. If the bubble moves off centre, bring it halfway back with the adjusting screws. Repeat the test and adjustment until all error has been removed.

Occurrence of instrumental errors

It is fortunate that the most frequently occurring instrumental errors are those which can be most easily checked and adjusted, namely:

- plate bubble error
- horizontal collimation error
- optical plummet error.

The plate bubble can be checked each time the theodolite is set up; the error can be removed or else the footscrews can be adjusted so that the bubble is a constant distance off centre in all positions of the alidade. The remaining adjustments should be checked monthly, as should the general soundness and mechanical operation of the theodolite and tripod.

Some electronic instruments provide automatic compensation for plate bubble error, horizontal collimation error, trunnion axis error and horizontal circle eccentricity once the checks have been carried out.

Accuracy of horizontal angle measurement

Measurements on each face will generate equal and opposite errors which will cancel out in the event of presence of horizontal collimation error, trunnion axis error, horizontal circle eccentricity or vertical index error. By commencing subsequent rounds of angles at different parts of the circle, eccentricity can be reduced. Mistakes (gross errors) should also be exposed. The effects of plate bubble error can be eliminated on setting up. The remaining sources of error are in sighting targets and in plumbing (centring) the theodolite. Targets should be set up accurately over stations, and coincident sighting achieved with the cross-hairs.

A centring error can have a considerable effect on the accuracy of angle measurement, and attention should be paid to eliminate, as far as possible, such an error. Accurate centring is often found difficult by inexperienced surveyors. They are also unaware of its importance, spending some time trying to eliminate a one division bubble error (effect negligible) while tolerating a centring error of several millimetres.

Example 3.2

Two points both 30.000 m from a theodolite subtend an angle of 90°. What will be the greatest possible error in measuring the angle if the theodolite is set up with a 5 mm centring error?

The greatest error will occur when the offset is along the internal bisector of the angle, as in Fig. 3.7.

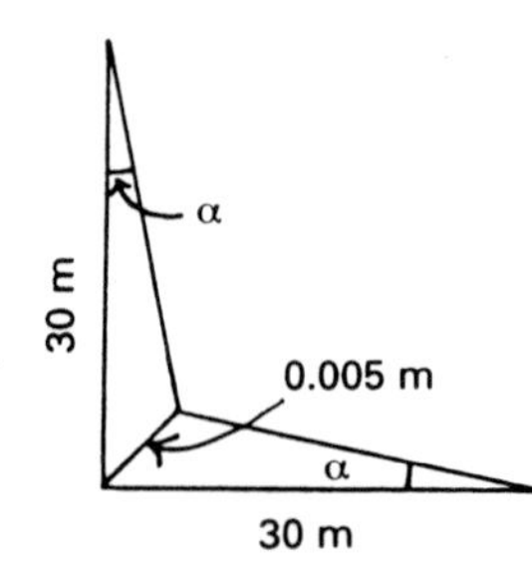

Fig. 3.7 Centring error.

$$\alpha = \tan^{-1}\frac{0.005\cos 45°}{30} = 24.3''$$

Angular error $= 2\alpha = 49''$

3.9 FURTHER READING

Cooper, M.A.R. (1982) *Modern Theodolites and Levels*. 2nd edn. Oxford: Blackwell Science.

3.10 EXERCISES

Exercise 3.1

Table 3.2 shows results from measuring horizontal rounds of angles at station I. Calculate the mean values of angles FÎG, GÎH and GÎJ. Comment on any inconsistencies, stating how they may be overcome.

Table 3.2

Instrument at	Sighting	Face	Reading
I	F (RO)	L	00°07′21″
	G	L	44°16′35″
	H	L	91°13′24″
	J	L	159°53′08″
	J	R	339°52′58″
	H	R	271°13′20″
	G	R	224°16′38″
	F	R	180°07′18″
	F	L	91°24′36″
	G	L	135°33′53″
	H	L	182°30′33″
	J	L	251°10′15″
	J	R	71°17′10″
	H	R	2°30′41″
	G	R	315°33′56″
	F	R	271°24′35″

Exercise 3.2

A check is to be made on marks X and Y on piling platforms in a river by taking theodolite observations from stations A and B, both on the same bank. A theodolite at A is used to sight targets (on tripods) at B, X and Y; the theodolite is then moved to B and a target is set up at A.

From the data in Table 3.3, calculate the plan length of XY and determine the reduced levels of X and Y.

Calculate the zenith angle that would be measured by the theodolite at B to sight the target at X.

Table 3.3

Reduced level	of B = 15.135 m AOD
Slope length	of AB = 103.270 m
Zenith angles	AB = 91°23′42″ AX = 92°42′36″ AY = 92°13′24″
Theodolite/target height	above B = 1.330 m
Target height	above X = 1.350 m above Y = 1.295 m
Horizontal clockwise angles	$X\hat{A}B$ = 39°33′24″ $A\hat{B}X$ = 47°52′06″ $Y\hat{A}B$ = 63°41′12″ $A\hat{B}Y$ = 61°37′15″

Exercise 3.3

As part of EDM measurements between two stations A and B, reciprocal trigonometric heighting was carried out with the following results:

Inst at	Height of theo axis above stn (m)	Station sighted	Slope measurement (m)	Height of target above station (m)	Zenith angle
A	1.270	B	1317.500	1.340	88°16′19″
B	1.423	A	1317.508	1.310	91°44′25″

The EDM is a fixed distance above the theodolite collimation axis, equal to the height of the prism above the target. The reduced level of A is 136.370 m AOD. Calculate:

(a) The reduced level of B.
(b) The coefficient of refraction.
(c) The reduced level of C from the following readings using the coefficient of refraction determined in (b):

Inst at	Height of theo axis above stn (m)	Station sighted	Slope measurement (m)	Height of target above station (m)	Zenith angle
B	1.423	C	1563.200	1.250	86°34′44″

4 Linear Measurement

4.1 TAPING: DESCRIPTION AND USE

Description

For engineering surveys, tapes should be made of steel. The most popular type is the 30 metre tape made of mild steel with a protective coating. An open-frame winder reduces the build-up of dirt on the tape, whereas the encased winder is more compact, allowing the tape to be carried easily in the pocket of a jacket. Markings are every five millimetres with one millimetre divisions for the first and last metre, or one millimetre divisions throughout. A three metre long spring loaded 'pocket tape' is also useful, particularly for setting out.

Fifty and 100 metre tapes are available, but longer measurements are more conveniently carried out electromagnetically.

Use

(1) Keep the tape taut so that it is straight in plan and elevation.
(2) Align the tape precisely with the points being measured. Sometimes it is difficult to measure from the start of the tape, particularly if tension is being applied. An additional handle with a spring clip can be used to allow a firm grip, or the 1.000 m mark can be used as a zero; at the other end of the tape 1.000 m should be deducted from the tape reading.
(3) Apply the correct tension. Figure 4.1(a) shows the zero of the tape; Figs 4.1(b) and (c) show both ends of the tape using a spring balance to check the tension and zeroing at the 1.000 m mark.
(4) Distances greater than the length of the tape must be measured in two or more bays. A precise, stable mark must be made at the end of each bay (peg with nail, sharp mark on concrete). It is

(a)
(b)
(c)

likely that different corrections for each bay must be applied (Section 4.2).

Care of tapes

Care should be taken of tapes:

(1) In unwinding, so that the tape is not pulled off the winder.
(2) In use, so that the tape does not become kinked, scratched, stretched or broken.
(3) In winding, so that the tape is not twisted or kinked, and dirt and moisture are not carried into the winder.

Periodically the tape should be removed from its winder and cleaned and, together with the winding mechanism, lightly oiled. Broken tapes can be repaired by riveting connecting strips across the break; such methods should be merely temporary, there being the risk of damage to hands from sharp edges. Important measurements should not be taken with a repaired tape.

4.2 TAPING: CORRECTIONS

Slope reduction

A measurement may be taken on the slope where a horizontal distance is required. Either the zenith angle (θ) of the slope measurement, or the difference in elevation (V) of the ends of the tape must be found. The former can be determined with a theodolite or, with sufficient accuracy on gentle slopes, with a hand-held clinometer or Abney level. The difference in elevation of the ends of the tape can be found by levelling.

Referring to Fig. 4.2, it can be seen that:

$$\text{Horizontal distance } (H) = (D^2 - V^2)^{1/2}$$
$$= D \sin \theta$$

Standardization correction

Occasionally on a construction site a tape may survive long enough to become permanently stretched. If stretching is suspected, the tape

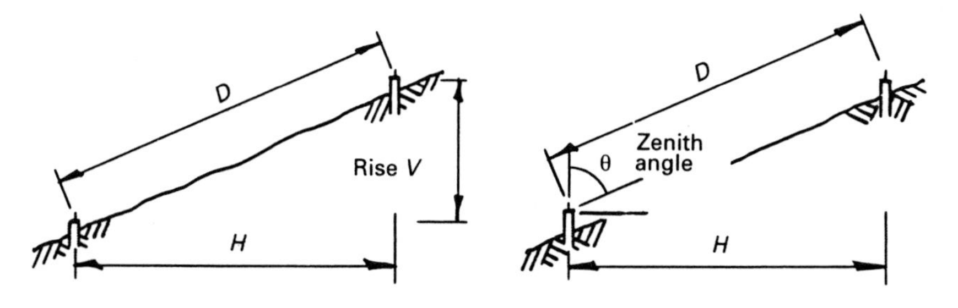

Fig. 4.2 Slope reduction.

should be compared with a known standard length, for example a new tape.

All measurements taken with the non-standard tape should be multiplied by:

$$\frac{\text{Tape actual length}}{\text{Tape nominal length}}$$

Tension correction

The corrections to compensate for tension effects (and for temperature, sag and altitude) are small, and may well be negligible if the tape is used at a tension close to the standard value (and 'on the flat' in any but extreme temperatures in a non-mountainous area).

A spring balance connected to the tape is used to measure its tension. It is convenient to attach it to the measuring end so that simultaneous readings of tension and distance can be noted. The balance can be connected by roller grips. Clearly the simplest method is to set the standard tension, usually 45 or 50 Newtons, the value being marked near the zero of the tape. If a non-standard tension is applied, the cross-section area of the tape must be determined either from the manufacturer's specification or by measuring; a micrometer screw gauge will be needed for the thickness. The Young's modulus of the tape material will also be required, the manufacturer's information again providing the value.

Sag correction

Tapes are standardized 'on the flat'. Where possible, measurements should be made on flat, or nearly flat, surfaces. Where the tape is suspended, it takes the form of a catenary and the measurement will be too long. To calculate the correction, the tension and the weight per unit length of the tape are required. The latter can be determined either by weighing the tape or from the manufacturer's specification.

Temperature correction

Tapes will expand or contract if the field temperature is above or below that at which they were standardized. The field temperature must be found with a thermometer. The coefficient of linear (thermal) expansion will be required (see manufacturer's specification).

Altitude correction

Measurements should be corrected to give the equivalent distance at sea level. The radius of the earth and the altitude of the measurements must be known.

Table 4.1 shows correction formulae and typical values for tension, sag, temperature and altitude effects.

Compromise tension

Where a steel tape is suspended, a compromise tension can be applied such that the effects of sag are balanced by a tension higher than the standard one.

The values are:

70 N for a tape 10 mm wide
105 N for a tape 13 mm wide

Accuracy of taping

Manufacturers quote a standard error of ±3 mm in a 30 m length for steel tapes. Lengths less than 30 m will have a proportionally lower error; lengths greater than 30 m measured in several stages with a 30 m tape will have a likely error determined as a 'root-mean-square' sum.

Table 4.1

Tape corrections

Typical physical data (values should be checked):
$P_s = 50\,N$; $t_s = 20°C$; $\alpha = 0.0000112/°C$; $E = 200\,000\,N/mm^2$; $R = 6370\,km$

Source of error	Tension variation	Sag	Temperature variation	Altitude effect
Correction	$\frac{(P - P_s)D}{AE}$	$\frac{-w^2D^3}{24P^2}$	$\alpha(t - t_s)D$	$\frac{-Dh}{R}$
Typical correction values for 30 m length with data as above	$P = 70$ $A = 3.25$ Correction $= +0.0009$	$P = 70$ $w = 0.240$ Correction $= -0.0132$	$t = 10°C$ Correction $= -0.0034$	$h = 1000\,m$ Correction $= -0.0047$

Symbols
P = Field tension (N)
P_s = Standard tension (N)
D = Measured distance (m)
A = Cross-section area (mm^2)
E = Young's modulus (N/mm^2)
w = Weight/unit length (N/m)
t = Field temperature (°C)
t_s = Standard temperature (°C)
α = Coefficient of linear expansion (/°C)
h = Height above sea level (km)
R = Radius of earth (km)

Example 4.1

What will be the standard error in a measurement of 73.455 m made with a 30 m tape which has a manufacturer's accuracy of ±3 mm?

Measurements will be 30 m, 30 m and 13.455 m
Standard errors will be ±3 mm, ±3 mm and ±3 × 13.455/30 000 (= ±1.35 mm)
Total error $= \pm(3^2 + 3^2 + 1.35^2)^{1/2} = \pm 4.5$ mm, say ±5 mm

Setting out distances

When setting out with a steel tape, the engineer must apply corrections, with the signs reversed, to the lengths to be set out.

4.3 EDM: DESCRIPTION AND OPERATION

Description and principle

Electromagnetic distance measurement (EDM) equipment consists of an aiming head/receiver unit set at one end of the line to be measured and pointed towards a reflective glass prism set up at the other end. An electromagnetic beam is emitted by the aiming head, projected towards the prism, reflected back and analysed to determine the distance.

The prism is of the 'corner cube' type, with three mutually perpendicular reflecting faces. It must be pointed towards the EDM outfit so that each reflecting face is at approximately 45° to the incoming signal which is then returned along a parallel path.

The aiming head is combined with a theodolite: originally 'add-on' units were produced to be mounted on the telescope or the standards. Manufacturers now also produce integral theodolite/distancers of the total station type so that the electromagnetic beam passes through the telescope optical system. The EDM unit can be set vertically over a station, while the prism (set in a tribrach on a tripod) can also be precisely centred over a station. For detailing work of lesser accuracy, the prism can be set on a hand-held 'plumbing' pole. The distance measured is a slope distance; the theodolite vertical angle allows the calculation of the horizontal distance and the vertical component.

Equipment is powered by nickel-cadmium rechargeable batteries, and for equipment used in engineering surveys, the electromagnetic signal is infra-red light produced by a gallium-arsenide diode. The wavelength of the carrier signal is too short for distance measurement, so it is amplitude modulated by a quartz crystal osciallator. The frequency is designed to give a suitable measuring wavelength (wavelength and frequency are inversely proportional), for example 10 m. The waveform may be depicted as sinusoidal and Fig. 4.3(a) shows the modulated signal.

Unless the double distance is a multiple of the wavelength, it will consist of a whole number of wavelengths plus a part of the wavelength.

$$2D = n\lambda + \Delta\lambda$$

where $2D$ is the double distance (path length of signal), λ is the wavelength, n is an integer and Δ is a fraction (Fig. 4.3(b)).

It is not possible to determine n, but the phase shift of the return

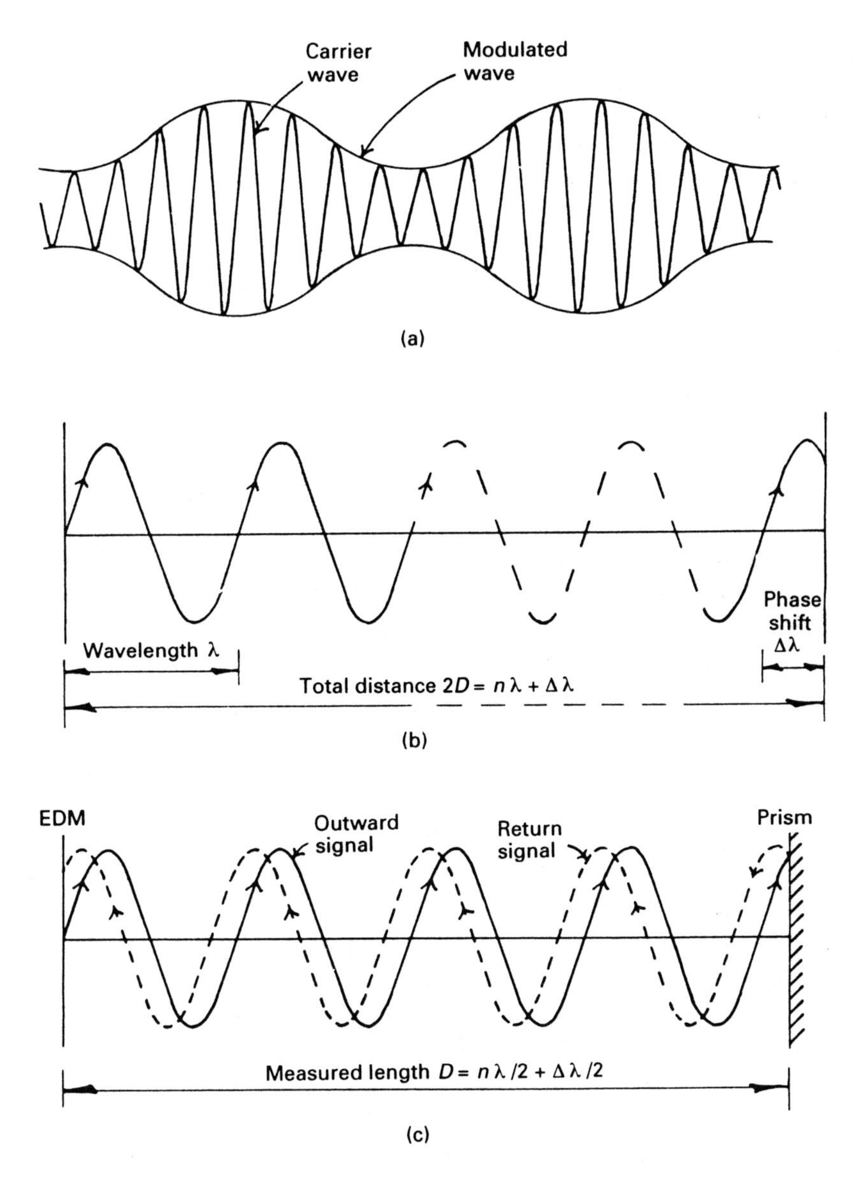

Fig. 4.3 EDM signal.

signal from the outgoing one can be found (Fig. 4.3(c)). The process is repeated for a different wavelength, producing a fresh value of the phase shift. The ratio between the frequencies is designed so that a unique determination of distance (over the instrument's range) can be made. Either a large change of frequency (×10, for example) is used to give fine and coarse measurements, or a small change (10%, for example) is employed so that there is no change in the number of whole wavelengths. Some EDMs use more than two frequencies to provide a sufficiently precise measurement.

The instrument automatically carries out the phase shift measurement at different frequencies, computes the single distance

$$D = n\lambda/2 + \Delta\lambda/2$$

and displays it. Close to the liquid crystal display, there is usually a keyboard, so that further data (vertical angle, prism constant, atmospheric constant) can be entered.

Operation

Figures 4.4(a) and (b) show add-on and integral outfits, while Fig. 4.4(c) shows some prism assemblies.

Operation of the equipment is simple, and is described in the manufacturers' handbooks. The EDM and prism must be carefully set up over the stations (similar to theodolite setting up – Chapter 3) to avoid centring errors. The aiming head must be precisely pointed at the prism, although the prism requires only approximately directing towards the EDM.

The range of EDM is very dependent on the weather and the number of prisms in use at a station. In clear conditions a range of 1000 m to 2000 m using one, two or three prisms is typical, though some instruments can measure much longer distances. In rain, mist or fog, the range may be cut to 100 m or less.

The signal is formed into a narrow beam by the lens system of the EDM, but diverges slightly as the distance increases. Over longer distances, a greater number of prisms must be used to ensure a return signal of adequate strength. If the outgoing signal is too strong, there is a risk of surfaces other than the prism returning the signal, which may affect measurements. On older instruments, there is a signal strength adjuster; newer machines provide automatic attenuation so that the intensity is reduced to suit conditions. Both types of instrument have a return signal strength indicator. In poor visibility, and when the beam is not aligned parallel with the theodolite optical axis, satisfactory pointing may have to be achieved electronically, the surveyor 'searching' by turning the horizontal and vertical tangent screws until a maximum return signal, as shown on the indicator, is obtained.

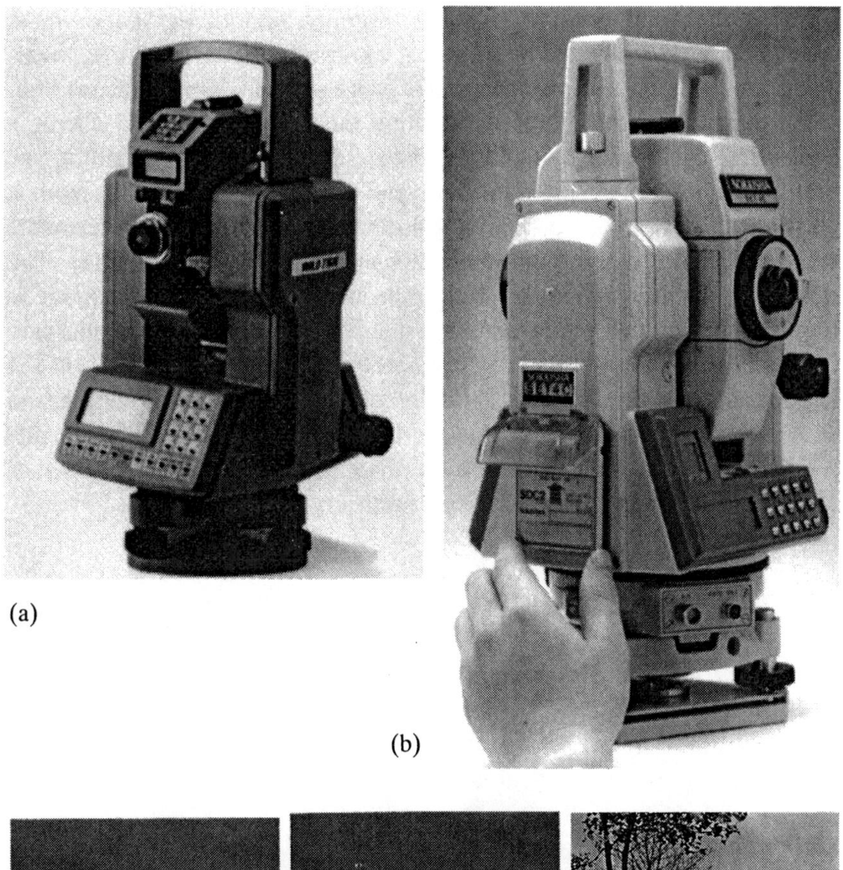

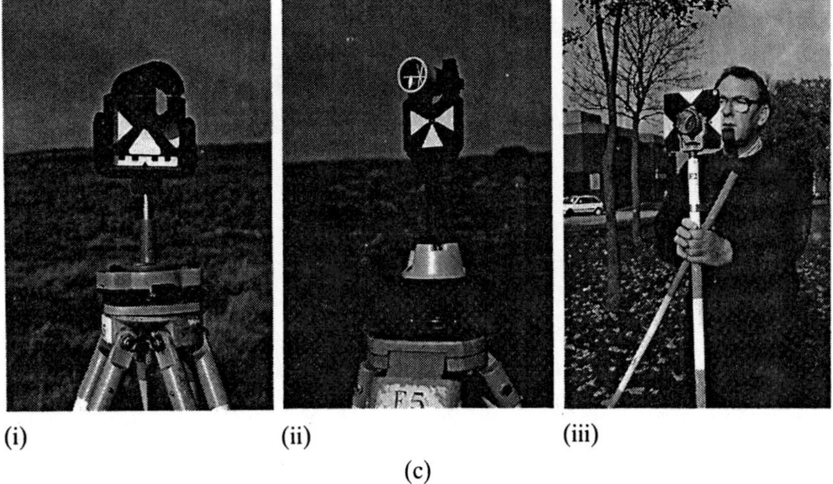

Fig. 4.4 EDM outfits: (a) Leica DI1600 EDM on T1610 electronic theodolite; (b) Sokkia SET 4C Total Station; (c) prisms: (i) Leica, (ii) Geodimeter, (iii) Sokkia.

More than one reading of each distance should be taken, most instruments providing repetition of measurements, tracking, averaging and staking out. Occasionally 'rogue results' are encountered, when the beam is interrupted or when the mode is changed during a measuring cycle. Each displayed measurement should be noted, so that rogue ones can be identified and rejected, particularly from a group for averaging. Tracking involves continuous measurements when the prism is being moved backwards or forwards (on line). The stake out facility is similar but indicates how much movement is required to set the prism at a specified distance. Where angles are manually read, the vertical angle can be observed and entered on the keyboard while the distance is being measured; horizontal length and vertical components will be displayed. Some instruments automatically read the vertical angle and directly display the horizontal distance, a useful feature when employing the stake out facility.

4.4 EDM: PROBLEMS, CORRECTIONS, ERRORS, ACCURACY

Problems – weak or absent return signal

There is little to go wrong with EDM equipment. Most problems occur when the return signal is absent or weak. Causes may be:

(1) *Prims not aligned*: Prisms must be approximately pointed towards the EDM.
(2) *Cable connections*: On older instruments, the battery is connected by cables to the aiming head. The connections, and the condition of the cable itself, should be checked. A spare cable (or set of cables) should be carried.
(3) *Batteries exhausted*: Battery problems are not unknown with EDM and steps must be taken to avoid failure through a flat battery. Nickel-cadmium (rechargeable) batteries are used to provide satisfactory signal strength over several hours. It is sound policy to have three batteries for each instrument: one in use, one (fully charged) spare and one on charge. Batteries should be used until low; a signal is displayed when the strength is low and the battery should be changed. Provided that there is not a long delay in battery changing, information stored in an EDM outfit will be retained.

Manufacturers recommend that batteries be fully discharged

before being recharged. Partially discharged batteries develop a resistance to being recharged fully (a so-called 'memory') so that apparently full batteries can discharged quite quickly. Such batteries should be fully charged then discharge several times to remove the memory effect. Full charging usually takes about 18 hours. To flatten a battery quickly continuous measurements in tracking mode can be made. Alternatively a discharger incorporating a suitable resistance can be assembled. Care must be taken not to charge a battery for too long (causing permanent damage); some chargers have a built-in timer, and some also incorporate a discharge facility.

(4) *Collimation error*: This occurs with add-on instruments when the axes of sight (or collimation) of the EDM and the theodolite are not parallel; the adjustment is usually simple, lining up the axes optically on a prism and target at first, then adjusting the EDM axis to give the maximum return signal. If the error is large, an initial adjustment will have to be carried out sighting a nearby prism; it can then be moved progressively further away – a single prism at 300 m will usually allow the EDM axis to be satisfactorily aligned. Telescope mounted apparatus will require aligning horizontally and vertically; standard mounted equipment horizontally only.

The collimation and adjustment will usually have to be carried out on new instruments.

Problems – prism centring errors

Prism centring errors, which may not be apparent, can give faulty results. They can occur when non-matching prisms are used for sloping sights. The EDM measuring centre of telescope mounted outfits and non-coaxial total stations will move backwards when an angle of elevation is set; a tilting target/prism assembly should be used and carefully aligned (Fig. 4.5(a)); alternatively the theodolite and EDM should in turn be pointed towards the prism, giving a negligible error on all but very short sights (Fig. 4.5(b)). Errors can also occur if a standard mounted EDM outfit (rare nowadays) is directed towards a tilting target/prism assembly (Fig. 4.5(c)).

Where EDM measurements are to be used for heighting, the vertical separation between target and prism should equal that between telescope and EDM objective lenses.

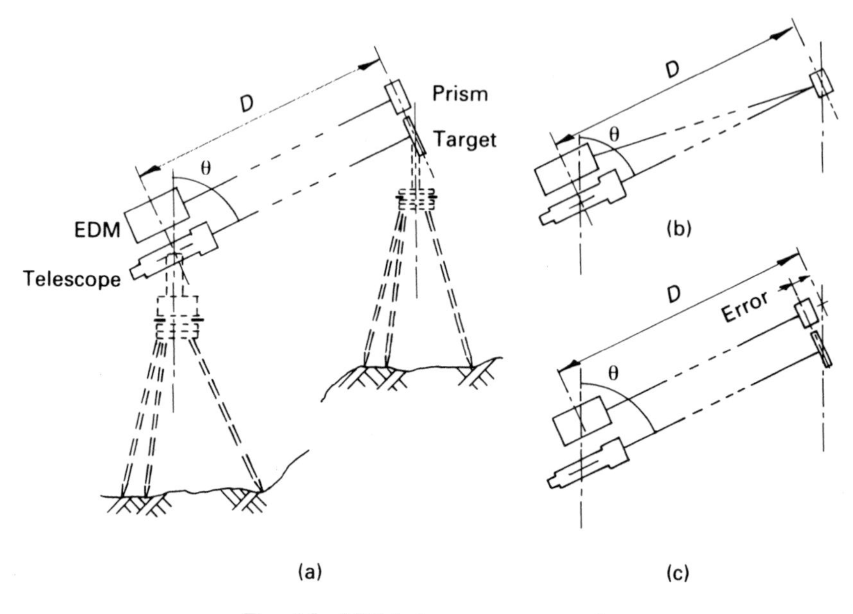

Fig. 4.5 EDM slope measurement.

A coaxial optics instrument (used with the matching prism) overcomes the above problems.

Corrections

Slope

Reductions of slope lengths can be carried out on the instrument keyboard, or may be performed automatically.

Horizontal distance $H = D \sin \theta$
Vertical component $V = D \cos \theta$

where D is the slope distance and θ is the zenith angle.

Atmospheric variations

Variations in atmospheric conditions affect the velocity of the signal. Corrections can be determined using a chart or nomogram showing a 'parts per million' correction for various temperatures and pressures/heights above sea level. Most sites have a thermometer, less frequently a barometer. Under normal conditions the corrections are small and may well be insignificant. In the UK, the greatest effects are usually in winter. A correction can be entered (via the keyboard) in its parts per million form; displayed values will then be corrected automatically.

Prism constant

The manufacturer usually calibrates the EDM to suit the corresponding prisms, and no constant need be added. For a non-matching prism the constant is likely to be different: a prism constant correction equal to the difference between the two prisms must be applied. A number of instruments allow the prism constant to be entered at the keyboard. If the constants are not known, they must be found by using the 'zero error' check described later.

Altitude

Measurements made above or below sea level must be corrected to give the corresponding sea level dimension.

$$\text{Correction} = \frac{-Dh}{R + h}$$

where D is the measured distance, h is the height above sea level and R is the radius of the earth. In engineering surveys the correction can be expressed in parts per million as $-1000\,h/R$ (h in m, R in km) and may be added to the atmospheric correction and entered at the keyboard.

Spheroidal/curvature

A distance measured as a chord will require correcting to give the corresponding spheroidal (earth's curvature) shape. Because of refraction, the path of the signal is curved, slightly reducing the amount of correction, which is usually insignificant in engineering surveying (see page 126).

Errors and checks

The simplest procedure for avoiding instrumental errors is to use matching prisms at all times and to have the equipment checked by the manufacturer or supplier once every twelve months. There are no calibration adjustments which can be carried out on site, although there are checks which will reveal the presence of instrumental errors, of which there are three categories.

Zero error

The zero error is a constant error caused by non-coincidence of mechanical and electronic centres or by use of non-matching prisms. The error can be detected by measuring a baseline whole and in two bays (taking care to minimize centring errors). The correction is the whole measurement minus the sum of the two bay measurements. Suppose a baseline has been measured whole as 216.327 m and in two bays as 73.975 m and 142.377 m. If each measurement has a zero error of $+x$ mm, then a correction of $-x$ mm must be applied to each quantity displayed.

Thus $73.975 - x + 142.377 - x = 216.327 - x$, and x is found to be 0.025 m; 25 mm must be subtracted from each measurement.

A more precise determination can be made by using a longer baseline divided into more bays. The correction is calculated using different combinations of measurements: either a mean value is taken, or a least squares method is applied (see example 7.13 in Chapter 7).

Cyclic error

A cyclic error can occur if there are variations in the phase shift measurement. A comparison with reliable measurements over the range of the main wavelength will expose cyclic errors. Zero errors must first be eliminated.

Proportional error

Atmospheric variations from standard conditions and changes in crystal frequency may cause proportional errors. A comparison with reliable measurements of short and long distances will expose proportional errors. Zero errors must first be eliminated.

It is unlikely that cyclic and proportional errors will be significant if equipment is regularly serviced. It will also be rare that time and facilities for checks of adequate precision are available on site.

Calibration tests

Multipillar baselines are used for calibration tests. Between six and ten pillars on a straight line of up to one kilometre can be set up. They should be arranged to give a variety of distances throughout the range of the main measuring wavelength (typically 10 m). Ideally they should be at the same level. They must be calibrated using a very precise EDM outfit.

Accuracy

Manufacturers quote a standard error in two parts: a constant part, typically ±5 mm, and a proportional part, typically ±5 ppm (parts per million, or mm per km).

The combined effect should be calculated as a 'root-mean-square' error. For example, for the errors given above, the standard error for a distance of 665.760 m would be:

$$\pm\left[(5)^2 + \left(5 \times \frac{665.760}{1000.000}\right)^2\right]^{1/2}$$

$$= \pm 6.0\,\text{mm}$$

The typical error values quoted apply to a number of instruments; some have smaller standard errors.

When an EDM is in tracking mode, the standard error is usually larger. For distances up to 50 m, careful use of a steel tape can give accuracy comparable with that of EDM; greater lengths are more precisely and more easily carried out electromagnetically.

4.5 EDM: APPLICATIONS

EDM is extensively used for linear measurement. The apparatus is not cheap, but given a reasonable amount of use, it will very quickly pay for itself. Modern instruments are of integral (total station) or modular construction combining an electronic theodolite with the distancer. Readings from both elements are combined and calculation facilities are incorporated allowing, for example, horizontal distances, remote object elevation and coordinates of points surveyed to be directly displayed.

A great deal of survey work can be carried out by EDM. For

control surveys involving direct measurement, the facility for simultaneous measurement of distances and angles, the speed and simplicity of operation, and accuracy achievable make it eminently suitable for such work. Much control work is carried out by traversing, or by measuring a network for which EDM is ideal. Redundant measures to check and strengthen the fieldwork can easily be carried out, and computer software is available to handle all the observations.

Detail surveying by polar (or radial) methods can also be carried out very efficiently using EDM. If instrument station and reference object coordinates are entered, a total station can display directly the coordinates of all points surveyed. Alternatively the observations can be recorded by a logging system, which can be linked subsequently to an office computer and plotter to produce a drawing automatically.

In setting out, EDM is useful in establishing and checking control points. In taped work, procedures were geared to short lengths, measuring along grid lines, avoiding obstacles and simplifying calculations; with EDM, polar methods are available where many points (road curve pegs, building column centres) may be located from a single instrument station.

One problem with EDM is the difficulty of setting out points with adequate accuracy. The process is one of trial and error, the chainman (assisting the surveyor) being guided in both distance (the stake out facility can be used to indicate the amount) and direction. The frailties of this procedure (requiring good communication between surveyor and chainman) are compounded by the difficulty in holding the pole vertical. Errors as large as 30 mm in distance and direction are possible in adverse conditions. One way of overcoming this problem is to measure the distance to a peg on line known to be slightly short of the required measurement. (The prism can be held directly over a nail in the peg, or a tripod can be used). The extra distance can then be taped by a reliable chainman and alignment performed with the instrument. The final point should be checked with a prism on the peg or on a tripod.

4.6 EDM: DEVELOPMENTS

The main recent developments in EDM have been concerned with making the equipment easier or more flexible to use in the field, and increasing the compatibility with computing facilities.

In detail surveys the selection of points to be surveyed can be quite critical, especially when an automatically contoured plot is required and the ground is uneven, perhaps with a number of breaks of gradient. One solution, developed by Geotronics, is the Geodimeter System 4000 (Fig. 4.6). This is a one person surveying outfit where the surveyor is at the prism. The pole, in addition to holding the prism, has a telescope, a display unit, a keyboard and a radio link. The total station has a servomotor actuating the horizontal and vertical rotation and it searches for the prism. When the prism has been located (i.e. the signal has been returned), measurement is undertaken and results are displayed at the prism keyboard where the surveyor can add or amend detailed information. The prism/pole/keyboard assembly is known as the Remote Positioning Unit (RPU); contact is made with the total station by radio. Geotronics also produce modular instruments, the 600 series; the equipment has a number of add-on or plug-in components so that the surveyor can choose the level of precision, type of keyboard and battery capacity. The keyboard acts as a data logger so that this can be detached, replaced

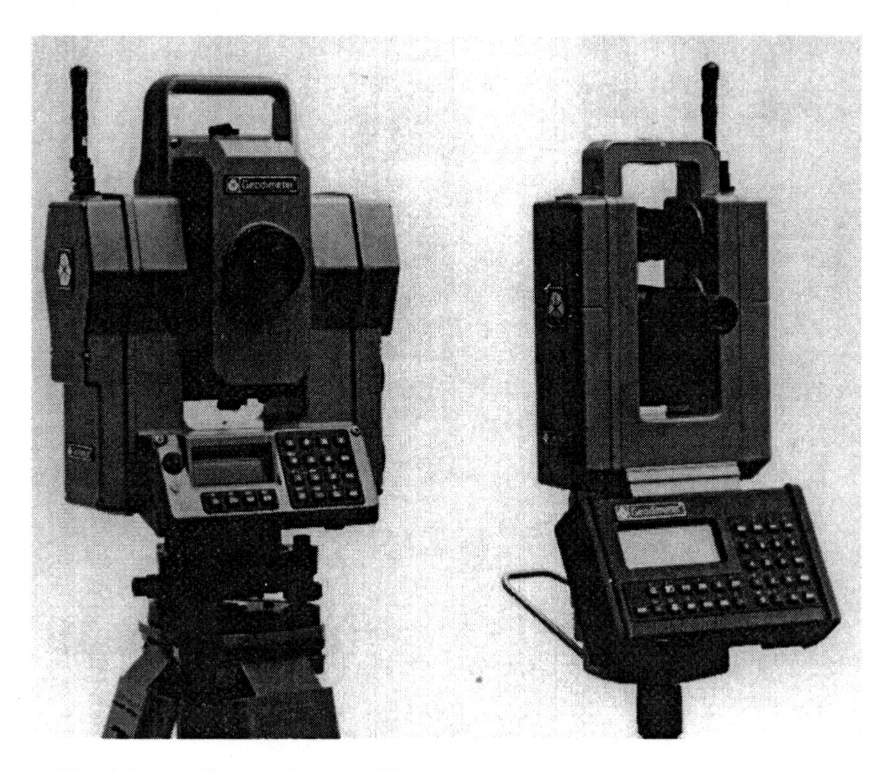

Fig. 4.6 Geodimeter System 4000: total station and remote positioning unit.

by another one for fresh field use, and taken to the office for processing of data (Fig. 4.7). A version of the Remote Positioning Unit is also available for this system.

Development work has also been done for reflectorless systems. The signal is reflected from a flat surface which must be at right angles to the outgoing beam. The distant point is located with a laser beam along the distancer collimation axis. The surface must have some reflective properties; the accuracy is lower (between ±5 and ±10 mm standard error) than for prism measurements, and the range is limited to a few hundred metres. Signals can also be returned from corners where three mutually perpendicular surfaces meet (the ceiling at the corner of a room) acting like a corner cube prism. The Leica DIOR models are of this type; Leica also produce the Disto hand-held distance meter using a similar principle.

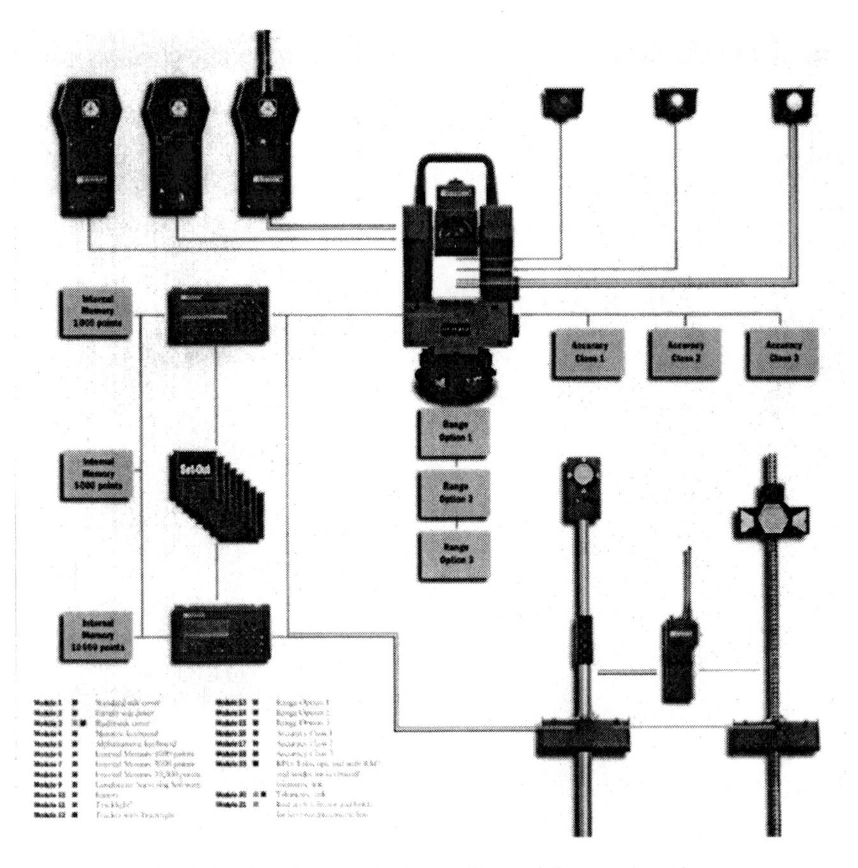

Fig. 4.7 Geodimeter System 600: modular total station.

The standard measuring accuracy of EDM is ±5 mm ±5 ppm, but an increasing number of instruments are being produced with improved specifications, typically giving ±2 mm ±2 ppm. The measuring range of equipment has been increased, and manufacturers are claiming distances measurable of 1000 m with a single prism, in good conditions. To simplify setting out, an optical diverging beam with red and green (off line) sectors and a white on line sector is available in Geodimeter instruments ('Tracklight').

Data logging and manipulation have received considerable attention in recent years. As an alternative to hand-held or tripod mountable loggers, a number of makers of total stations equip their instruments with memory cards. These are inserted in a slot or space in the instrument and observations are automatically logged. The card can be removed from the machine, taken to an office and put in a card

Fig. 4.8 Nikon DTM750 field station.

reader which is coupled to a computer for data handling, storage and plotting. Nikon have brought out the DTM 700 series total stations incorporating a program card in addition to a memory card; this should further reduce delays while cards are changed and data is handled (Fig. 4.8). Two-way communication between the cards is possible making the instrument more responsive in handling design and setting out information. Either card can be exchanged, so that fresh data can be collected, or different system functions can be imposed in conjunction with the unchanged card.

4.7 FURTHER READING

Burnside, C.D. (1991) *Electromagnetic Distance Measurement*. 3rd edn. Oxford: Blackwell Science.

4.8 EXERCISES

Exercise 4.1

A, B and C are three stations used in a control survey. Line AB was measured with a steel tape, line BC by EDM and the angles by theodolite. From the readings given, calculate:

(a) The corrected plan lengths AB and BC.
(b) The plan length of CA.

Length AB (two bays)

Bay	Slope length (m)	Flat or suspended and field tension	Field temp	Rise (m)
1	29.789	Flat 60 N	15°C	+1.105
2	28.666	Catenary 50 N	21°C	+1.675

'30 m' tape was actually 30.013 m at 50 N and 20°C on the flat
Mass = 0.011 kg/m Cross section = 2.5 mm^2
$E = 20 \times 10^4$ MN/m^2 Coeff^t of linear expan^n = 11×10^{-6}/°C

Length BC (two bays)

Bay	Slope length (m)	Zenith angle	Atmospheric correction (ppm)
1	42.353	88°30′20″	−50
2	69.240	91°15′00″	+60

Angles
$A\hat{B}C = 31°22'36''$ $B\hat{C}A = 26°14'18''$ $C\hat{A}B = 122°23'30''$

5 Control Surveys

5.1 INTRODUCTION

In surveying and setting out, a control survey covering the area should be carried out in the early stages. The object will be the establishment of permanent or semi-permanent points for plan control and levelling so that a unified, self-checking survey of the required precision can be produced. Whereas an error in detailing will often have no effect on the location of subsequent points and, if small, may be insignificant, an error at the control stage may affect the whole survey. Most control surveys must be tied in to existing fixed points.

Traditionally, control surveys have been carried out using linear, angular and levelling measurements which connect the stations. Control surveys should incorporate checks (for closure) at several stages. Large discrepancies indicate that the work (in part or whole) must be repeated. Small discrepancies (after the correction of systematic errors) within the specified tolerances may be left, or, more commonly, may be 'removed' by the adjustment of observations to produce mathematical consistency.

Before the development of computer programs for survey adjustment, field methods were used that provided few redundant measures (e.g. traversing), allowing a manual adjustment to be performed. Computer programs can now adjust complex networks, often in three dimensions, with many redundant measures.

Such procedures are in no way designed to compensate for poor precision in the field. There is still a school of thought believing that measurements requiring adjustment are not of a satisfactory standard. However the use of EDM facilitates the rapid taking of extra measurements which can provide further checks, and if consistent will strengthen the survey.

Satellite surveying (GPS) is now being used to establish control of stations. All surveying operations use two receivers, one of which must be at a point of known corrdinates so that corrections can be determined and applied in the location of new stations. Redundant measures can be obtained by receiving signals from more than the minimum number of four satellites, and by taking paired observations from opposite ends of lines forming a network.

Planning

(1) A specification linking precision with permitted tolerances must be drawn up.
(2) The whole area must be reconnoitred.
(3) Suitable survey methods must be selected.
(4) Equipment of the appropriate type and adequate precision must be selected.
(5) Control stations must be suitably positioned.
(6) The survey must be self-checking so that its validity can be assessed.
(7) Acceptably small errors of closure can be neglected or eliminated by adjustment, as appropriate (large errors will require some or all of the work to be repeated).
(8) For complex projects an error analysis will indicate whether the specified precision has been achieved.
(9) Periodical checks must be made when the stations are in use over a long time.

Station selection (plan control)

(1) Stations should define a control figure of suitable shape.
(2) Stations should permit the convenient surveying of detail.
(3) Stations should be intervisible for linear and angular measurements.
(4) Stations should be accessible so that instruments can be set up over them.
(5) For GPS fixing, stations should be in an open location.
(6) Stations should be free from the risk of disturbance.
(7) Stations should be easily referenced.

Bench mark establishment

If there is not an Ordnance bench mark nearby, TBMs must be established. They must be:

(1) Centrally located.
(2) Free from the risk of subsidence or ground heave.
(3) Sharply defined so that a staff repeatedly held on one will be at the same level.

Where GPS is to be used, a number of points must be surveyed by satellite and by spirit-levelling to enable correspondence (between reference ellipsoid and geoid) to be achieved.

Linear measurement

The principles described in Chapter 4 should be followed, in particular:

(1) Equipment should be regularly calibrated.
(2) All necessary corrections should be applied.
(3) Accurate indexing of tapes and centring of tripods at stations should be carried out.

Angular measurement

The principles described in Chapter 3 should be followed, in particular:

(1) Theodolite permanent adjustments should be regularly checked.
(2) Instruments and targets should be accurately centred.
(3) Several rounds of angles should be taken and mean values calculated.

Levelling

The principles described in Chapter 2 should be followed, in particular:

(1) The instrument should be free from collimation error.
(2) Closure checks and cross runs should be carried out.
(3) Effects of errors from staff tilt, curvature and refraction must be eliminated.

5.2 COORDINATES

Rectangular coordinates are used for the plotting of control surveys. There are several reasons.

(1) Coordinate plotting eliminates angle (or bearing) plotting, a source of error.
(2) Each point is plotted directly from the axes, thus eliminating cumulative plotting errors.
(3) The arithmetic for survey adjustments can be carried out more easily on coordinates than on field measurements.
(4) Information can be stored in coordinate form.
(5) Surveys can be tied to existing coordinate systems.
(6) Land areas within traverses can easily be evaluated.
(7) On many construction sites setting out is specified in coordinate form.

Axes are taken north–south and east–west, the distances of a point from them being easting (east positive) and northing (north positive). The easting is always quoted before the northing. A coordinate system may be purely local with the axes arbitrarily chosen (on industrial construction sites often to suit the orientation of roads and structures) or the national system may be used. It is customary to position the origin south and west of the survey area to make all coordinates positive. In Great Britain, the National Grid has a false origin off the southwest coast. Surveys of limited areas may be taken as plane, assuming the earth to the flat. Where these surveys have to be related to the National Grid, and where surveys of larger areas are carried out, corrections will be required to allow for the earth's curvature. The connection of local surveys to the National Grid is explained later in this chapter.

Bearings

The coordinates of a point are determined by finding the differences in easting and northing from a point already coordinated. These differences are calculated from the polar coordinates of the joining line. While the length can be measured in the field, the bearing must be deduced from angle measurements.

Bearings are measured clockwise from a meridian or north line. (This contrasts with mathematical notation.) They were formerly called whole circle bearings to distinguish them from the now obsolete reduced (or quadrantal) bearings. Bearings are measured in degrees, minutes and seconds, values being in the range 0° to 360°; 90° represents due east, 180° due south and 270° due west (Fig. 5.1(a)).

A bearing refers to a line where a start and an end are specified. Alternatively, it can refer to a point (corresponding to line end) relating it to another point (= line start). Occasionally the term back bearing may be encountered; this is the bearing of a line taking the clockwise angle measured from a meridian at the end of the line. The bearing at the start may then be further described as a forward bearing. Thus in Fig. 5.1(b), the bearings of lines AB, CD, EF and GH are 80°, 212°, 320° and 121° respectively, while the back bearing of AB is 260°. The (forward) bearing and back bearing always differ by 180°. The back bearings of the other lines are 32°, 140° and 301° respectively.

It is sometimes useful in conversion calculations between polar and rectangular coordinates to know the quadrant of a line. The quadrant does not refer to the position of the line relative to the coordinate axes, but is determined from the line bearing. The 1st, 2nd, 3rd and 4th quadrants lie between 0° and 90°, 90° and 180°, 180° and 270° and between 270° and 360°/0°. In Fig. 5.1(b), line EF with a bearing of 320° is in the 4th quadrant.

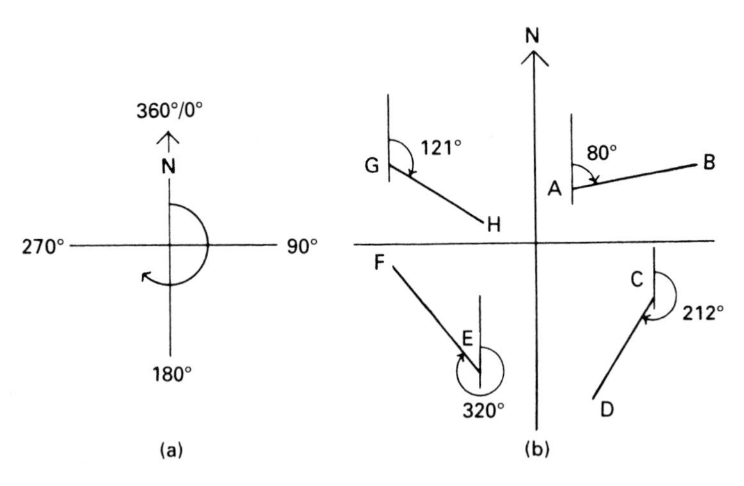

Fig. 5.1 Bearings.

Orienting bearing

Before any bearings can be evaluated, an initial orienting bearing must be determined. Where the survey is to be tied in to an existing coordinate system, two known points must be observed. The simplest method involves using one of the known points as a control station and taking sightings of the other with a theodolite. The bearing of the line formed can be found by a rectangular to polar conversion.

If an approximate orientation is adequate, the magnetic bearing of a line can be found using a prismatic compass. Although a correction for magnetic declination (difference between magnetic and true north) can be applied using data on an Ordnance Survey map, errors can arise from the lack of precision of bearing measurement (within 30 minutes) and the possible presence of local magnetic attraction.

If there is no requirement for precise or even approximate orientation, it can be assigned arbitrarily, perhaps giving one of the lines a suitable value. It is usually convenient for an arbitrary north to have some approximation to true north.

Polar to rectangular conversion: coordinate differences

For a line AB as in Fig. 5.2:

Easting difference = easting of B minus easting of A

$$\text{i.e. } \Delta E_{AB} = E_B - E_A$$

Northing difference = northing of B minus northing of A

$$\text{i.e. } \Delta N_{AB} = N_B - N_A$$

From an inspection of Fig. 5.2 it can be seen that:

ΔE = plan length of line × sine of bearing

ΔN = plan length of line × cosine of bearing

If a calculator is used, the correct sign should be given as in Table 5.1.

Probably the simplest way of determining coordinate differences is to use the polar to rectangular facility of a calculator. The exact sequence of operations will depend on the model of instrument, but the length should be entered before the bearing and the northing difference will be displayed before the easting difference.

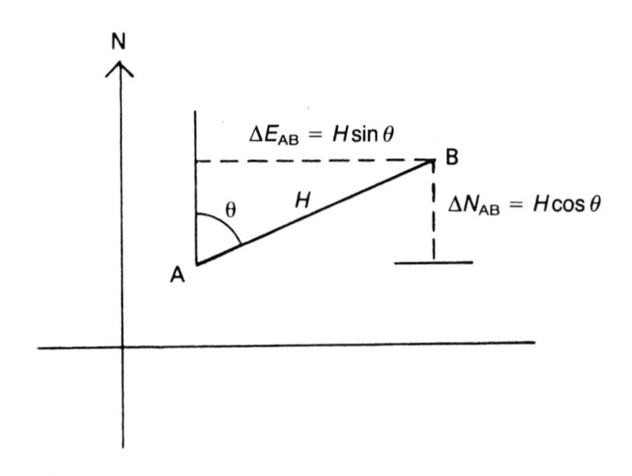

Fig. 5.2 Coordinate differences.

Table 5.1

Bearing	Quadrant	Easting difference	Northing difference
0° to 90°	1	+	+
90° to 180°	2	+	−
180° to 270°	3	−	−
270° to 360°	4	−	+

Example 5.1

Find the coordinate differences of a line of horizontal length 114.325 m and bearing 123°31′25″.

Solution

Either:

$$\Delta E = 114.325 \sin 123°31'25'' = 95.308\,\text{m}$$
$$\Delta N = 114.325 \cos 123°31'25'' = -63.139\,\text{m}$$

or:

Type 1 Calculator		Type 2 Calculator	
Press	$P \rightarrow R$		
Enter	114.325	Enter	114.325

Press	$X \leftrightarrow Y$	Press	$x \leftrightarrow y$
Enter	123°31′25″ (enter as a decimal)	Enter	123°31′25″
Press	=	Press	R
Read	−63.139477 (ΔN)	Read	−63.13947683 (ΔN)
Press	$X \leftrightarrow Y$	Press	$x \leftrightarrow y$
Read	95.307986 (ΔE)	Read	95.30798545 (ΔE)

The values should be rounded off to 95.308 m and −63.139 m.

Most scientific calculators have a sexagesimal conversion, so that angles can be entered as degrees, minutes and seconds in turn, pressing the sexagesimal button after each entry. The angle is then displayed as a decimal. If the conversion facility is not provided, the angle must be manually converted by entering the seconds and dividing by sixty, adding the minutes and dividing by sixty and finally adding the degrees. This process must be carried out before the conversion, the decimal value then stored in the memory and recalled when the bearing is to be entered.

Some older calculators will not give the negative sign for coordinate differences when the bearing is greater than 180°. A check in accordance with Table 5.1 should be carried out.

After the coordinate differences have been calculated, a check should be carried out either:

(1) By repeating the calculation, preferably by a different method. Or:
(2) By performing a rectangular to polar conversion.

Rectangular to polar conversion

Either calculate:

$$\text{Length} = [(\Delta E)^2 + (\Delta N)^2]^{1/2}$$

$$\text{Bearing} = \tan^{-1}\frac{\Delta E}{\Delta N}$$

The bearing value will be between −90° and +90°. To convert to a (whole circle) bearing:

if ΔN is positive, add 360° to a negative value
if ΔN is negative, add 180°.

Or use rectangular to polar conversion on a calculator. The procedure is similar to that for polar to rectangular conversion, the northing difference being entered before the easting difference and the length being read before the bearing. To enter a negative difference, enter the value, then press the +/− button.

The bearing will be quoted in one of three ways; the surveyor must ascertain which one.

(1) Value between −90° and +90°; use the method previously described to find the bearing.
(2) Value between −180° and +180°; 360° should be added to negative values.
(3) Value between 0° and 360°.

On some calculators the bearing must be stored and recalled before 180° or 360° can be added to it. An inspection should always be carried out to see that the correct bearing has been evaluated.

Example 5.2

Stations P and Q have rectangular coordinates of (1357.275 m E; 1482.360 m N) and (825.980 m E; 1035.715 m N) respectively. Find the length and bearing of PQ.

Solution

Referring to Fig. 5.3:

$$\Delta E = E_Q - E_P = -531.295\,\text{m}$$
$$\Delta N = N_Q - N_P = -446.645\,\text{m}$$

Either:

$$\text{PQ length} = [(-531.295)^2 + (-446.645)^2]^{1/2} = 694.094\,\text{m}$$
$$\text{bearing} = \tan^{-1}\frac{-531.295}{-446.645} = 49°56'50''$$

As N is negative, add 180° to give 229°56′50″.

Or:

Press $R \rightarrow P$
Enter −446.645
Press $X \leftrightarrow Y$
Enter −531.295
Press =

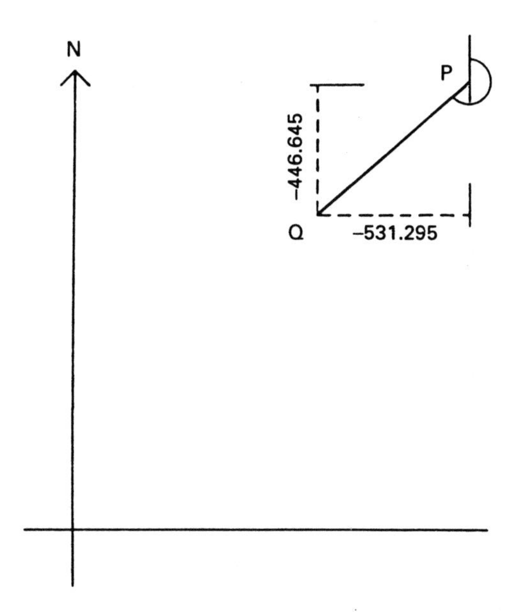

Fig. 5.3 Rectangular/polar conversion.

Read 694.094
Press $X \leftrightarrow Y$
Read −130°03′10″
Add 360° to bearing to give 229°56′50″.

5.3 TRAVERSING

Traversing is one of the traditional methods of carrying out a control survey in plan. Stations are set out to define a series of traverse lines or legs, the plan lengths of which can be measured as can the angles between pairs of lines at each station. There are three types of traverse (Fig. 5.4):

(1) Closed loop traverse, where the legs form a closed polygon.
(2) Closed tied (or connecting or link) traverse, where the traverse runs between two stations of known position.
(3) Open traverse, where the lines, although starting from a known position, do not finish at one.

Closed traverses provide a check on the validity and accuracy of field measurements. The loop traverse is suitable for many engineering

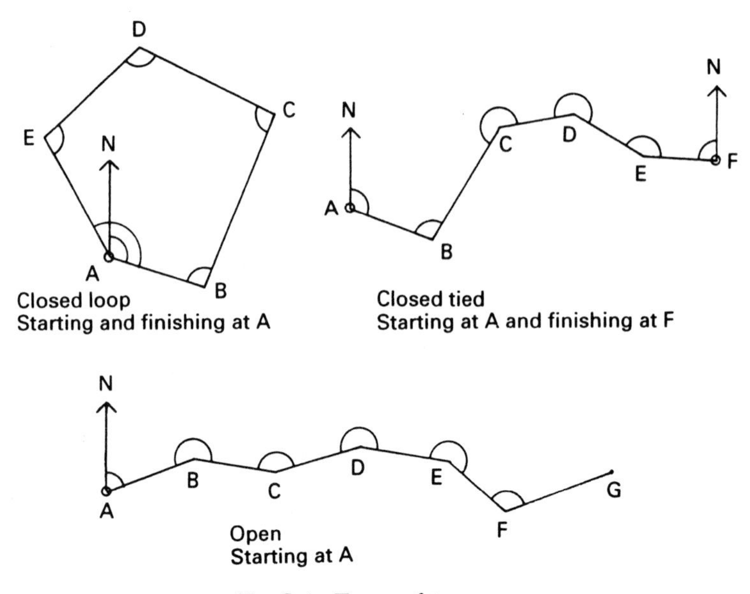

Fig. 5.4 Types of traverse.

surveys, although for road, railway and pipeline projects the tied traverse is more appropriate. Open traverses, providing no check, are not recommended; they may find application in underground surveying although it is better practice to return the traverse along the tunnel (preferably using different stations) to the starting point.

Angles are measured by theodolite, distances being determined by taping or by EDM. The development of EDM has contributed to the prominence of traversing; in addition to the greater ease in measuring distances, measurements of distance and angle can be carried out almost concurrently and with comparable precision.

Coordinate plotting is used, checks and adjustments being made during the calculation process. A loop traverse will include a station of known (or assumed) coordinates with an orienting bearing available. In a tied traverse two stations of known coordinates are necessary and, for an angle check to be possible, orienting bearings at both stations should be available.

Station selection

The criteria listed in Section 5.1 apply, with the following additional requirements:

(1) The stations should form a traverse of suitable shape.
(2) Only neighbouring stations along traverse lines need be intervisible.
(3) Where traverse legs are to be taped, the ground should be accessible.
(4) Traverse legs should be approximately equal in length.
(5) Existing stations and reference objects should be incorporated.

Stations should be firmly and clearly marked out and strongly referenced.

Distance measurement

The principles mentioned in Section 5.1 and described in Chapter 4 should be followed. If EDM is being used, two legs may be measured from one station.

Angle measurement

There are some aspects additional to those covered in Section 5.1 and in Chapter 3 that should be considered.

Ideally, angles should be measured in sequence progressing around the traverse from the start station. Frequently this is not convenient and care must be taken to measure all angles (not forgetting the one at the closing station). Angles should be properly recorded, so that it is known whether an angle is internal or external.

It is common to use 'three tripod' equipment when traversing. The theodolite can be detached from its tribrach and exchanged with a target, similarly detachable. While the theodolite is set up at the 'instrument station' (B), at the 'back station' (A) and 'forward station' (C) a tripod, tribrach, optical plummet and target are set up. After observation of the angle $A\hat{B}C$, the complete target assembly at A is transferred to station D, and the theodolite at B and target at C are detached from their tribrachs and exchanged. Angle $B\hat{C}D$ is then measured (Fig. 5.5(a)).

Three tripod equipment is always used when distances are measured electromagnetically. A combined prism and target is set up at each station sighted. Three tripod equipment eliminates centring errors to the extent that, at each station, measurements are taken to and from the same point in plan. There can, of course, still be errors in station

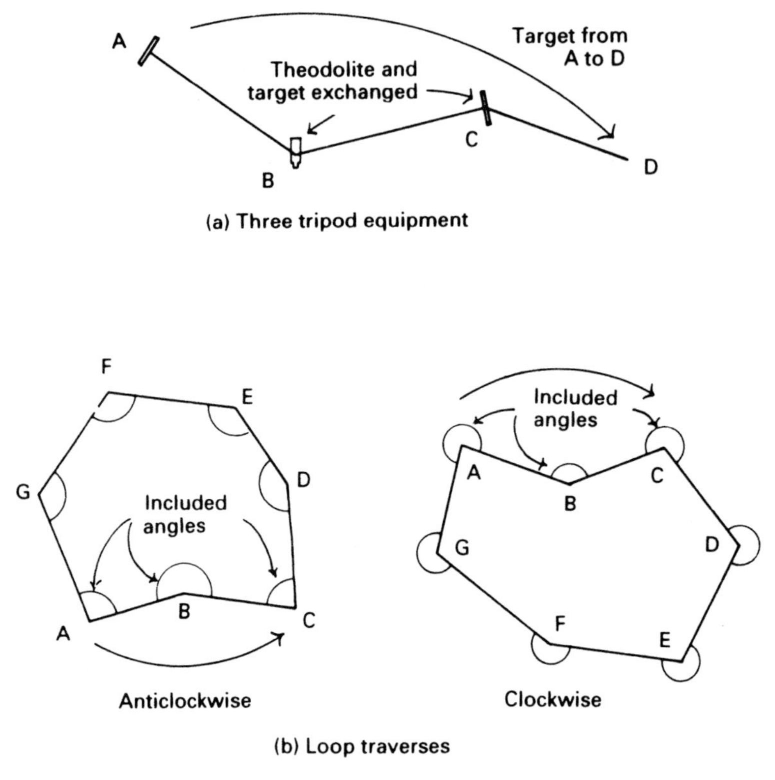

Fig. 5.5 Traversing.

coordinates if the targets have not been accurately plumbed over the stations.

Included angles

For consistency and ease of calculating bearings, the included angle should be measured at each station. This angle is the clockwise angle swept out when turning from the back to the forward station, i.e. at B sighting A then turning clockwise to sight C. (It does not matter whether the theodolite is actually turned clockwise or not, the circle being graduated in a clockwise sense.)

In a loop traverse carried out in an anticlockwise sense, the included angles will be internal; a clockwise one will produce external angles (Fig. 5.5(b)).

Angle closure check

Once all the angles of a closed traverse have been measured, a check should be carried out. This can often be performed in the field, so that in the event of an unacceptable misclosure, remeasurement can be commenced without delay.

Loop traverse check

Anticlockwise traverse:

Sum of included angles should equal $(n - 2) \times 180°$ (internal)

Clockwise traverse:

Sum of included angles should equal $(n + 2) \times 180°$ (external)

for a traverse round n stations. (The orienting angle should not be used.) In the infrequent event of the traverse lines crossing, the sum should equal $n \times 180°$ for an *odd* number of crossings.

Tied traverse check

Sum of included angles (counting orienting angles) should equal final orienting bearing minus initial orienting bearing plus

$(n + 1) \times 180°$ usually
$(n - 1) \times 180°$ sometimes
$(n + 3) \times 180°$ infrequently

for a traverse of n stations.

Alternatively, the bearings of the traverse legs can be successively calculated from the unadjusted angles and the deduced value of the final orienting bearing compared with the known value.

Inevitably there will be some misclosure. Provided that the amount is acceptably small, the angles can be adjusted so that geometrical consistency can be achieved. (Otherwise two different sets of bearings could be assigned to lines depending on the direction of calculation.)

Acceptable misclosure

The closing errors will be affected by the precision of the equipment and the skill and care of the surveyor. Acceptable errors will depend

Table 5.2

Precision of survey	Acceptable angular misclosure	Acceptable linear misclosure
Geodetic	$2(n)^{1/2''}$	1 in 20 000
Secondary	$10(n)^{1/2''}$	1 in 10 000
Tertiary	$20(n)^{1/2''}$	1 in 5 000
Fourth order	$60(n)^{1/2''}$	1 in 2 000

on the precision required of the survey. Recommended values for n stations are shown in Table 5.2.

Second order precision is appropriate for many engineering surveys. Geodetic precision may be necessary in tunnelling work, while on a construction site of limited area tertiary precision may suffice.

If the angular misclosure is acceptable, the error is distributed before the calculation of the bearings, unless a least squares adjustment is to be performed. The adjustment is usually applied equally at each station, unless there are reasons for apportioning greater adjustments at some stations than at others.

Bearing calculation

The bearing of one line, preferably at the start station, must be known. Taking the lines in sequence (AB, BC, CD, ...):

$$\begin{aligned} \text{Bearing of line} = {} & \text{Bearing of previous line} + \text{included angle} \\ & +180^\circ \text{ if sum} < 180^\circ \\ & -180^\circ \text{ if } 180^\circ < \text{sum} < 540^\circ \\ & -540^\circ \text{ if sum} > 540^\circ \end{aligned}$$

In a closed traverse a check on the arithmetic should be carried out. Bearings should be successively calculated around the traverse. In a loop traverse the bearing of the first line, calculated successively, should equal its initial value. In a tied traverse the known and deduced values of the final orienting sight should be equal.

Example 5.3

A closed loop traverse ABCDEFA starts at A where X, the reference object, has a bearing of 52°32′00″. Find the angular misclosure, adjust the angles equally and calculate bearings of the traverse legs.

Included angle	Value
XÂB	87°41′30″
FÂB	71°22′45″
AB̂C	131°36′35″
BĈD	147°24′20″
CD̂E	171°16′55″
DÊF	91°22′15″
EF̂A	106°57′40″

Solution

Figure 5.6 is a plan.

Angle sum:

71°22′45″	→	71°22′40″
131°36′35″	→	131°36′30″
147°24′20″	→	147°24′15″
171°16′55″	→	171°16′50″
91°22′15″	→	91°22′10″
106°57′40″	→	106°57′35″
Σ 720°00′30″	→	720°00′00″

Closing error = 30″, i.e. $12.2(6)^{1/2}$, in between secondary and tertiary precision.

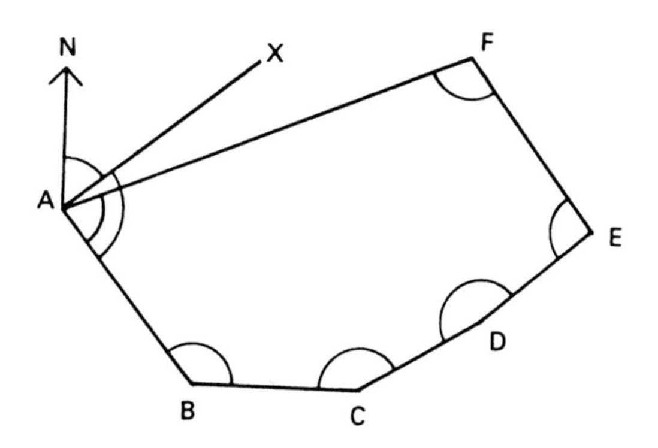

Fig. 5.6 Loop traverse example.

The error has been distributed by subtracting 5″ from each angle.

Bearing of AB = 52°32′00″ + 87°41′30″ = 140°13′30″
Bearing of BC = 140°13′30″ + 131°36′30″ (= 271°50′00″) − 180° = 91°50′00″
Bearing of CD = 91°50′00″ + 147°24′15″ (= 239°14′15″) − 180° = 59°14′15″
Bearing of DE = 59°14′15″ + 171°16′50″ (= 230°31′05″) − 180° = 50°31′05″
Bearing of EF = 50°31′05″ + 91°22′10″ (= 141°53′15″) + 180° = 321°53′15″
Bearing of FA = 321°53′15″ + 106°57′35″ (= 428°50′50″) − 180° = 248°50′50″
Bearing of AB = 248°50′50″ + 71°22′40″ (= 320°13′30″) − 180° = 140°13′30″ (check)

Example 5.4

A tied traverse PQRST has been run between stations P and T where orienting sights have been taken to X and Y. The coordinates of X, P, T and Y have previously been determined. Find the angular misclosure, adjust the angles and calculate bearings of the traverse legs.

Station	Easting	Northing
X	1736.265	2227.865
P	1974.870	1906.550
T	843.350	2068.485
Y	1039.220	2171.715

Included angle	Value
X$\hat{P}$Q	241°52′15″
P$\hat{Q}$R	243°21′20″
Q$\hat{R}$S	219°47′35″
R$\hat{S}$T	238°22′50″
S$\hat{T}$Y	235°24′05″

Solution

Figure 5.7 is a plan.

For PX, ΔE = 1736.265 − 1974.870 = −238.605 m
ΔN = 2227.865 − 1906.550 = 321.315 m
From rectangular to polar conversion, bearing = −36°35′50″
Adding 360° gives 323°24′10″
For TY, ΔE = 1039.220 − 843.350 = 195.870 m
ΔN = 2171.715 − 2068.485 = 103.230 m
From $R \rightarrow P$, bearing = 62°12′33″

Either:

Bearing of PQ = 323°24′10″ + 241°52′15″ − 360° = 205°16′25″
Bearing of QR = 205°16′25″ + 243°21′20″ (= 448°37′45″) − 180° = 268°37′45″
Bearing of RS = 268°37′45″ + 219°47′35″ (= 448°25′20″) − 180° = 308°25′20″
Bearing of ST = 308°25′20″ + 238°22′50″ (= 546°48′10″) − 540° = 6°48′10″
Bearing of TY = 6°48′10″ + 235°24′05″ (= 242°12′15″) − 180° = 62°12′15″

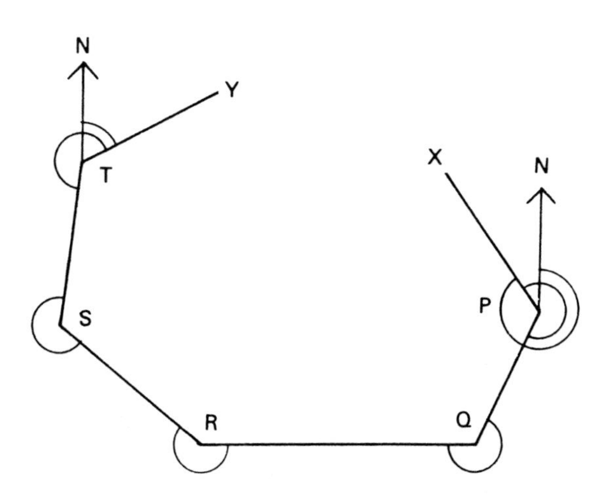

Fig. 5.7 Tied traverse example.

Actual value of bearing of TY = 62°12′33″; error = −18″, i.e. $8.0(5)^{1/2}$, in between geodetic and secondary precision.

A +18″ adjustment may be divided between the five stations +4″, +3″, +4″, +3″, +4″ to keep integral values of seconds.

Applying the adjustments,

Adjusted angles		Adjusted bearings
$X\hat{P}Q$ = 241°52′19″	recalculating the bearings as before	PQ = 205°16′29″
$P\hat{Q}R$ = 243°21′23″		QR = 268°37′52″
$Q\hat{R}S$ = 219°47′39″		RS = 308°25′31″
$R\hat{S}T$ = 238°22′53″		ST = 6°48′24″
$S\hat{T}Y$ = 235°24′09″		TY = 62°12′33″ (check)

Or:

Sum of included angles 241°52′15″
243°21′20″
219°47′35″
238°22′50″
235°24′05″

Σ 1178°48′05″

Final bearing − initial bearing + $(n + 1) \times 180°$
= 62°12′33″ − 323°24′10″ + 1080°
= 818°48′23″

Clearly this is nowhere near to 1178°48′05″! However, it may be observed that the difference is almost 360°; therefore this is an occasion when $(n + 3) \times 180°$ must be added.

818°48′23″ + 360° = 1178°48′23″

Sum of measured angles = 1178°48′05″

The error can now be seen to be −18″, and may be distributed as before.

Station coordinates

Station coordinates are calculated by successively adding the line coordinate differences to the coordinates of the start station. In closed traverses, the sums of the easting differences and of the northing differences are first calculated to assess the linear closing error; if acceptably small, it is distributed by adjusting the coordinate differences. For an open traverse, the station coordinates can be calculated as soon as the coordinate differences have been evaluated. Before the adjustment of a closed traverse is considered, an example of an open traverse will be given.

Example 5.5

An open traverse GHJKLM has been carried out. The coordinates of G are (1523.635 m E; 720.316 m N) and an orienting sight has been taken to A, which has a bearing of 149°23′00″ from G.

Included angle	Value	Leg	Horiz length
$A\hat{G}H$	63°27′15″	GH	151.360
$G\hat{H}J$	201°22′43″	HJ	175.227
$H\hat{J}K$	153°30′45″	JK	186.105
$J\hat{K}L$	177°51′47″	KL	150.885
$K\hat{L}M$	183°15′22″	LM	198.432

Calculate the coordinates of H, J, K, L and M, and find the distance and bearing of M from G.

Solution

There is no angle check, nor check on coordinate differences.

Figure 5.8 is a plan. Typical calculations:

$$\text{Bearing of GH} = 149°23'00'' + 63°27'15'' = 212°50'15''$$

$$\text{Bearing of HJ} = 212°50'15'' + 201°22'43'' - 180° = 234°12'58''$$

$$\text{For GH,} \quad \Delta E = 151.360 \sin 212°50'15'' = -82.076$$

$$\Delta N = 151.360 \cos 212°50'15'' = -127.174$$

(or use $P \rightarrow R$ function)

$$\text{Station coordinates of H } E_{\text{H}} = 1523.625 - 82.076 = 1441.549$$

$$N_{\text{H}} = 720.316 - 127.174 = 593.142$$

The calculations are entered in Table 5.3.

The distance and bearing of M from G may be deduced from the coordinate differences.

Table 5.3

Stn	Horizontal line	length	Included angle	Bearing	Coord diff ΔE	ΔN	Station coords E	N
G			63°27′15″				1523.625	720.316
	GH	151.360		212°50′15″	−82.076	−127.174		
H			201°22′43″				1441.549	593.142
	HJ	175.227		234°12′58″	−142.149	−102.460		
J			153°30′45″				1299.400	490.682
	JK	186.105		207°43′43″	−86.592	−164.733		
K			177°51′47″				1212.808	325.949
	KL	150.885		205°35′30″	−65.176	−136.082		
L			183°15′22″				1147.632	189.867
	LM	198.432		208°50′52″	−95.740	−173.808		
M							1051.892	16.059
					Σ −417.733	−704.257		

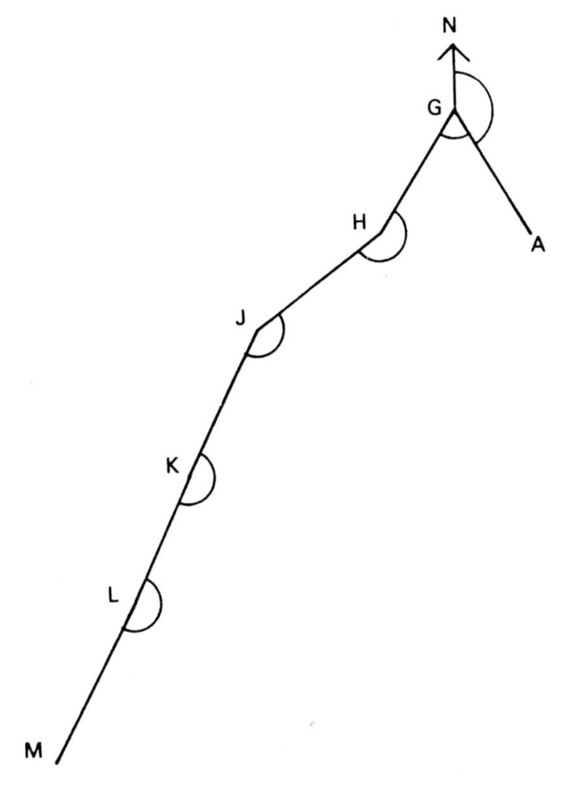

Fig. 5.8 Open traverse example.

$\Delta E = -471.733$ Using rectangular → polar conversion

$\Delta N = -704.257$

$$\text{Length} = 847.650\,\text{m}$$

$$\text{Bearing} = -146°11'05'' \rightarrow 213°48'55''$$

Misclosure and adjustments: Bowditch's method

In a closed traverse, a check is made on the linear misclosure when the coordinate differences are determined. The misclosures in eastings e_E and northings e_N are calculated, the linear misclosure being:

$$e = [(e_E)^2 + (e_N)^2]^{1/2}$$

The (fractional) linear misclosure = 1 in $\Sigma H/e$ where ΣH is the sum of the line lengths.

For a loop traverse, $\Sigma\Delta E$ and $\Sigma\Delta N$ should equal zero, thus:

$$e_E = \Sigma\Delta E \text{ and } e_N = \Sigma\Delta N$$

For a tied traverse with start station coordinates (E_A; N_A) and end station coordinates (E_Z; N_Z):

$$e_E = \Sigma\Delta E - (E_Z - E_A) \text{ and } e_N = \Sigma\Delta N - (N_Z - N_A)$$

Provided that the linear misclosure is within the acceptable limit (Table 5.2), adjustments may be applied to remove the error, thus making all derived coordinates consistent.

The best known traverse adjustment in engineering surveys is the Bowditch method. It assumes that the closing error has been generated at a constant rate around the traverse. Hence, at any point, the direction of adjustment will have a bearing 180° from the line of misclosure and the magnitude of the adjustment will be in proportion to the distance around the traverse from the start. The adjustment is applied to the coordinate differences in proportion to the respective line lengths, the sign being opposite to that of the error, thus:

$$\text{Adjustment to } \Delta E_{AB} = -e_E \times \frac{H_{AB}}{\Sigma H}$$

$$\text{Adjustment to } \Delta N_{AB} = -e_N \times \frac{H_{AB}}{\Sigma H}$$

H_{AB} being the (horizontal) line length of AB.

It can be seen that the total adjustments are $-e_E$ and $-e_N$, removing all closing error. An integral part of the Bowditch method is the initial adjustment of angles, as previously described.

Example 5.6

Find the fractional linear misclosure of the closed loop traverse ABCDA, and calculate coordinates of B, C and D using the Bowditch method of adjustment.

Coordinates of A are (1739.565 m E; 2375.652 m N)
Bearing of H (R.O.) from A is 333°21′47″

Included angle	Value	Leg	Horizontal Length (m)
HÂB	147°27′36″		
		AB	135.365
AB̂C	261°22′37″		
		BC	201.781
BĈD	263°31′03″		
		CD	208.553
CD̂A	293°06′56″		
		DA	256.417
DÂB	261°59′10″		

Solution

Figure 5.9 is a plan.

Bearing of AB = 333°21′47″ + 147°27′36″ − 360°00′00″ = 120°49′23″. The calculations are entered in Table 5.4.

Typical calculations:

Angle check: $(n + 2)180° = 1080°$; actual sum = 1079°59′46″
Error = −14″. $10(n)^{1/2} = 20''$; ∴ error is acceptable.
Adjust by +3″, +4″, +3″, +4″. Note check of adjusted angles.

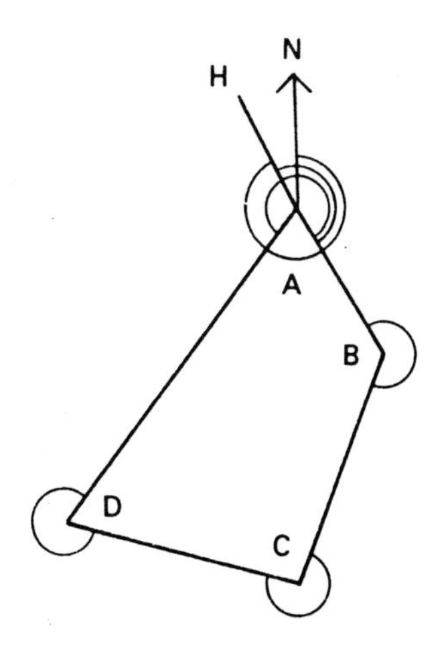

Fig. 5.9 Loop traverse example.

Table 5.4

Line and plan length (m)	Included angles measured	Included angles adjusted	Bearing	Coord diffs ΔE	Coord diffs ΔN	Adjustment to ΔE	Adjustment to ΔN	Adjd coord diffs ΔE	Adjd coord diffs ΔN	Station coords E	Station coords N	Stn
										1739.565	2375.652	A
AB 135.365			120°49′23″	116.245	−69.359	−0.008	−0.009	116.237	−69.368			
	261°22′37″	261°22′40″								1855.802	2306.284	B
BC 201.781			202°12′03″	−76.244	−186.822	−0.011	0.014	−76.255	186.836			
	263°31′03″	263°31′09″								1779.547	2119.448	C
CD 208.553			285°43′10″	−200.753	56.503	−0.012	−0.015	−200.765	56.488			
	293°06′59″	293°06′59″								1578.782	2175.936	D
DA 256.417			38°50′09″	160.797	199.734	−0.014	−0.018	160.783	199.716			
	261°59′14″	261°59′14″								1739.565	2375.652	A
Σ 802.116	1079°59′46″	1080°00′00″	120°49′23″ AB check	0.045	0.056	−0.045	−0.056	0.000	0.000			

Bearing of BC: 120°49′23″ + 261°22′40″ − 180°00′00″ = 202°12′03″
Note check on final bearing.

For AB $\Delta E = 116.245\,\text{m}$, $\Delta N = -69.359$ using $P \rightarrow R$ conversion.

Linear closing error: $\Sigma\Delta E = 0.045\,\text{m} = e_E$, $\Sigma\Delta N = 0.056\,\text{m} = e_N$

$$\therefore e = [(0.045)^2 + (0.056)^2]^{1/2} = 0.0718\,\text{m}$$

Fractional linear misclosure: $= 1 \text{ in } \dfrac{802.116}{0.0718} = 1 \text{ in } 11\,165$
which is acceptable.

$$\text{Adjustments: AB } \delta E_{AB} = \frac{-0.045 \times 135.365}{802.116} = -0.008$$

$$\delta N_{AB} = \frac{-0.056 \times 135.365}{802.116} = -0.009$$

The sum of the adjustments is checked before the adjusted coordinate differences are calculated; rounding off errors of ±0.001 m may occur and can be neglected.

$$\text{Adjusted } \Delta E_{AB} = 116.245 - 0.008 = 116.237;$$
$$\Delta N_{AB} = -69.359 - 0.009 = -69.368$$

The sums of the adjusted coordinate differences should be checked for arithmetic errors.

$$\text{Station coordinates: } E_B = 1739.565 + 116.237 = 1855.802\,\text{m}$$
$$N_B = 2375.652 - 69.368 = 2306.284\,\text{m}$$

Example 5.7

Applying the Bowditch method of adjustment, calculate the coordinates of stations H, J and K, forming a tied traverse between stations G and L.

Coordinates of G (2675.320 m E; 3072.779 m N)
L (3289.563 m E; 2936.421 m N)
Orienting bearing GX 341°37′30″
LY 201°52′27″

Included angle	Value	Leg	Horizontal length (m)
XĜH	88°52′29″		
		GH	173.922
GĤJ	206°52′57″		
		HJ	179.531
HĴK	149°44′05″		
		JK	124.246
JK̂L	257°12′03″		
		KL	270.540
KL̂Y	237°33′50″		

Solution

Figure 5.10 is a plan.

Angle check:

Bearing GH = 341°37′30″ + 88°52′29″ − 360° = 70°29′59″
Bearing HJ = 70°29′59″ + 206°52′57″ − 180° = 97°22′56″
Bearing JK = 97°22′56″ + 149°44′05″ − 180° = 67°07′01″
Bearing KL = 67°07′01″ + 257°12′03″ − 180° = 144°19′14″
Bearing LY = 144°19′04″ + 237°33′50″ − 180° = 201°52′54″
Error = 201°52′54″ − 201°52′27″ = +27″; $10(n)^{1/2}$ = 22.4″.

The error is just outside the limit; in most instances it would be cousidered acceptable.

∴ Adjust angles by −5″, −6″, −5″, −6″, −5″

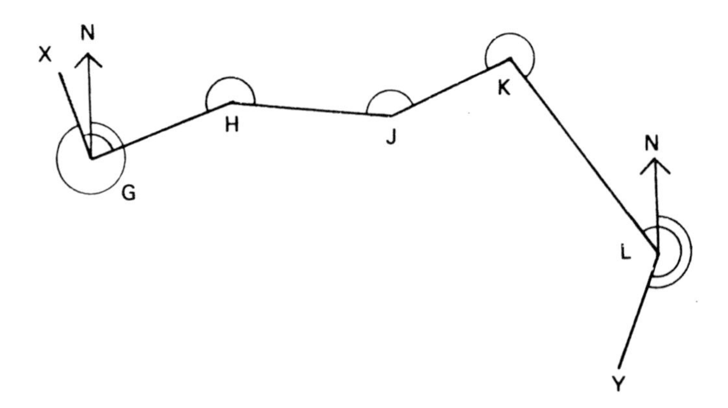

Fig. 5.10 Tied traverse example.

Table 5.5

Line and plan length (m)		Included angles			Coord diffs		Adjustment		Adjd coord diffs		Station coords		
		measured	adjusted	Bearing	ΔE	ΔN	to ΔE	to ΔN	ΔE	ΔN	E	N	Stn
GX				341°37′30″									
		88°52′29″	88°52′24″								2675.320	3072.779	G
GH	173.922			70°29′54″	163.944	58.061	−0.008	+0.012	163.936	58.073			
		206°52′57″	206°52′51″								2839.256	3130.852	H
HJ	179.531			97°22′45″	178.044	−23.058	−0.009	+0.012	178.035	−23.046			
		149°44′05″	149°44′00″								3017.291	3107.806	J
JK	124.246			67°06′45″	114.464	48.322	−0.006	+0.008	114.458	48.330			
		257°12′03″	257°11′57″								3131.749	3156.136	K
KL	270.540			144°18′42″	157.827	−219.733	−0.013	+0.018	157.814	−219.715			
		237°33′50″	237°33′45″								3289.563	2936.421	L
Σ	748.239		940°14′57″	201°52′27″	614.279	−136.408	−0.036	+0.050	614.243	−136.358			

Coordinate difference GL:

$$\Delta E = 3289.563 - 2675.320 = 614.243$$
$$\Delta N = 2936.421 - 3072.779 = -136.358$$

The calculations are entered in Table 5.5. Calculations are performed in the same manner as for the loop traverse.

Angle check:

$$201°52'27'' - 341°37'30'' + 6 \times 180° = 940°14'57''$$

which checks

Linear misclosure:

$$e_E = 614.279 - 614.243 = 0.036$$
$$e_N = -136.408 + 136.358 = -0.050$$

(coordinate differences minus known differences)

$$e = [(0.036)^2 + (0.050)^2]^{1/2} = 0.0616$$

Fractional linear misclosure:

$$= 1 \text{ in } \frac{748.239}{0.0616} = 1 \text{ in } 12144$$

Traverse adjustments: other methods

The Bowditch method was evolved when traversing was carried out by compass and chain. It makes assumption that the effects of errors in bearing and linear measurement are similar. Nowadays bearings are deduced from angles measured with a theodolite. By applying the Bowditch method, the angles are adjusted twice: initially to satisfy the summation condition, and again in the main part of the adjustment. With the adjustments sometimes having a cumulative effect and sometimes having a compensating effect, the adjustments to some angles may be disproportionately large. Tied traverses running more or less in a straight line suffer this effect, the opening and closing angles being adjusted much more than the others.

A further disadvantage is that angle and distance measurements are not usually of comparable precision. In a tape and theodolite traverse, angles may well be determined with more precision than distances; in an EDM traverse, the opposite may occur. An ideal procedure would adjust accordingly.

When closing errors are small, the method of adjustment (provided that it has some logical foundation) is often immaterial. Large errors indicate a mistake in the field through poor technique or faulty equipment, and some or all of the fieldwork should be repeated.

If a more rigorous adjustment is required, the '*method of least squares*' should be used. Features of this method are that angles are not initially adjusted, and that comparative weightings of angle and distance measurement can be applied. The principles are covered in Chapter 7. In practice, a computer adjustment would be performed, using a *variation of coordinates* program suitable for all types of control survey.

Gross errors

If a large angular or linear misclosure is found in a traverse, a gross error in fieldwork or calculations has occurred. The first step in locating such an error is to check the calculations. Gross errors in fieldwork may be revealed if cross measurements have been taken in loop traverses. It will, however, be difficult to incorporate cross measurements in adjustment arithmetic unless a computer least squares solution is possible.

Gross angular errors

A single gross angular error can be located by calculating the coordinates in anticlockwise and clockwise directions for a loop traverse and from each end for a tied traverse. The station where the error has occurred will have the same coordinates. Confusion between internal and external angles close to 180° is a not infrequent cause of angular discrepancy.

Gross linear errors

If a single gross linear error has occurred, the faulty line will be parallel to the linear misclosure.

Two or more gross errors

Two or more gross errors may be difficult to locate. Calculations and approximate plotting may enable a surveyor to estimate where the errors lie.

5.4 INTERSECTION

A point may be located in plan by two intersecting sights by theodolite from the ends of a baseline of accepted length. The triangle formed should be well-conditioned. The unmeasured horizontal angle is deduced so that the sine rule can be applied to determine one or both distances from the fixed stations at the end of the baseline. Provided that coordinates of one station and the baseline bearing are known, coordinates of the intersected point can be calculated (Fig. 5.11(a)).

There is no check on the fieldwork, and horizontal angles should be measured with care. A better procedure is to measure angles for intersecting sights from three (or more) stations (Fig. 5.11(b)). With the incorporation of redundant measures, a *variation of coordinates* adjustment should be carried out. Vertical angles from two stations will also provide a redundant measure.

The method is employed for precise measurement by a non-contact method of industrial components. Theodolites at the ends of a short baseline measure horizontal angles; the results are analysed by a computer linked to both instruments. Such methods are known as

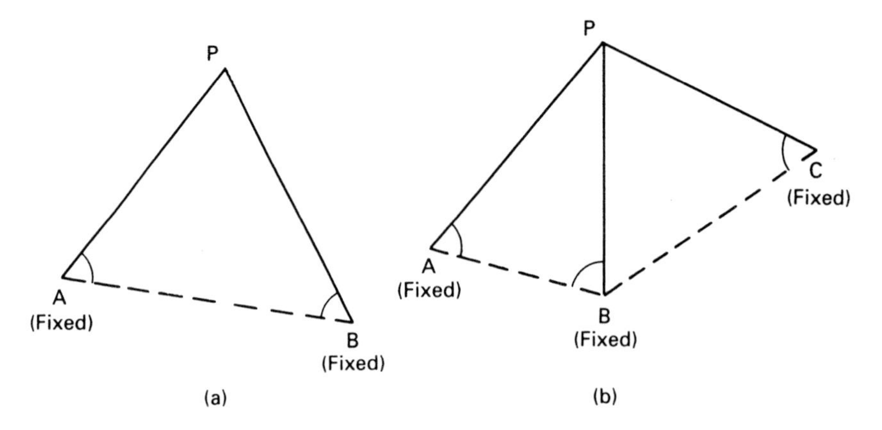

Fig. 5.11 Intersection.

Coordinate Measuring Systems, or Industrial Measuring Systems and are described further in Chapter 13.

Terrestrial Photogrammetry with converging axes uses the principles of intersection in determining the positions of points surveyed.

5.5 RESECTION

An instrument station can be located by measuring angles subtended by three distant stations whose positions are known. The method has been useful for relating a station to triangulated points and for locating a station on an offshore structure. Both operations can now often be more easily performed using EDM.

If the four stations involved are concyclic, there can be no solution; if they are nearly concyclic, the 'fix' of the instrument station will be weak. Accordingly, either the unknown station should lie within the fixed stations, or the fixed three should present a convex formation to the unknown point.

In Fig. 5.12(a), A, B, and C are three known stations observed from P. Lengths l_1 and l_2, and angle c will be known (or can be deduced from coordinates). Angles α and β are measured at P.

Let unknown angle $\text{C}\hat{\text{A}}\text{P} = \theta$. Then $\text{P}\hat{\text{B}}\text{C} = 360° - (\alpha + \beta + \theta + c)$

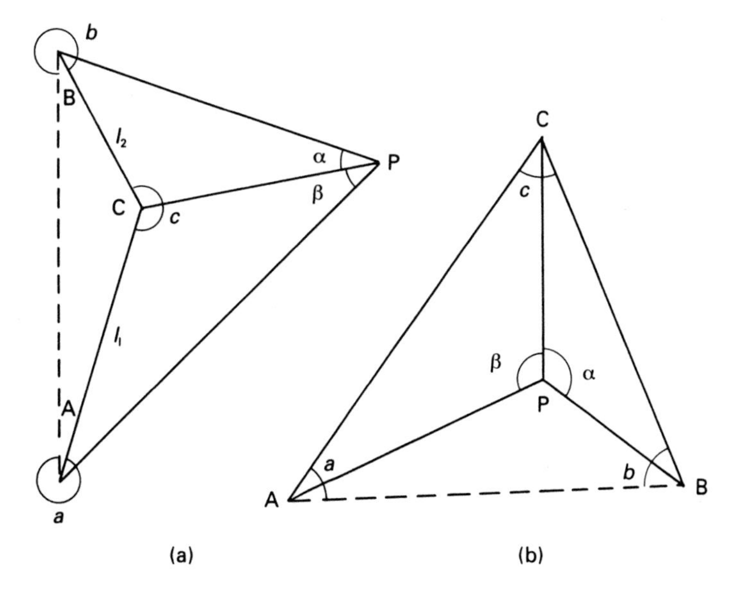

Fig. 5.12 Resection.

$$\text{Then } \frac{\text{CP}}{\sin\theta} = \frac{l_1}{\sin\beta} \text{ and } \frac{\text{CP}}{\sin(360 - \alpha - \beta - \theta - c)} = \frac{l_2}{\sin\alpha}$$

$$\therefore \frac{l_1 \sin\theta}{\sin\beta} = \frac{-l_2 \sin(\alpha + \beta + \theta + c)}{\sin\alpha}$$

$$l_1 \sin\theta \sin\alpha = -(l_2 \sin\beta)\,[\sin(\alpha + \beta + c)\cos\theta + \cos(\alpha + \beta + c)\sin\theta]$$

$$\text{Hence } \theta = \tan^{-1}\frac{-l_2 \sin\beta \sin(\alpha + \beta + c)}{l_1 \sin\alpha + l_2 \sin\beta \cos(\alpha + \beta + c)}$$

Triangle APC can be solved and P related to A, B, and C.

Example 5.8

A theodolite station P is to be located from stations A, B, and C by resection. From the following data, determine the coordinates of P.

Station	Coordinates	Angle	Value
A	(1036.279 m E; 742.645 m N)		
		AP̂C	28°13′55″
B	(987.550 m E; 897.042 m N)		
		CP̂B	26°42′35″
C	(1047.389 m E; 856.691 m N)		

Solution

Figure 5.12(a) applies. The lengths and bearings of CA and CB must be found using a rectangular – polar conversion.

For CA $\Delta E = -\ 11.110$ m Length = 114.586 m
$\Delta N = -114.046$ m Bearing = 185°33′50″

For CB $\Delta E = -\ 59.839$ m Length = 72.173 m
$\Delta N = 40.351$ m Bearing = 303°59′34″

Angle BĈA = 360° + 185°33′50″ − 303°59′34″
= 241°34′16″

Then tan CÂP =

$$\frac{-72.173 \sin 28°13'55'' \sin(+\ 28°13'55' + 26°42'35'' + 241°34'16'')}{114.586 \sin 26°42'35'' + 72.173 \sin 28°13'55'' \cos(+\ 28°13'55'' + 26°42'35'' + 241°34'16'')}$$

$\therefore$ CÂP = 24°35′42″

Then $\quad P\hat{C}A = 180°00'00'' - (24°35'42'' + 28°13'55'') = 127°20'23''$

Bearing $\quad AP = 05°33'50'' + 24°35'42'' = 30°09'32''$

$$\text{Length} \quad AP = \frac{AC \sin P\hat{C}A}{\sin A\hat{P}C} = \frac{114.586 \sin 127°10'23''}{\sin 28°13'55''} = 193.041\,\text{m}$$

For AP, $\Delta E = 96.970\,\text{m};\ E_P = 1036.279 + 96.970$

$\Delta N = 116.887\,\text{m};\ N_P = 742.645 + 166.887$

Coordinates of P are (1133.249 m E; 909.532 m N).

An alternative solution working in coordinates is provided by the Tienstra formula:

$$E_P = \frac{K_1 E_A + K_2 E_B + K_3 E_C}{K_1 + K_2 + K_3};\ N_P = \frac{K_1 N_A + K_2 N_B + K_3 N_C}{K_1 + K_2 + K_3}$$

$$\text{where} \quad K_1 = \frac{1}{\cot \alpha - \cot a},\ K_2 = \frac{1}{\cot \beta - \cot b},\ K_3 = \frac{1}{\cot \gamma - \cot c}$$

The labelling of the angles is important; Fig. 5.12(a) and (b) shows the labelling for point P without and within the triangle ABC.

A check on the accuracy can be obtained by taking sights onto four known stations. The instrument coordinates can be calculated from two sets of three stations and compared; if good correspondence is obtained, a mean set of coordinates can be taken. Alternatively a (computer) least squares solution can be found.

5.6 NETWORKS

A control network involves linear and angular measurements between stations. There is no restriction on which distances and angles should be measured in contrast with the methods of traversing, triangulation and trilateration. Provided that measurements of sufficient precision can be taken, a control network can establish the levels as well as the positions of all stations.

Before the development of computer operated least squares adjusment processes, control methods, although designed to incorporate checks on the field measurements, were geared to manual calculations and adjustments. Such adjustments were often carried out sequentially to satisfy geometric conditions in turn, and to adjust a small part of the whole survey at a time. The inclusion of redundant measurements, while providing additional checks on the fieldwork, increased the complexity of the adjustment process.

The adoption of EDM for fieldwork (facilitating simultaneous angle and distance measurement) and the use of computer programs for calculations, adjustments and strength analysis have meant that a network process allows many more field observations to be taken and incorporated in calculations than formerly was possible.

In the past, much control surveying was carried out by triangulation, and this method, and trilateration, will be considered briefly.

Triangulation

Stations are established to form a framework of linked triangles. Figure 5.13 shows connecting triangles, a centre-point polygon and two braced quadrilaterals. The side of one triangle and all angles of all triangles must be measured; the sine rule can then be applied successively to find the side lengths of all triangles. Provided start station coordinates and orientation are known, coordinates can be calculated for all stations.

An angle check is provided at each station (this should be carried out in the field) and for each triangle. A triangle of large area may require angle corrections for spherical excess so that calculations are

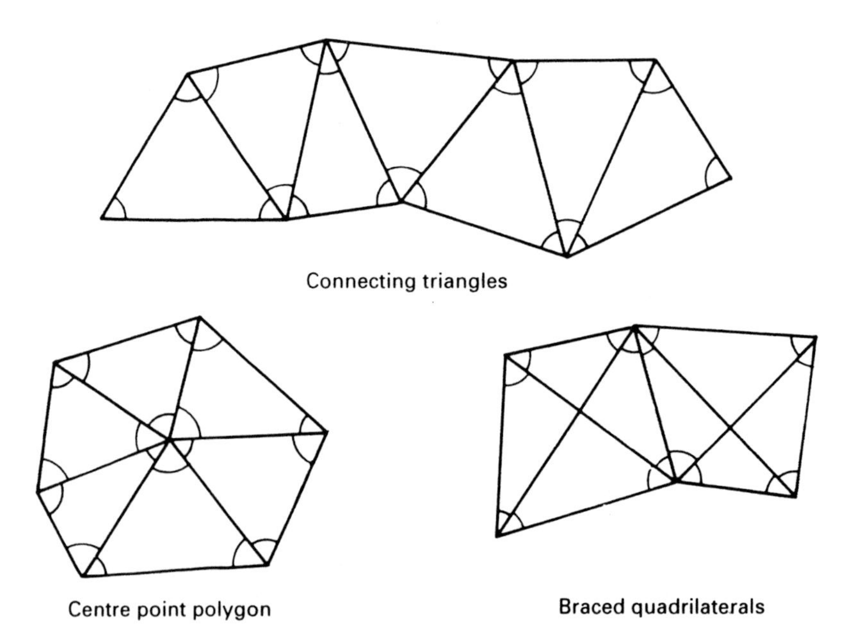

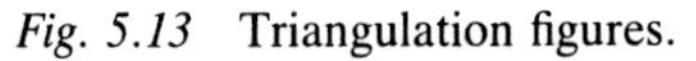
Fig. 5.13 Triangulation figures.

performed for an equivalent plane figure. In addition a trigonometric check (the side condition) is available. This requires two (distant) sides to be measured (and accepted) for connecting triangles, but can be applied to a centrepoint polygon and a braced quadrilateral without any distance measurements. However, to prevent cumulative, or scale, errors from creeping in, two or more sides for all configurations are normally measured. This makes a sequential adjustment process virtually impossible. The figure (triangle) and side checks (and adjustments) are automatically carried out in a least squares process: care must be taken to select the minimum number of independent unknowns in a manual solution; usually a computer program would be run.

For primary triangulation, stations (triangulation points or trig points) were established on pillars at hill or mountain summits to provide intervisibility for angle measurement. Long baselines of several kilometres length were precisely taped over level ground and linked to the primary triangulation by further triangles. In Great Britain, now that control can be maintained by GPS (satellite) methods not requiring intervisibility, many trig points are surplus to requirements and will no longer be maintained (either physically or by measurement) by the Ordnance Survey, to whom enquiries for details of those that are being maintained should be addressed.

Trilateration

In trilateration, stations are again established to form linking triangles. The side lengths of all triangles are measured. The traditional linear detail survey employed a chain or tape for such measurements; for control work EDM is used. Triangle angles can be deduced from the cosine formula and station coordinates can be calculated. There are no checks corresponding to the figure and side checks in triangulation. A check for gross errors exists in that no triangle side should exceed the sum of the other two! The only redundancy which can be built in is to measure distances between stations in non-adjacent triangles. A least squares method must then be used for calculation and adjustment.

Network methods

The measurement process for a network can be imagined as a combined triangulation and trilateration. (It has also been called tri-angulateration). It could also be seen as a loop traverse with additional linear and angular measurements between non-adjacent stations and to additional stations within the perimeter.

The station selection criteria at the start of the chapter should be applied. Additional considerations are to make triangles well-conditioned, to avoid the incorporation of 'grazing rays' (where a sight line could be refracted through being close to a surface) and to consider how many, and which, quantities to observe. It will not normally be feasible (or desirable in terms of time spent) to take every theoretically possible measurement within a network. There must be a sufficient number of redundant measurements within the whole network to provide checks, and also enough measurements so that individual points are not weakly fixed in terms of geometry or number of measurements. Figure 5.14 shows a plan of a typical network.

There are two fixed stations and ten to be located in plan. This means that there are $10 \times 2 = 20$ independent unknowns. Also 45 angles and 26 distances are shown to be measured, resulting in $45 + 26 - 20 = 51$ redundant measures. Although there is nothing wrong with this number of redundants, it could be reduced, so that there are fewer single (angle only) observations compared with double (angle

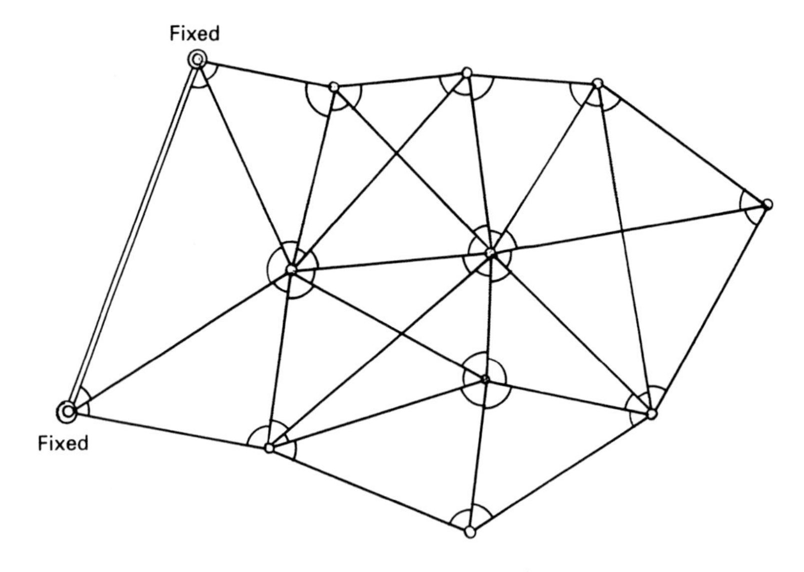

Fig. 5.14 Control network.

and distance simultaneously) ones. The amount of re-occupation of stations can be reduced. All stations should have at least one redundant measure to them.

Fieldwork

A theodolite/EDM outfit (or total station) should be used. At least two tripod mounted prism outfits should be employed. It may be helpful to have further tripods and tribrachs available so that setting up ahead of measurements can be performed. The sequence of measurements should be determined before work starts – one of the problems can be communication between the surveyor and the prism team. Two-way radio may be helpful.

Care must be taken in setting up equipment to minimize centring errors. The optical plummets should be regularly checked and adjusted if necessary. Members of the prism team must be made aware of the importance of accurate centring.

For distance measurement a recently calibrated EDM should be employed. Attention must be given to entering the correct prism and atmospheric constants. Horizontal distances should be booked; repeat measurements may be taken to provide checks. A distance should be measured in both directions and a mean taken; this is because a derived horizontal distance of more than a few hundred metres will contain errors from curvature and refraction effects on the zenith angle. For distances exceeding 1 km, a further correction is required to compensate for the convergence of plumb lines; this is described in the next section ('Calculations').

If angles are to be measured with an optical theodolite, change of face and change of zero must be carried out. An electronic theodolite may have axis compensation and horizontal circle scanning simplifying the measurement process. The surveyor should carry out regular checks for measurement on both faces, so that any change of compensation factor can be applied.

Calculations

Before field results are entered into a computer for adjustment and coordinate evaluation, some manual manipulation may be required. Checks for gross errors should be carried out (angle summation check in each triangle) and mean values for duplicated (consistent) distance observations should be calculated.

Distance measurements may require corrections. For elevated or depressed EDM sights of about 500 m or more, the effect of normals to the ellipsoid at the stations not being parallel becomes significant. To derive a correction it is necessary to consider a mean elevation angle for reciprocal measurements (Fig. 5.15).

At A,
Elevation angle $B\hat{A}B_1 = 90° - \alpha + \hat{c} - k\hat{c}$
At B,
Depression angle $A\hat{B}A_1 = \beta - 90° - \hat{c} + k\hat{c}$
(α and β are zenith angles)

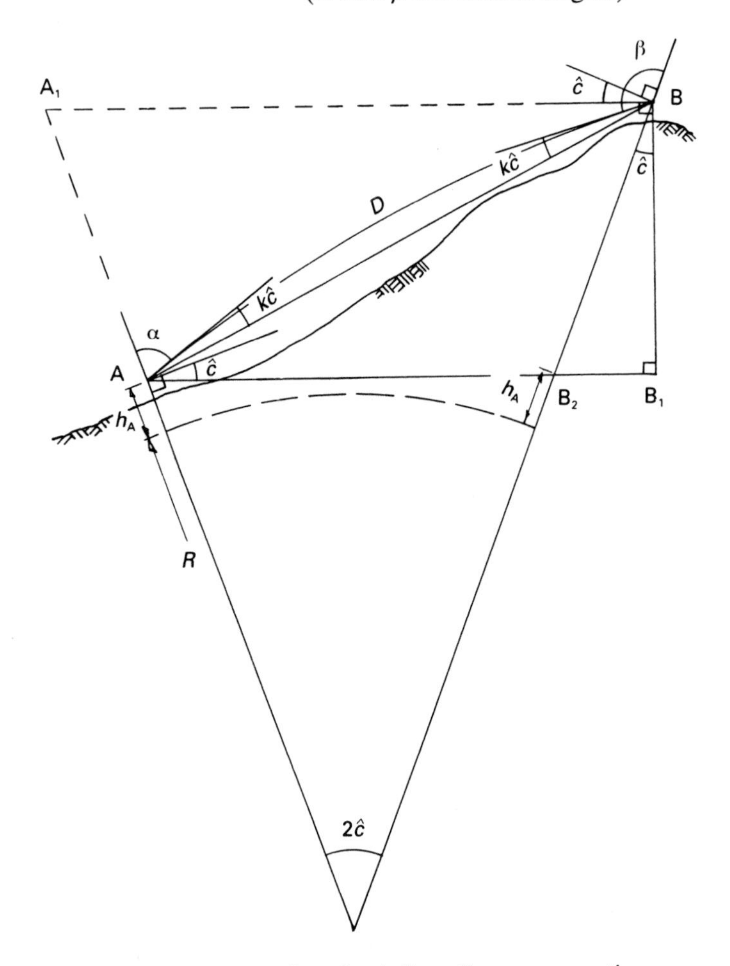

Fig. 5.15 Converging plumb line: distance correction.

$\therefore$ Mean elevation angle (at A) $= \dfrac{\beta - \alpha}{2}$

($\hat{c}$ and k are curvature correction angle and coefficient of refraction, defined in Chapter 3.)

The horizontal distance component at the axis level at A is

$$AB_1 = D\cos\left(\frac{\beta - \alpha}{2}\right)$$

Because normals to the ellipsoid at A and B are not parallel, a correction must be subtracted.

$$\text{Correction } B_1B_2 = BB_1 \tan \hat{c} \text{ and } \hat{c} = \frac{180D}{2\pi R} \text{ degrees}$$

(with negligible error)

$$\therefore \text{Correction } B_1B_2 = D\sin\left(\frac{\beta - \alpha}{2}\right)\tan\frac{180\,D}{2\pi R} \quad (R \text{ is earth's radius.})$$

Thus corrected horizontal distance

$$= D\left[\cos\left(\frac{\beta - \alpha}{2}\right) - \sin\left(\frac{\beta - \alpha}{2}\right)\tan\frac{180\,D}{2\pi R}\right]$$

Now

$$\sin\alpha + \sin\beta = 2\sin\left(\frac{\alpha + \beta}{2}\right)\cos\left(\frac{\beta - \alpha}{2}\right)$$

$$\text{and } \cos\alpha - \cos\beta = 2\sin\left(\frac{\alpha + \beta}{2}\right)\sin\left(\frac{\beta - \alpha}{2}\right)$$

$\dfrac{\alpha + \beta}{2} \simeq 90^\circ$ and $\sin\left(\dfrac{\alpha + \beta}{2}\right)$ therefore approximates to 1.

$\therefore \cos\left(\dfrac{\beta - \alpha}{2}\right)$ may be taken as $(\sin\alpha + \sin\beta)/2$,

and $\sin\left(\dfrac{\beta - \alpha}{2}\right)$ may be taken as $(\cos\alpha - \cos\beta)/2$

$$\text{Thus } H_{MEAN} = D\left(\frac{\sin\alpha + \sin\beta}{2} - \frac{\cos\alpha - \cos\beta}{2}\tan\frac{180\,D}{2\pi R}\right)$$

with negligible error

$$\text{i.e. } H_{MEAN} = \frac{H_{AB} + H_{BA}}{2} - \left(\frac{V_{AB} - V_{BA}}{2}\right)\tan\frac{180\,(H_{AB} + H_{BA})}{4\pi R}$$

Thus the mean displayed H and V values from reciprocal observations can be used. Only in the cases of extreme values of k (coefficient of refraction) and long distances (10 km or more) is the approximation insufficiently accurate, when the mean elevation angle, taken from the zenith angles and equal to $(\beta - \alpha)/2$, and the slope distance D should be used.

The straight distance AB_2 should, theoretically, be corrected further by adding $D^3/24R^2$ to obtain a spheroidal distance. This amount is small (1 mm for 10 km distance). For EDM measurement the signal path is curved with refraction coefficients of approximately 1.5 for light waves (including infra-red) and 1.65 for radio waves (coefficient relative to the earth's radius); the combined correction is then reduced to $D^3/43R^2$ for light waves and $D^3/38R^2$ for radio waves and is often neglected.

Horizontal distances will need reducing to mean sea (or other datum) level by applying the altitude correction for EDM (Chapter 4).

Spherical excess

A triangle covering a large area of ground will not be plane, and the sum of its angles should exceed 180°.

In geodetic work, angles should be adjusted so that the sum in a triangle equals 180° plus the spherical excess. Where calculations for plane figures are to be carried out, a correction, equal to one third of the spherical excess must be subtracted from each angle.

$$\text{Excess} = \frac{\text{Plan area of triangle}}{R^2} \times 206\,265 \text{ seconds}$$

For an equilaterla triangle, the sides must be 21 km long for an excess of 1″. The area can normally be scaled or calculated from unadjusted observations.

Least squares adjustment

A computer 'variation of coordinates' program should be used for adjustment of observations and calculation of adjusted coordinates. The theory is covered in Chapter 7. Standard errors for all observations are required. Some configurations of network and some programs require provisional coordinates to be entered; these can be calculated manually from some of the unadjusted observations. A program will evaluate standard errors of the calculated coordinates and of the adjusted observations. From these values and from the displayed error ellipses, the surveyor can assess the precision of individual observations and the quality of the network as a whole.

5.7 GEODETIC SURVEYS

Surveys of large areas, for new town development for example, may have to be connected to a reference system which takes account of the shape of the earth. Over a large area there will be discrepancies as a curved surface is represented as a plane. Projections of the earth's surface will result in distortions of areas, or directions or both. Once a grid is superimposed (for coordinate reference of points), differences between ground and grid bearings and ground and grid distances become apparent.

For surveys carried out by the Global Positioning System (GPS) transformation of coordinates between reference systems is necessary.

Reference systems

The earth approximates to an oblate spheroid, the solid obtained by rotating an ellipse about its minor axis; the polar diameter is less than the equatorial diameter. For position surveys, a number of oblate spheroids (called ellipsoids) of slightly different dimensions have been used to provide good correspondence with the earth's surface for each region being surveyed. Traditionally for Great Britain the Airy spheroid with an equatorial semi-axis of 6 377 563.396 m and polar semi-axis of 6 356 256.909 m has been used. This is the basis for the OSGB36 datum used by the Ordnance Survey in establishing the National Grid for Great Britain. Other ellipsoids have been used for other parts of the world. WGS84 (World Geodetic System of 1984) is increasingly being used for world-wide surveying. The corresponding semi-axes are 6 378 137.000 m and 6 356 752.314 m. Spheroidal coordinates are used to define the position of points relative to the reference ellipsoid, these being latitudes and longitudes.

The latitude of a point on the ellipsoid's surface is the angle between the normal to the ellipsoid at the point and the equatorial plane (and quoted as north or south). The intersection of a polar plane (i.e. through the ellipsoid's minor axis) and the surface defines two meridians 180° apart in longitude. For reference purposes, one such meridian is termed the central meridian. Conventionally this is taken to pass through Greenwich (England). A longitude is the angle about the minor axis between the central meridian and the meridian through the point (and quoted as east or west) (Fig. 5.16(a)).

Level surveys carried out by spirit-levelling or trigonometric heighting, have established level surfaces normal to the direction of gravity at the location of the survey. Such directions diverge slightly from normals to the ellipsoid because of the irregularity of land masses. Thus a shape differing from an ellipsoid is defined. This is the geoid, a solid resembling an ellipsoid with surface irregularities. The geoid is

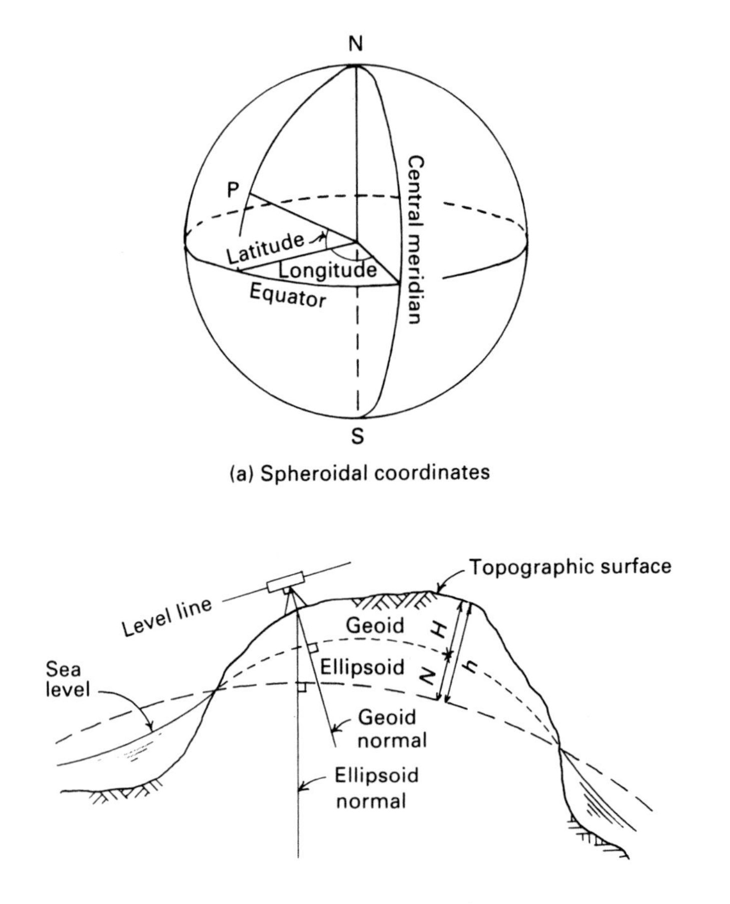

(a) Spheroidal coordinates

(b) Geoid-ellipsoid separation

Fig. 5.16 Geodetic reference systems.

represented by mean sea level for oceans and seas, but for land masses the surface lies between the topographic surface and an idealized ellipsoid (Fig. 5.16(b)). If narrow canals were cut through the land masses to connect the oceans and seas, the mean water level (free from tidal or frictional effects) would correspond to the geoid in landlocked areas. The geoid cannot be represented by any mathematically defined solid. Levels related to it are known as orthometric heights.

Three-dimensional coordinates can be referred to the ellipsoid as latitude, longitude and height (ϕ, λ, h). The height, h, is the height of a point above the ellipsoid's surface, measured along the normal to the surface. The difference between the ellipsoidal height and the orthometric height (H) is a variable N, such that

$$h = H + N.$$

The results from satellite surveys (GPS) are produced in a three-dimensional Cartesian (X, Y, Z) form related to the WGS84 ellipsoid. The Z axis is the minor (polar) axis, the X axis in the direction of Greenwich meridian with the Y axis at right angles to the X and Z axes. For surveying purposes, conversion to ellipsoidal coordinates is required; this may even involve conversion from one reference ellipsoid to another, requiring knowledge of the ellipsoids' dimensions and the relationships between the axes.

To convert between ellipsoidal heights and orthometric heights, GPS observations of points orthometrically levelled within the survey area must be taken to allow interpolation to be carried out.

Projections

For points to be represented on a flat plan, a projection from an ellipsoid is required. For survey purposes, a projection in which the scale at a point is constant in any direction is convenient. Such a projection is called conformal (or orthomorphic). The Mercator projection has often been used to depict the whole world in plan. This is a conformal projection achieved by projecting surface features onto a cylinder wrapped around the ellipsoid, touching the equator, its axis coinciding with the ellipsoid minor axis. The cylinder is then unwrapped to produce a flat surface. Areas and distances are increasingly exaggerated for zones further away from the equator.

Transverse Mercator projection

For an area with a predominantly north–south land mass a transverse Mercator projection is preferable with the cylinder axis lying in the equatorial plane. For Great Britain, the Ordnance Survey National Grid is based on a transverse Mercator projection from the Airy spheroid. The central meridian is taken as 2° west of Greenwich, this being more central to the land mass. If the cylinder, of elliptical section, matched the ellipsoid at the central meridian, distances along the central meridian when projected would be unchanged, but all others would be enlarged. In fact a cylinder smaller in section is employed so that it is in contact with the ellipsoid 180 km (along the cylinder) from the central meridian. Hence, on the projection, at meridians 180 km from the central meridian, the scale is correct; between them it is reduced, and outside them it is increased (Fig. 5.17(a)). At the central meridian the reduction is 9996/10 000 approximately, while at the edges of the projection the increase is approximately 10 004/10 000.

The origin for the projection is at 2° W and 49° N. To keep all Great Britain coordinates positive, and (on the mainland) less than 1000 km, this origin is labelled (400 km E; −100 km N), so that there is a 'false origin' off the southwest coast of England.

When conversions between ground measurements and grid measurements (and between ground and grid bearings) are carried out, certain corrections must be applied. For Great Britain these are detailed, for precise work, in *Transverse Mercator Projection – Constants, Formulae and Methods*, an Information Sheet produced by the Ordnance Survey.

The following methods described are sufficiently accurate for almost all surveying work for construction.

Local scale factor

A ground distance at sea level must be multiplied by a scale factor to determine the corresponding grid distance. In conformal projections (the transverse Mercator being one), the scale factor at a point is independent of the direction of the line.

The factor will clearly depend on the distance east or west of the meridian of the line. In engineering surveys, lines are sufficiently short for a single scale factor to apply throughout their lengths (Fig. 5.17(b)).

The scale factor is calculated at the middle of the line, and the grid

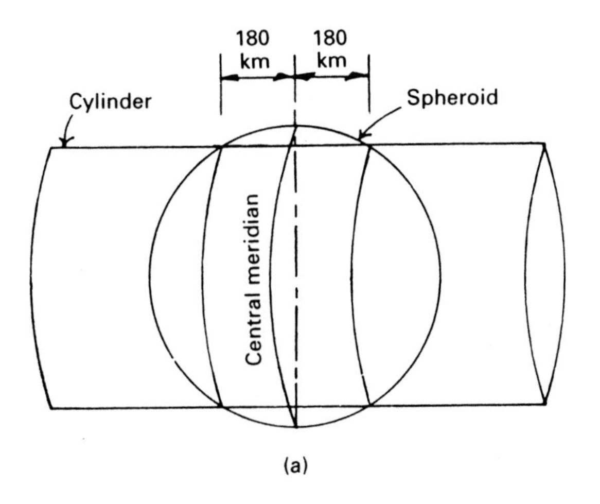

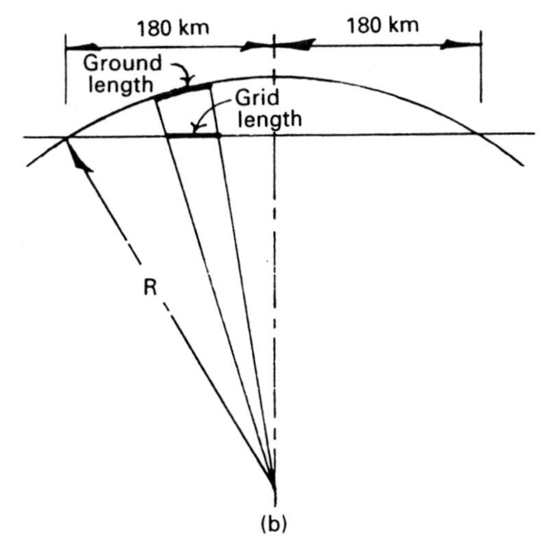

Fig. 5.17 Transverse Mercator projection.

easting is required, the mean of the eastings at each end of the line being taken.

$$\text{Local scale factor, } F \simeq 0.99960127 \times \left[1 + \frac{(E_M - 400\,000)^2}{2R^2}\right]$$

which, taking a mean value of the earth's radius (R) and E_M in metres,

$$= 0.99960127 + [1.228 \times 10^{-14} \times (E_M - 400\,000)^2]$$

Note that the ground distance must be a sea level length; distances measured above or below sea level must be converted using the altitude correction (Chapter 4). Because the LSF formula contains approximations, the altitude correction cannot be applied by adjusting R in the formula.

Example 5.9

Determine the grid length of a line of length 1405.365 m at sea level with an easting of the mid-point of 350 000.000 m.

Solution

$$F = 0.99960127 + [1.228 \times 10^{-14} \times (-50\,000)^2]$$
$$= 0.99963197$$

$$\text{Grid length} = 1405.365 \times 0.99963197 = 1404.848\,\text{m}.$$

Convergence of meridians

Meridians (lines of longitude) converge towards the poles (Fig. 5.18(a) and (b)). Only at the central meridian will bearings have identical true and grid values. At points west of the central meridian, true north will be east of grid north, so that a correction must be subtracted from a grid bearing to give a true bearing. East of the central meridian, the corresponding correction must be added (Fig. 5.18(c)).

$$\text{Convergence, } C, \simeq \Delta\lambda \sin\phi$$

where ϕ is the latitude of the point in question and $\Delta\lambda$ is the difference in longitude from the central meridian (spheroidal coordinates). The latitude and longitude can be estimated by interpolation from Ordnance Survey Maps (1:50 000) where intersection points at 5′ intervals are shown. Because the grid meridian is 2° west of Greenwich, west longitudes must be subtracted from 2°00′00″ and east longitudes must have 2°00′00″ added to them to give $\Delta\lambda$.

Hence:

$$\text{Convergence (degrees)} = (2°00'00'' - \lambda)\sin\phi \text{ (for points west of Greenwich)}$$

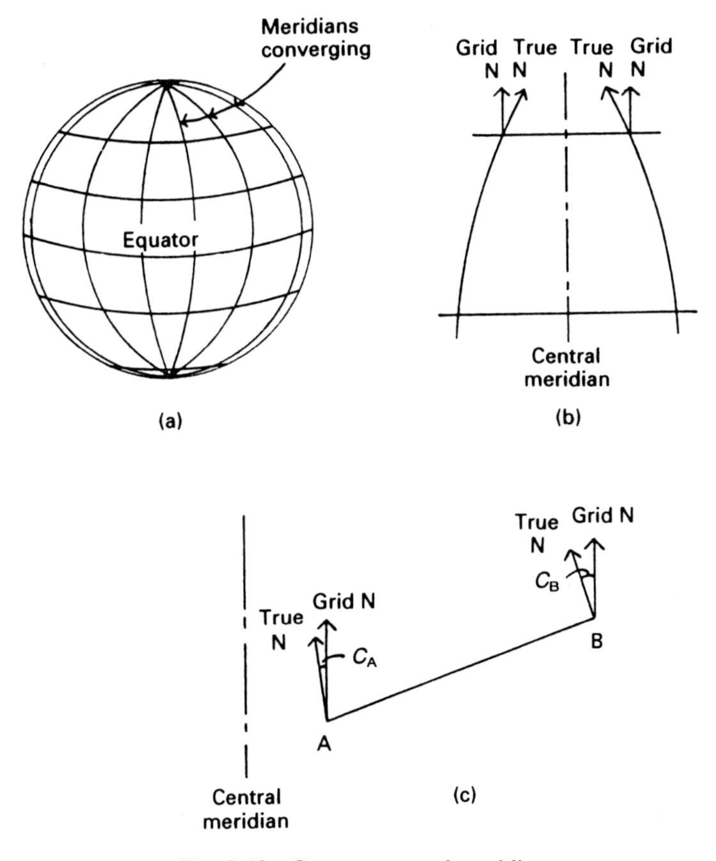

Fig. 5.18 Convergence of meridians.

or

$$= (2°00'00'' + \lambda) \sin \phi \text{ (for points east of Greenwich)}$$

The sign of correction from grid to true bearing is given automatically.

Example 5.10

Determine the true bearing of the line AB with a grid bearing of 123°41′23″ with A having National Grid coordinates of (259 314.760 m E; 361 796.320 m N).

Solution

From a 1:50 000 map, $\phi = 53°08.1'\,N$, $\lambda = 4°06.2'\,W$;
 so $\Delta\lambda = -2°06.2'$
Convergence $= (-2°06.2')\sin 53°08.1' = -1°40'58''$
True bearing of AB $= 123°41'23'' - 1°40'58'' = 122°00'25''$.

Arc-chord or $(t-T)$ correction

A further small correction is sometimes applicable when carrying out conversions between grid and true bearings. The shortest line joining two points on the surface of the ellipsoid (a 'geodesic', corresponding to a great circle route on a sphere) will appear curved when projected onto a flat plan. The direction 'T' (a tangent to the curved line) from one point will differ from the straight line direction 't' drawn on the plan. Thus a grid bearing (straight) will require a correction to produce the bearing related to north on the ellipsoid. The correction will be different at both ends of a line. The curved line will always be concave towards the central meridian, allowing the sign of the correction to be checked, see Fig. 5.19.

For a line AB,

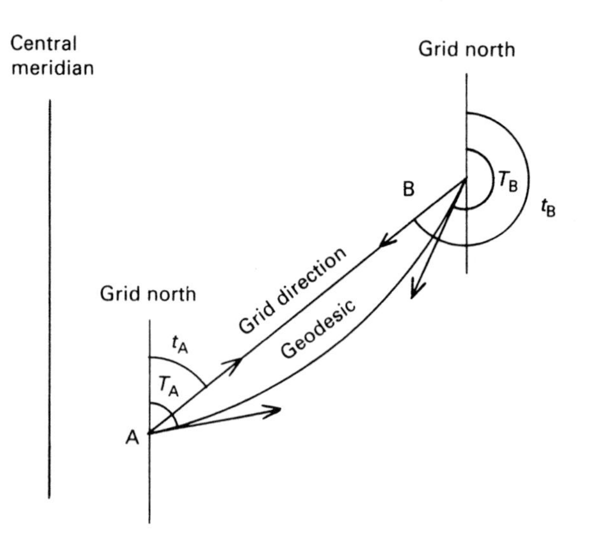

Fig. 5.19 Arc–chord correction.

Correction, $(t-T)_{AB} \simeq (2y_A + y_B)(N_A - N_B) \times 0.845 \times 10^{-9}$ seconds

$(t-T)_{BA} \simeq (2y_B + y_A)(N_B - N_A) \times 0.845 \times 10^{-9}$ seconds

where y_A and y_B are true eastings (related to the central meridian) of A and B and N_A and N_B are their grid northings.

(0.845×10^{-9} is a factor appropriate for Great Britain and based on the curvature of the ellipsoid).

True bearing = Grid bearing + $C - (t-T)$

The largest values for $(t-T)$ will be obtained for lines that have a large northing difference and that are remote from the central meridian. For a line 10 km long, the correction cannot exceed 7″ in the worst case at the edge of the projection of Great Britain. The correction may well be of a similar magnitude to the error involved when applying the approximate correction given above for Convergence. Thus if the $(t-T)$ correction is likely to be significant, it is recommended that a more precise method of determining C is used.

5.8 HEIGHT CONTROL

At the start of a surveying or engineering project, height control stations must be established. Usually these will relate to mean sea level, or to other national or local datum. It should be determined whether GPS will be used for levelling, or whether conventional spirit-levelling will be employed.

If GPS is being contemplated, the likely accuracy of it must be evaluated; relative heighting (with one receiver over a reference bench mark) may be sufficiently precise. It must be remembered that in absolute terms, GPS levels are referred to the reference ellipsoid, whereas spirit-levelling refers to the geoid (see Fig. 5.16(b)). It may be necessary to provide a network of points correlated between ellipsoid and geoid altitudes.

For spirit-levelling, an engineer's level or a precise level should be used to carry levelling from an Ordnance bench mark on to site and to establish site temporary bench marks. The levelling must be closed to an acceptably small amount, and an additional check is provided if two or more OBMs are involved. OBM levels should be verified: current values can be obtained from the Ordnance Survey. It should be remembered that values are quoted to the nearest 10 millimetres.

The principles in Chapter 2 should be followed; an example of a least squares adjustment of a simple level network is given in Chapter 7.

If trigonometric heighting is carried out (possibly as part of a three-dimensional survey using a 3D variation of coordinates adjustment), attention must be paid to the precision of zenith angle measurement, and reciprocal observations should be employed to eliminate the uncertainties of refraction effects. An error analysis to investigate the precision of trigonometric heighting is recommended.

Throughout a construction project, the levels of temporary bench marks should be checked in case they have been affected by disturbance, subsidence or ground heave.

5.9 GLOBAL POSITIONING SYSTEM (GPS)

Introduction

Surveying by the Global Positioning System involves information from orbiting satellites being received by a ground station at the point to be surveyed. The method is analogous to the determination of three-dimensional position by measuring distances from fixed stations of known coordinates. In GPS, the instantaneous positions of satellites correspond to these fixed stations, while the time taken for a signal to travel from satellite to ground receiver is the equivalent of the distance.

Although not originally designed for surveying, GPS has been refined so that survey accuracies can be achieved.

Advantages are:

(1) Measurements can be taken relating points that are not intervisible.
(2) Measurements can be taken for all open locations at night and in poor weather.
(3) Each point is individually located, so that cumulative errors do not occur.
(4) Measurements can be made continuously.

Disadvantages are:

(1) An open position with a wide field of view of the sky is required for the GPS receiver; in particular, points close to buildings, overhung by trees and underground cannot be surveyed.

(2) Instantaneous observations do not always yield the required accuracy; measurements for a period of up to 30 minutes with 'post processing' of results are often needed (1994).
(3) Levels are related to the ellipsoid, not the geoid.

Satellites

The satellites are part of the 'NAVSTAR' system of the United States Navy, designed for navigation purposes. There are 21 operational satellites with 3 spare having 4 in each of 6 orbital planes at 55° to the equatorial plane. Each satellite is at an altitude of 20 200 km with an orbit time of 12 hours. Between five and nine are 'in view' at any ground point at any time. The satellites' positions are constantly monitored from Earth; this information is continually transmitted to each satellite which in turn relays the information with the time signal to the ground receiver. The satellites contain very precise atomic clocks so that the transmitted time is compared with the time of a synchronized clock at the receiver. From these times, distance measurements (ranges) can be evaluated. Measurements of this kind from three satellites give two alternative locations for the receiver; inevitably one of these locations is absurd (e.g. nowhere near the earth's surface), so that a unique determination is possible. In practice, a fourth satellite is employed to help to remove systematic errors; this also provides confirmation of the position (Fig. 5.20).

While the atomic clocks in the satellites are extremely accurate (and constantly monitored), the receiver clock (in an economically priced system) is liable to small errors. The measurements made (with such errors present) are called pseudo-ranges. If measurements are made from four satellites, the same clock error is present in all measurements, and a corrected position can be found. (For a two-dimensional fix, where two measurements locate a point, a third is needed if all measurements have a constant error).

Transmitted signal

Two frequencies of carrier signal are used, the L1 frequency at 1575.42 MHz and the L2 frequency at 1227.60 MHz. These signals are phase modulated by two binary coded signals, a Coarse Acquisition signal at 1.023 MHz for general use, and a Precise signal at 10.23 MHz

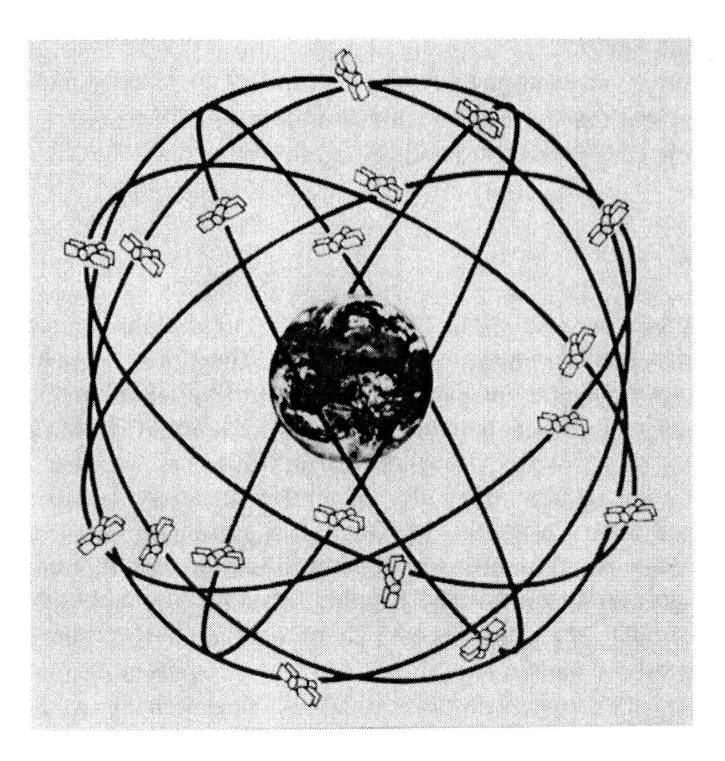

Fig. 5.20 Satellite constellation.

restricted in use for military purposes. Pseudo-random codes are used, the L1 signal carrying both, the L2 signal carrying just the Precise coded signal.

Carrier signal		Modulated signal			
Identity	Frequency	Code identity	Type	Frequency	Use
L1	1575.42 MHz	C/A	Coarse Acquisition	1.023 MHz	General
		P	Precise	10.23 MHz	Restricted
L2	1227.60 MHz	P			

The position parameters of each satellite are also transmitted. For navigation purposes, the C/A code (also known as the S – Standard code) with a wavelength of approximately 300 m may be acceptable for general use; the P code for restricted use has a wavelength of approximately 30 m.

Each satellite has its own distinctive modulated signal(s). The signal, pseudo-randomly generated with a square shaped wave form, is compared with a similar signal at the receiver to give the elapsed time, and hence the distance. Synchronization can only be carried out to within a few per cent, resulting in an error in the range of about 0.5 m at best (Fig. 5.21).

Except for reconnaissance, such accuracy is not adequate for survey work. Accordingly, the carrier wave itself is analysed, the L1 frequency allowing range determination within 3 mm, if phase measurement can be carried out to within five degrees. However, in measurement of the carrier wave, an ambiguity, the number of complete cycles, exists, just as happens in EDM measurement. In EDM the frequency is continuously swept up, allowing an unambiguous measurement (over a given range) to be calculated. For GPS, a number of solutions to the ambiguity problem have been devised. One method is to use the modulated wave to give a coarse measurement, another is to take continuous measurements over a prolonged period so that a significant movement of the same satellites occurs, and a third is to use receivers at two ground stations, the position of one already being established.

Because other errors may also be present (atmospheric effects, satellite position errors, satellite clock errors), this third method is usually employed in surveying. Compared with the satellite altitudes, measurements within a survey area are small so that the effects of the errors stated will be constant at any one time. These errors can be established by measurements to the known station and applied to measurements to subsequent points. All calculations are carried out

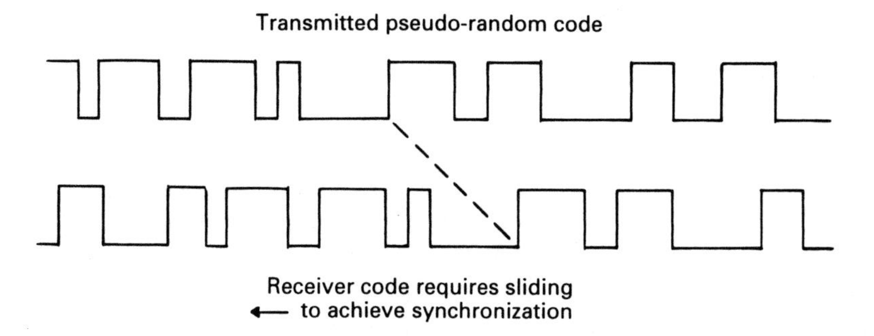

Fig. 5.21 Pseudo random code synchronization.

by computer, and the selection of the exact field procedure will be determined by the available software.

Ground receiver

The signal is received by an antenna. At the control station it is mounted on a tripod or fixed directly to a survey pillar. For detailing work, the antenna may be mounted on a hand-held pole. The antenna is connected to an analysing unit incorporating a computer, display screen and data storage (Fig. 5.22). Surveying accuracy requires simultaneous reception of signals from four (or more) satellites, and multi-channel receivers are required.

Two or more receivers are always used in survey work. For the most precise applications, observations must be made continuously for about 30 minutes. The computer records all the information, and can then resolve the integer ambiguity and reduce errors by averaging processes. In 'post processing', all data is analysed after measurement. All information received must be logged together with the precise time of its reception. Alternatively a radio link between the processing units can be used to allow simultaneous analysis of all information gathered.

Transmitted information is related to the WGS84 rectangular coordinate system. The software must be capable of transformation into the local system in use; for Great Britain this may involve conversion to ellipsoidal coordinates (WGS84 and then Airy) and further transformation to OSGB36 projection coordinates. The software is being improved so that 'real time' analysis can be carried out, of considerable benefit in detailing work, volume measurement and setting out.

Dilution of precision

As in all branches of surveying, the geometry of measurements is important. A weaker fix will be obtained from closely clustered satellites than from well dispersed ones. The phenomenon is called Geometric Dilution of Position (GDOP). If the satellite orbits and position within these orbits (the ephemeris) are known, it is possible to predict superior times for conducting the survey, and to carry it out accordingly. However, usually the receiver has software enabling data

Fig. 5.22 GPS systems: (a) Leica Wild SR299; (b) Trimble Site Surveyor; (c) Geotracer System 2000; (d) Sokkia GSS1.

from the best four out of all the available satellites to be used. The variable effect from the disposition of the satellites gives rise to Position Dilution of Precision (PDOP), a factor indicating whether suitable accuracy can be achieved. A low value is required for precise work:

less than 7. GPS software can analyse the PDOP. This can be displayed on the screen against a time-scale, as can satellite availability and elevations. Sky plots of the satellites are also available on screen (Fig. 5.23).

Atmospheric effects

The speed of the signal is affected by the ionosphere (a blanket of electrically charged particles between altitudes of approximately 80 km and 400 km) and the troposphere (the region of the earth's weather effects, i.e. water vapour, up to an altitude of 11 km approximately). In addition to the signal being slowed down, refraction can occur. Simultaneous measurements at two receivers ('Differential GPS') will enable compensation to be applied for atmospheric

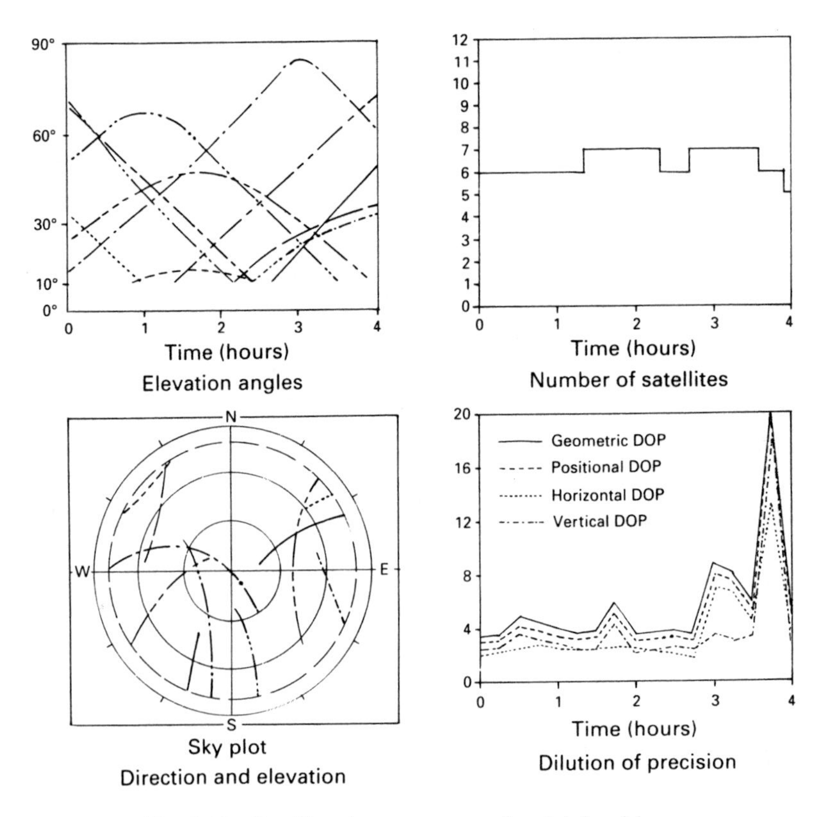

Fig. 5.23 Satellite data at a ground point for 4 hours.

effects. However, the effects are not constant, so different compensation will be required after a time interval.

Multipath errors

Receivers should be set up in open areas so that signals can be received directly. A receiver close to a reflective surface (a heavily glazed building, for example) may receive the signal bounced off the surface, so that the signal arrives via two (or more) paths. The delayed (reflected) signal can interfere with the direct signal, reducing the accuracy. Receivers should be positioned with care to be well clear of possible reflective surfaces.

Selective Availability (S/A)

The US Department of Defense, for security reasons, can introduce 'noise' into the satellite clock signal, which can reduce the accuracy. Differential GPS will allow these errors to be counteracted.

Ephemeris errors and satellite clock errors

The satellite orbits are continuously checked, and updated ephemeris (position prediction) information transmitted. Small errors can occur, but Differential GPS will overcome these and satellite clock errors.

Differential GPS

Differential GPS uses two ground receivers. They are set over different ground points and receive information simultaneously from the same satellites. One is positioned at a known location, the other at the opposite end of the baseline to be measured, or at a series of detail points to be located with reference to a fixed baseline.

For survey work the distance between the two receivers (up to 50 km; much less for most projects) is insignificant compared with satellite altitudes. Thus errors of transmission (orbit, satellite clock, atmospheric, selective availability) are constant for both ground positions. These errors can be determined for the known point, and used to provide corrections for the determined position of the unknown point or points. Several different techniques of Differential GPS have been developed; in all of them there is a procedure for resolving (or eliminating) the integer ambiguity in the carrier phase measurement. This procedure is known as initialization.

During subsequent measurements it is usually necessary for the receivers to maintain 'lock' on the four (or more) satellites, i.e. continuous reception of signals. Interruptions cause cycle slips, necessitating fresh initialization.

Differential GPS can produce accuracies of between 5 and 10 mm in eastings and northings and between 25 and 40 mm in heighting. As GPS is a three-dimensional survey method, the precise determination of the antenna height above each station is important. Usually a built-in tape or measuring rod is supplied.

The techniques of Differential GPS are either Static Differential or Kinematic Differential (also known as Dynamic) surveying. Faster Kinematic techniques are being developed as analysis algorithms and computer solutions are improved.

Vertical control

There are two particular problems associated with vertical control by GPS: reduced accuracy and geoid – ellipsoid separation.

Because of the geometry between satellites and survey points, a weaker fix applies to levelling than to plan positioning. Atmospheric delays also affect height measurements more than plan location.

Heights are related to the reference ellipsoid, whereas conventional levelling relates heights to the geoid. No mathematical correspondence can be achieved as the geoid is an irregular 'surface' defined by gravity. Correlation must be established for a number of points in the survey area between GPS heights and orthometric levelling (related to the geoid). Interpolation is then carried out to deduce the orthometric heights of points subsequently surveyed by GPS. A greater density of vertical control points is required in an area of topographic irregularity where the geoid is likely also to be irregular.

Static GPS surveying

This is the basic method for the establishment of survey control points, although faster methods are being developed. One receiver is set up over a station with known coordinates at one end of a baseline. The other receiver is set up at the unknown end. Observations for a period of about 30 minutes are required. This period provides enough time for the satellites to change position significantly to allow resolution of the integer ambiguity.

The process can be repeated to survey other points required for control. If the receivers' lock on the satellites can be maintained, the station occupation time can be greatly reduced. For a precise control survey, distances (and bearings) between many combinations of pairs of points can be determined. If lock is lost, the integer ambiguity can be resolved by occupying a station where coordinates are known to within a few centimetres. Additional control is provided by using more than one reference station (Fig 5.24(a)).

When sufficient control measurements have been obtained, a least squares computation can be carried out to provide a best estimate of coordinates and analysis of results. Achievable accuracy of range is between 1/100 000 and 1/5 000 000.

Rapid static surveying

In operation this is similar to static surveying but faster. Initialization (i.e. resolution of ambiguity) is achieved either by using pseudo-range (P code) measurements, or by utilizing carrier wave measurements from additional satellites. Either a receiver capable of analysing the P code (usually the code is removed during carrier phase measurements) is required, or there must be more than four satellites visible throughout the initialization process. The number and disposition of the satellites is critical. Occupation times are between 5 minutes (6 or more satellites, PDOP less than 5) and 20 minutes (5 satellites, PDOP between 5 and 7) (Fig. 5.24(b)). Achievable accuracy of range is between 1/100 000 and 1/1 000 000.

Pseudo-kinematic surveying

Kinematic methods of GPS involve a fixed receiver and a moving receiver. Pseudo-kinematic surveying (also known as pseudo-static surveying and as reoccupation surveying) uses some of the principles of static GPS within a kinematic framework.

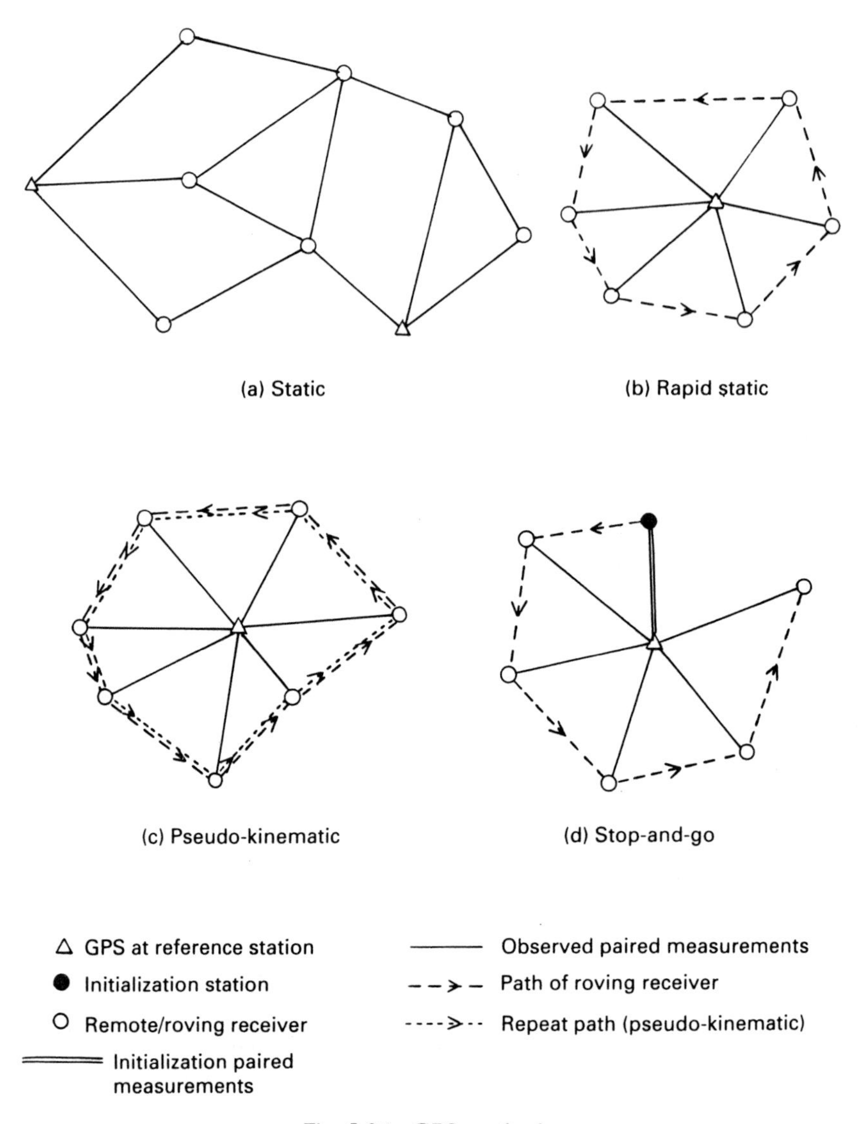

Fig. 5.24 GPS methods.

One receiver is set up and kept at a reference station. The roving receiver occupies remote stations in turn. It does not need to keep lock on the satellites. Stations must be occupied for a few minutes. After all stations have been surveyed, the process must be repeated, each station again being occupied in turn so that the integer ambiguity is resolved for each station. There must be at least one hour (to allow significant change of satellite geometry) and not more than four hours between occupation times. The method is suitable for regions where overhead obstructions prevent lock being maintained. It is convenient for there to be enough points to survey to fill the time between occupations (Fig. 5.24(c)). Achievable accuracy of range is between 1/50 000 and 1/5 000 000.

Stop-and-go surveying

One of the standard kinematic methods of surveying, stop-and-go surveying is also known as semi-kinematic surveying. A fixed receiver and a roving receiver are used. The survey must commence with initialization measurements to resolve integer ambiguities. There are two methods of achieving this.

In the first method both receivers are set up at the ends of a baseline whose length and bearing are known either from previous measurements, or determined at the time of the survey by static or rapid static methods. After observations, the roving receiver can be moved to points to be surveyed.

The second involves establishing a short baseline of 5 to 10 metres length incorporating the reference station. The two receivers are set on tripods over the stations and observations are made for about 2 minutes. The antennae are then exchanged (the tripods being kept fixed) and the process is repeated. The antennae are restored to their original positions, and observations made for a further 2 minutes. During the exchanges lock must not be lost. The integer ambiguity is resolved and kinematic surveying then commences.

The roving receiver must maintain lock on four satellites at all times. If lock is lost (the receiver gives warning of this), either a previously surveyed station must be reoccupied, or one of the initialization procedures described must be used. Points must be occupied during 2 distinct periods of measurement (2 epochs); according to conditions this may take between a few seconds and one minute (Fig. 5.24(d)). Achievable accuracy of range is between 1/100 000 and 1/1 000 000.

Kinematic surveying

In true kinematic surveying, the roving receiver is continuously moving, i.e. in a boat, or a road or rail vehicle. Initialization is carried out similarly to that for stop-and-go surveying. Because instantaneous observations are taken, the accuracy is about half that for the stop-and-go mode.

Kinematic surveying is useful where the position of individual points is unimportant, in marine work, in road kerb location, and in ground levelling for volume calculations. Observations are taken at fixed time intervals, which can be as short as one second. In assessing the accuracy, consideration must be given to the offsets, vertically from the surface, and horizontally from the vehicle axis, of the receiver. Lock on the satellites must be maintained (Fig. 5.24(d)).

General operation

A certain amount of planning is necessary before carrying out a GPS survey. Although there should be a sufficient number of satellites visible throughout the survey period, this, and the 'health' of the satellites should be checked. Software is designed so that a screen display of satellite availability can be provided. Also a display of the PDOP can be obtained; the value should always be less than 7, and for best quality work less than 5.

The method to be used will depend on the accuracy required, the availability and disposition of control stations, the feasibility of the receivers maintaining lock on the satellites and the capability of the available software. It is possible to combine different methods (e.g. stop-and-go and pseudo-kinematic methods) to achieve the required survey results. Receivers should be positioned so that lock is maintained and so that they are clear of reflective surfaces which could cause multipath errors.

A programme for the occupation of control stations must be drawn up. Precise timing for synchronization of observation periods is necessary. Accurate watches must be used, or radio communication between operators must be maintained.

Developments

The applications of GPS are considerably wider than the military use for which it was developed. Marine navigation, dredging, airborne applications, road vehicle tracking and personal navigation for walkers can all employ the system. The widespread general use of GPS will surely complement the adoption of it for many specialist surveying purposes.

Algorithms (and the associated computer programming) are being developed to allow 'real time' surveys of acceptable accuracy to be performed. The main limitation is that of obstructed signal paths for congested, overhung and underground areas. GPS is excellent for control work, detailing, volume measurement, surveys of movement and deformation and relative heighting. GPS is insufficiently precise for absolute heighting. Initial control of setting out points can be carried out with GPS; it is more difficult to envisage its employment for day-to-day setting out, bearing in mind obstruction risks and the ease and cheapness of conventional techniques.

5.10 FURTHER READING

Allan, A.L. (1993) *Practical Field Surveying and Computations*. 2nd edn. London: Butterworth-Heinemann.

Cooper, M.A.R. (1987) *Control Surveys in Civil Engineering*. Oxford: Blackwell Science.

Jackson, J.E. (1987) *Sphere, Spheroid and Projections for Surveyors*. Oxford: Blackwell Science.

Ordnance Survey (1983) *Transverse Mercator Projection: Constants, Formulae and Methods*. London: HMSO.

Schofield, W. (1993) *Engineering Surveying*. 4th edn. London: Butterworth-Heinemann.

Shepherd, F.A. (1981) *Advanced Engineering Surveying*. London: Arnold.

5.11 EXERCISES

Exercise 5.1

A two span bridge is to have beams supported by bankseats at each end and by a central pier. Points X and Y, each on bearing plates on the central pier, are to be checked for position and level by a theodolite set in turn at points A and B on the north and south bankseats. The bearing plates can be sighted directly.

From the following information, determine the coordinates and reduced levels of X and Y.

Coordinates of A (689.932 m E; 458.997 m N)
B (679.235 m E; 407.330 m N)
Theodolite axis level of A 127.523 m AOD

Theodolite at A

Sighting	Horizontal reading (mean)	Zenith angle
X	00°07′30″	91°07′36″
B	05°40′03″	91°06′19″
Y	08°35′52″	–

Theodolite at B

Sighting	Horizontal reading (mean)
Y	00°13′23″
A	03°18′39″
X	08°53′20″

Exercise 5.2

A closed loop traverse has been carried out between stations P, Q, R and S. Using the following information, calculate adjusted coordinates of Q, R and S after applying Bowditch's rule.

Bearing of reference object from P = 273°50′00″
Coordinates of P = (400.000 m E; 200.000 m N)

Plan lengths (m)	Included angles
	RO–P̂–Q = 138°15′00″
PQ = 251.23	
	P–Q̂–R = 225°36′00″
QR = 213.67	
	Q–R̂–S = 272°42′00″
RS = 273.44	
	R–Ŝ–P = 281°14′00″
SP = 388.00	
	S–P̂–Q = 300°24′00″

Exercise 5.3

A closed tied traverse has been conducted between A and E via stations B, C and D. From A, reference object 1 (RO1) with a bearing of 64°30′00″ was sighted; from E, reference object 2 (RO2) with a bearing of 118°46′00″ was sighted.

From the following data, determine the adjusted coordinates of B, C and D applying Bowditch's rule.

Fixed station coordinates:
A (150.000 m E; 407.000 m N)
E (487.020 m E; 151.470 m N)

Line	Horizontal Length	Angle	Value
		RO1–Â–B	85°27′38″
AB	113.630		
		A–B̂–C	141°41′20″
BC	110.570		
		B–Ĉ–D	152°28′44″
CD	129.380		
		C–D̂–E	255°18′02″
DE	138.400		
		D–Ê–RO2	139°20′26″

Exercise 5.4

Angles have been measured from an offshore station P sighting shore stations X, Y and Z whose National Grid coordinates are known. From the following data calculate the (National Grid) coordinates of P.

Coordinates:
X (260 853.2 m E; 376 183.7 m N)
Y (260 985.6 m E; 376 812.3 m N)
Z (260 955.0 m E; 377 387.6 m N)

Horizontal angles:
XP̂Y = 38°41′ YP̂Z = 23°16′

Exercise 5.5

Total station measurements were taken between two stations A and B.

(a) Using the figures in Table 5.6, determine the mean plan distance between the stations at the altitude of station A, allowing for the effects of curvature, refraction and convergence of plumb lines (normals) at the two stations. The radius of the earth may be taken as 6380 km.
(b) Perform the same calculation using the figures in Table 5.7.

Table 5.6

	Measurements at A	Measurements at B
Slope distance (m)	5165.427	5165.427
Zenith angle	87°27′30″	92°35′00″

Table 5.7

	Measurements at A	Measurements at B
Horizontal distance (m)	5160.345	5160.177
Height difference (m)	229.066	−232.818

Exercise 5.6

National Grid coordinates of two points A and B are (363 783.771 E; 418 238.305 N) and (362 658.638 E; 417 177.425 N). The latitudes of the points are 53°39′35″(A) and 53°39′00″(B) and the appropriate longitudes are 2°33′55″W(A) and 2°32′55″W(B). The mean altitude is 297 m AOD and the radius of the earth may be taken as 6380 km.

Calculate:

(a) The grid length of AB.
(b) The grid bearing of AB.
(c) The geographical bearings (azimuths) of AB and BA.
(d) The plan length of AB.

6 Detail Surveys

6.1 INTRODUCTION

Detail surveys are carried out within areas covered by control surveying, being tied in to the stations established. Detailing often follows control work, although the two stages can be carried out concurrently when electronic methods are used for both types of work.

Objects of a detail survey

(1) To produce a plan (to scale) of an area of land showing all detail, natural and man made.
(2) To produce a contoured plan (with detail).
(3) To provide a record (in coordinate form) of detail within an area.
(4) To produce a longitudinal ground section.
(5) To produce ground cross-sections.
(6) To take measurements to allow earthwork areas and volumes to be calculated.

Principles

The object of the survey will determine the precision with which the survey is to be carried out. If the survey is to be used for the subsequent setting out of construction, and particularly the extending of or connecting to existing works, an accurate plan will be required.

It is, however, unlikely that natural features (vegetation, watercourses) can or need be accurately plotted. Perhaps the greatest problem in a detail survey is the sheer volume of information to be measured and recorded, and a systematic method of booking is essential.

Factors to be considered in planning a detail survey are:

(1) Frequency of measurements.
(2) Precision of measurements.
(3) Method of booking.
(4) Time available.
(5) Equipment available.
(6) Cost.

Of these factors, (3) to (6) may well suggest particular methods.

Frequency of measurements

The frequency will be dictated by the amount and accuracy of detail required, the shape and nature of the detail and the scale of plotting. For example:

(a) A straight length of road of constant width. Three measurements are required: location of one side of the road at each end, and the width of the road.
(b) A house in plan (Fig. 6.1).
(c) A river (Fig. 6.2).

Precision of measurement

Although construction drawings often bear an inscription 'Do not

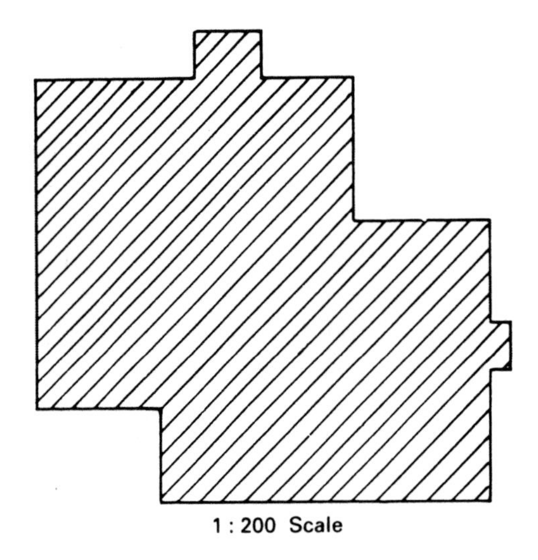

Fig. 6.1 Plot of house.

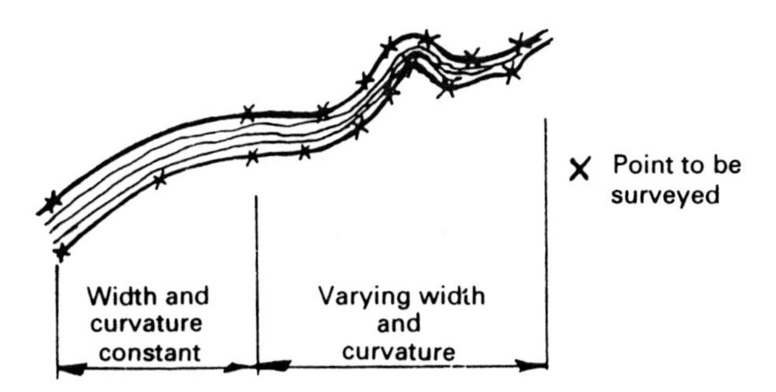

Fig. 6.2 Plot of river.

scale' (drawings can expand or shrink with age), it will be desirable to record construction detail reasonably precisely. In addition to a plan, dimensions for designing may be required; measurements should be taken to the nearest 5 mm, and care taken to prevent errors from accumulating.

Where detail is required solely for plotting, an awareness of the plotting accuracy will help to determine the field precision, for example 0.1 m on the ground will become 1 mm at 1:100 scale and will be 0.1 mm at 1:1000 scale.

6.2 LINEAR SURVEYING

The traditional linear (or chain) survey uses a control framework of chain lines forming (in plan) a rigid figure. (Chains, which gave their name to the method, are now obsolete.) The figure is the equivalent of a structural pin-jointed but rigid frame. Additional lines between the stations are measured as checklines, corresponding to redundant members in a structural frame (Fig. 6.3). Detail is then measured by tape from the lines.

Linear surveying is simple and the equipment is cheap. The work, however, is slow and laborious, all booking and plotting being done manually. Thus the method of linear surveying for carrying out a complete survey is now obsolete; nevertheless, linear measurements with a tape frequently form a significant part of detail surveys.

Where a linear detail survey is to be undertaken, it is likely that the stations will be located by a theodolite and tape (or EDM) traverse. Stations should be sited accordingly.

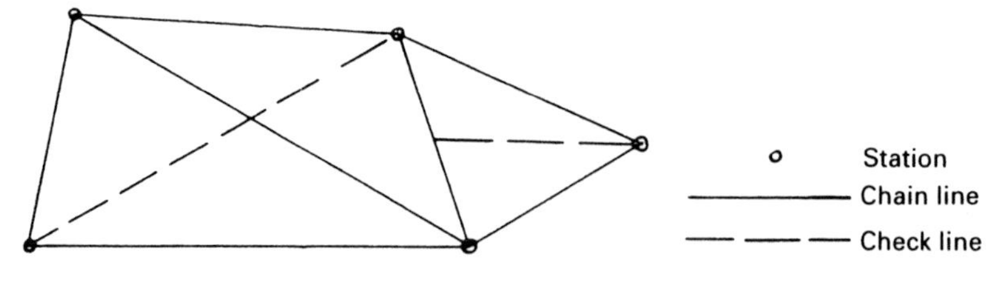

Fig. 6.3 Layout of chainlines.

Station siting criteria for linear detailing

(1) Lines should be close to detail.
(2) End stations of each line should be intervisible.
(3) Lines should pass over accessible ground.

Field procedure

Equipment: two tapes, several ranging rods, ten arrows, optical square.

Method: With the zero of the tape at a station, lay out a tape along a chain line, using ranging rods at the stations and one at the end of the tape to line (range) it in. Pull the tape tight. Record detail by measuring with the second tape from the one forming the chain line. A single taping at a right angle to the chain line is called an *offset*; perpendicularity can be estimated or established using an optical prism (a small hand-held instrument for setting out right angles). Avoid long offsets where poor perpendicularity can cause errors.

Alternatively, take two oblique measurements (tie lines) from different parts of the chain line ('tying in'). Figure 6.4(a) shows both methods.

Record the lengths of offsets or tie lines, and the distances along the chain line of the points from which the offsets or tie lines are measured ('chainages').

Booking

A plain page in a field book with two lines down the middle is used. The two lines represent the chain line, and chainages are entered here; detail is sketched either side of the double line and offset

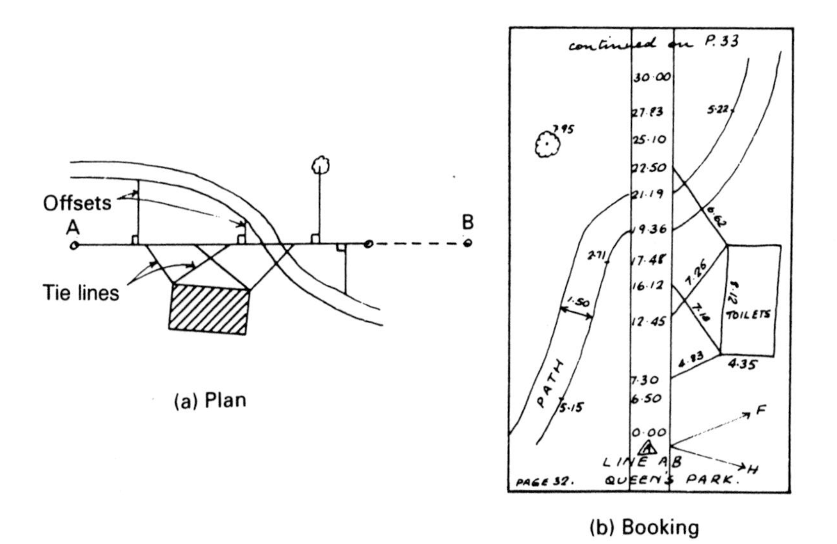

Fig. 6.4 Linear surveying.

distances or tie lines and dimensions drawn in. The chainage is shown increasing up the page. Figure 6.4(b) illustrates.

Notes

(1) Detail is drawn freehand either side of the two lines but not between them; plenty of space should be taken.
(2) An offset value is written beside the detail to which it refers.
(3) Tie lines are drawn.
(4) Dimensions are added.
(5) Descriptions are added.

Slope corrections

Chainages

If the chainages are measured on the slope, they must be converted to plan measurements using levels or vertical angles obtained. Care must be taken when using a theodolite to keep the heights of instrument and target equally above the ground. Alternatively, the tape may be stepped. This involves the distance being measured as a series of horizontal lengths, plumbing up or down at each step (Fig. 6.5).

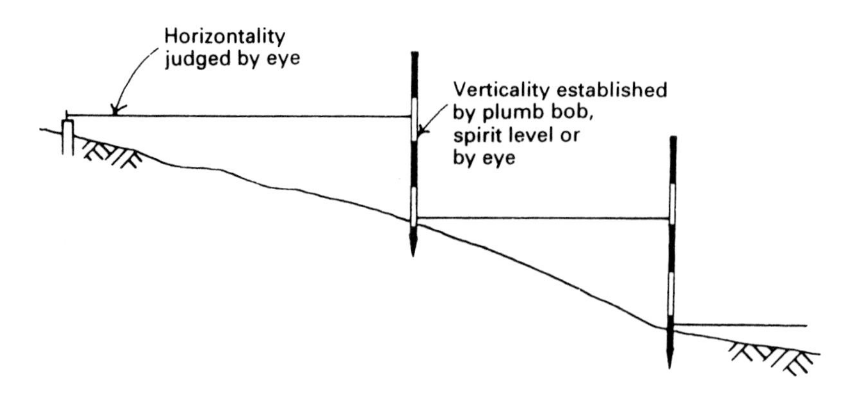

Fig. 6.5 Tape stepping.

Notes

(1) The work is more conveniently carried out downhill than uphill.
(2) Beginners tend to hold the downhill end of a stepped length too low. A third person standing to the side of the tape can assist in achieving horizontality; nearby buildings can often be used to judge horizontal and vertical lines.

Offsets/tie lines

Unless the detail lines are long, the tape can usually be held level. Long lines, which should be avoided (especially for offsets), must be corrected where they pass over sloping ground.

Obstructions

Although routines have been devised for measuring round obstructions (using simple geometric principles, for example similar triangles), other methods of detailing, involving angular measurement, will usually be preferable. A traverse can be conducted around the obstacle, or additional stations can be positioned by intersection or resection (Chapter 5) from which the linear survey can be continued.

Plotting

Plotting is carried out as a reversal of the field procedure. Distances are measured with a scale rule, right angles are set out with a set square and points which have been located by tying in are plotted from intersecting compass arcs.

On a plan, detail is drawn to scale although some features are represented by conventional symbols. A map, smaller in scale than a plan (i.e. the number in the representative fraction being larger), cannot show all detail to scale (widths of roads, building details) and further symbols are required. Figure 6.6 shows symbols for 1:500 plans. Further symbols for 1:500, 1:1000 and 1:1250 plans may be found in Department of Transport *Model Contract Document for Topographical Survey Contracts*. For smaller scales, Ordnance Survey symbols may be used.

In plotting a survey, some initial planning prior to setting pencil to paper is advisable. The scale, size of paper and orientation must be determined. A trial plot of the control framework on rough paper will enable the surveyor to be sure that the survey can be conveniently fitted on the paper.

Plotting should be done in pencil; when the surveyor is satisfied, the detail can be inked in and construction lines erased. The title, scale, drawing number and plotter's name should be entered in a title block. A north point should be shown and a drawn scale should be added; this is necessary for a check on whether the drawing has expanded or shrunk before readings are scaled from it. Written descriptions may be added where they provide further explanation.

6.3 EDM DETAILING

A great deal of detailing is now carried out with EDM. A theodolite/ EDM outfit set up over a control station can be used to determine the horizontal distance, bearing and rise or fall of a detail point. (A second fixed station is required to provide a reference bearing.) A hand-held pole-mounted prism is used at each detail point.

In comparison with linear detailing, EDM detailing is quick and accurate. It is possible to record information automatically with a data logging system and to obtain a computer aided drawing. Nevertheless the sheer volume and variety of detail to be surveyed may cause problems. It may be convenient to use EDM for the location and

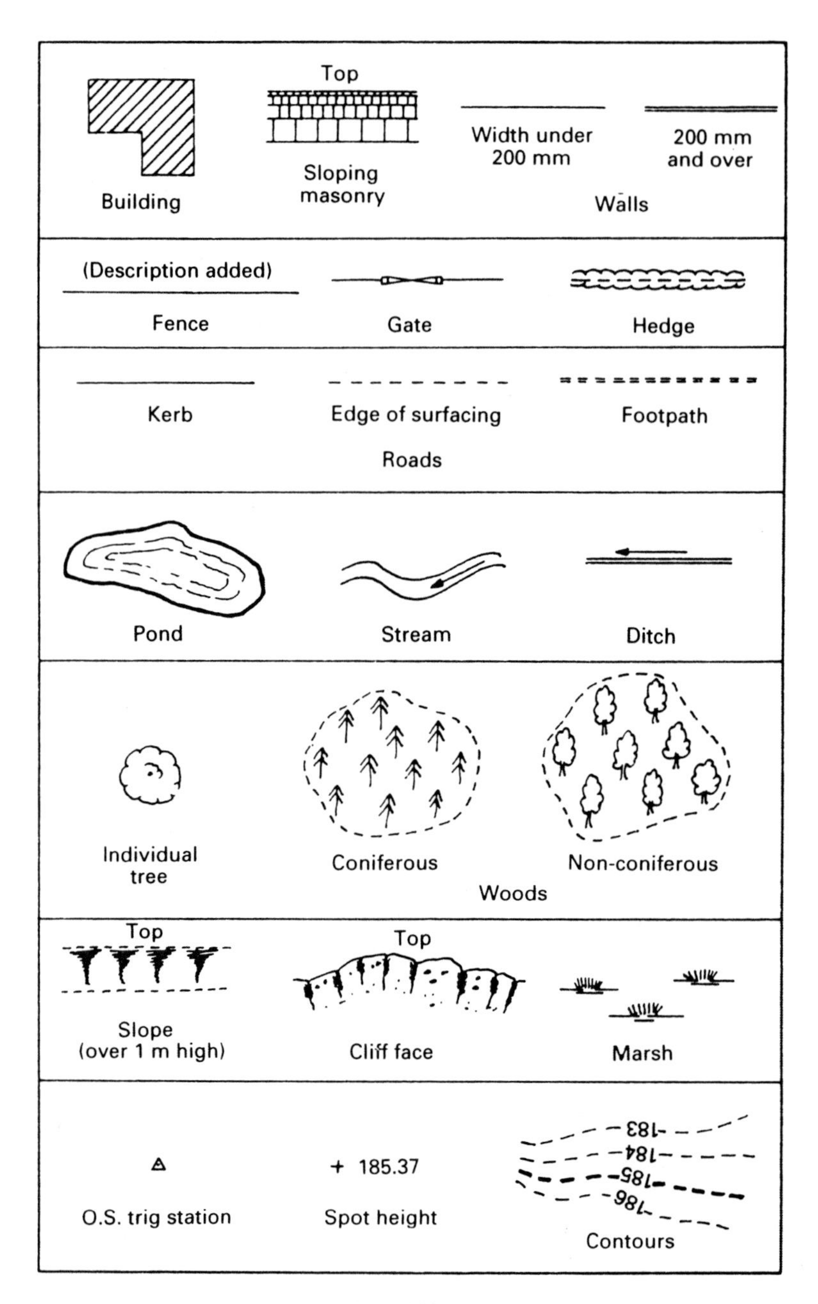

Fig. 6.6 1:500 symbols.

plotting of features, but to use a steel tape to determine sizes and to measure fine detail.

The theodolite/EDM outfit measures horizontal angle, zenith angle and slope distance. Modern instruments will display the calculated horizontal distance and rise or fall. Most (of the total station type) can also display the three-dimensional coordinates (easting, northing and reduced level) of surveyed points provided that sufficient information is entered at the start (3D coordinates of instrument station, bearing of reference object and heights of instrument and prism).

Field procedure

(1) Set up a theodolite mounted EDM outfit over a control station, or at a convenient location. An atmospheric correction can be entered at this stage, although the effects of such variations are usually negligible in detailing. The prism constant value should be checked.

(2) Sight a known reference object, setting the horizontal circle to zero or to other required value. If the equipment is not over a station, take EDM measurements to two known stations. To provide a check, a second reference object should be used, or an extra distance or angle should be measured. Data logging software will display the effects of a check.

(3) Level calculations will be simplified if the length of the prism pole is adjusted so that the height of the prism above the ground will equal the height of the theodolite trunnion axis above the station. (Prisms to match the instrument should be used so that for a non-coaxial optics outfit the target/prism vertical separation equals that of theodolite/EDM and there is no tilt error.)

If the prism height is not set to match that of the instrument, measure and record both heights. (Obstacles to visibility may require the prism height to be altered.)

For accurate heighting in a detail survey, the errors incurred in taping or staffing an instrument height may be excessive; the collimation level should be deduced from readings to a detail point of known level.

(4) If coordinates are to be displayed, or if automatic data logging is to be employed, ensure that enough current information is entered. There is always a danger that control information refer-

ring to a previous station (or previous survey!) may still be registered.

(5) Direct the chainman to the first point to be detailed. The pole should be held vertically, a check being made with the bullseye bubble on it. In windy conditions a ranging rod can be used as a diagonal brace for the pole. The target and prism should be pointed towards the EDM.

(6) Point the telescope/distancer towards the target/prism. For a non-coaxial optics outfit, the target of the (matching) prism assembly should be sighted through the theodolite.

(7) Follow the manufacturer's instructions for taking measurements and obtaining the required display. Usually readings on one face only are taken. The instrument should be set to take repeated measurements, so that a check on work is possible, although automatic logging allows only a single reading.

    Where coordinates are to be booked, or where a data logger is to be used, a display of the horizontal or slope distance should be incorporated so that a check for gross errors can be carried out.

    Usually a single keystroke activates data recording; detail point numbers are automatically incremented. Point information (identification and description) must be manually entered, although descriptive coding can be employed.

(8) If levelling accuracy of a few millimetres is required, vertical index and collimation errors must be eliminated (or double face readings should be taken) and curvature and refraction corrections should be applied (deduce the refraction coefficient from reciprocal measurements).

(9) A sketch is essential for manual booking of distance and angles, and highly recommend for coordinate and automatic booking. Record any change in prism height.

(10) Direct the chainman to the next point and continue the survey.

Planning of detailing

Before embarking on an EDM detail survey of any consequence, the surveyor and chainman must plan the work. Distinctions must be drawn between points to be surveyed by EDM, and those subsequently to be taped. The spacing of points to be surveyed should be agreed: this will be influenced by the irregularity in plan of features, the irregularity of the ground and the scale of drawing required.

Where a computer plot is to be produced, points to be joined up will have to be surveyed in strings, i.e. continuously and in sequence, and appropriately coded.

In an area of irregular topography, points to be levelled must be carefully chosen. Contour lines are often drawn connecting integer values obtained by linear interpolation (often within a triangular framework) from spot heights. Humps, hollows and break lines (where gradients or levels change suddenly) must be adequately surveyed.

Booking, calculations and plotting

Figure 6.7 is an elevation of EDM detailing. If the equipment is not set up over a station, the station coordinates should be determined from the measurements to the two reference object stations.

(1) EDM on optical theodolite (zenith vertical angles)

Use face left.
Read and book horizontal angle, D and θ.
Calculate: $H = D \sin \theta$
$V = D \cos \theta$
Rise $= h_I + V - h_T$
Bearing

Plotting may be by protractor and scale, or for more precision by rectangular coordinates calculated from H and the bearing.

(2) EDM connected to electronic theodolite

Either: Read horizontal angle, H and V
Calculate: Rise $= h_I + V - h_T$
Bearing
Plot by protractor and scale, or by rectangular coordinates

or: Enter initial bearing, station level, h_I and h_T
Read bearing, H and reduced level
Plot by protractor and scale, or by rectangular coordinates

or: Enter station 3D coordinates, initial bearing, h_I and h_T
Read 3D coordinates
Plot by rectangular coordinates

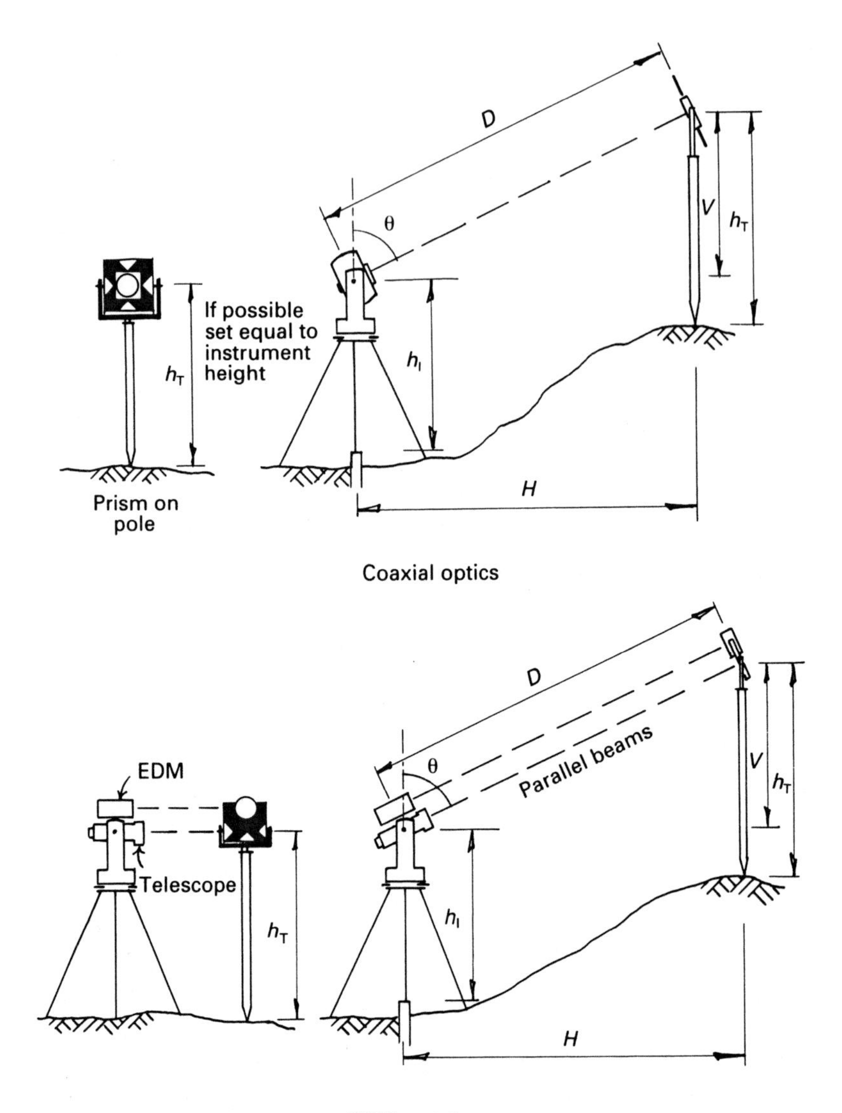

Fig. 6.7 EDM detailing.

or: Enter station 3D coordinates, initial bearing, h_I and h_T (on instrument or on data logger).
Record data on logger.
Plot by rectangular coordinates, or automatically by computer.

(3) *Integral EDM/theodolite of total station type.*

All display formats for an EDM added to an electronic theodolite are normally available. Using the full coordinate display or automatic data logging saves time and eliminates calculation errors.

For all plotting, the general principles for plotting linear surveys should be followed.

Accuracy of EDM detailing

In tracking mode, a typical distance error is ±10 mm. To this must be added centring errors at instrument and prism. For a pole-mounted prism, errors within ±30 mm may occur. For precise detailing, 1″ angular resolution, ±3 mm ±3 ppm distance error equipment and a tripod mounted prism should be used.

Example 6.1

EDM detailing was carried out from instrument stations X and Y, the coordinates of the points being (300.000 m E; 450.000 m N) and (500.000 m E; 650.000 m N) respectively. The reduced level at station X was 106.250 m AOD. Calculate the 3D coordinates of the points surveyed. Tables 6.1 and 6.2 show the bookings at stations X and Y.

Bookings

Table 6.1

At station	X	Station coordinates 300.000 E; 450.000 N	
Station level	106.250	Survey	Newtonfields
Ht of inst	1.530	Surveyor	F Bloggs
		Date	29.7.94

Point	Horizontal angle	H	V	h_T	Remarks/sketch
Y	00°00′00″	282.845	1.272	1.530	
A	328°30′35″	102.037	0.145	1.530	
B	71°24′20″	85.283	4.361	1.530	
C	16°49′05″	177.312	−6.723	1.530	

Table 6.2

At station	Y		Station coordinates 500.000 E; 650.000 N		
Station level			Survey		Newtonfields
Ht of inst	1.490		Surveyor		F Bloggs
			Date		29.7.94

Point	Horizontal angle	H	V	h_T	Remarks/sketch
X	00°00′00″				
C	335°36′10″	124.209	−7.963	1.530	
D	139°00′50″	157.674	5.235	1.530	

Solution

Figure 6.8 is a plan.

Coordinate calculations

(Coordinates are further explained in Chapter 5.)

At X sighting RO (Y)

$\Delta E = 500.000 - 300.000 \qquad = 200.000\,\text{m}$

$\Delta N = 650.000 - 450.000 \qquad = 200.000\,\text{m}$

$\text{Bearing of XY} = \tan^{-1}\dfrac{200.00}{200.00} \qquad = 45°00'00''$

$\text{Bearing of XA} = 45°00'00'' + 328°30'35'' = 373°30'35'' - 360°00'00'' = 13°30'35''$

XA Length = 102.037 $\therefore \Delta E = 102.037 \sin 13°30'35'' = 23.837$

Bearing = 13°30′15″ $\Delta N = 102.037 \cos 13°30'35'' = 99.214$

Add to coordinates of X (300.000 m E; 450.000 m N)
to derive coordinates of A (323.837 m E; 549.214 m N).

Bearing of XB = 45°00′00″ + 71°24′20″ = 116°24′20″

XB Length = 85.283 $\therefore \Delta E = 85.283 \sin 116°24'20'' = 76.385$

Bearing = 116°24′20″ $\therefore \Delta N = 85.283 \cos 116°24'20'' = -37.927$

$\therefore$ Coordinates of B are (376.385 m E; 412.073 m N).

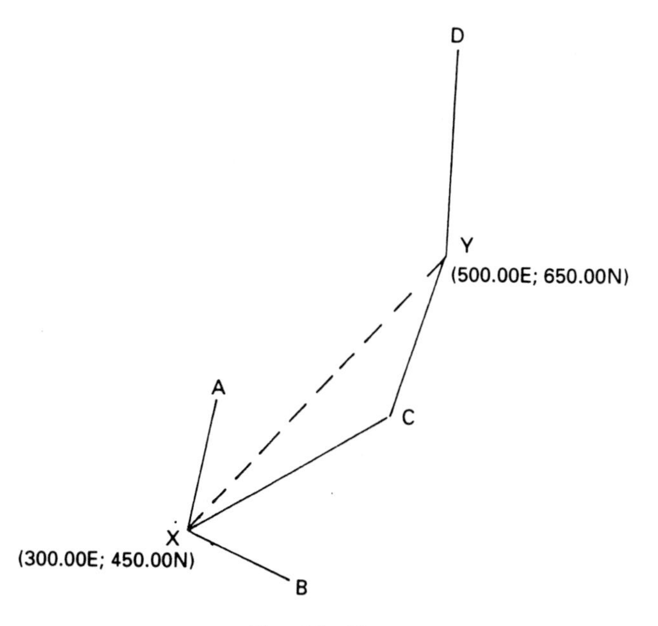

Fig. 6.8 Plan.

Bearing of XC = 45°00′00″ + 16°49′05″ = 61°49′05″

XC Length = 177.312 ∴ ΔE = 177.312 sin 61°49′05″ = 156.292

Bearing = 61°49′05″ ΔN = 177.312 cos 61°49′05″ = 83.740

∴ Coordinates of C are (456.292 m E; 533.740 m N).

Note: There is a check on the horizontal distance XY:

From coordinates XY = $[(200.00)^2 + (200.00)^2]^{1/2}$ = 282.843 m

By EDM XY = 282.845 m

The error of 2 mm is acceptable.

At Y sighting RO (X)

YX Length = 282.843

Bearing = 225°00′00″

Bearing of YC = 225°00′00″ + 335°36′10″ − 360°00′00″
= 200°36′10″

YC Length = 124.209 ΔE= 124.209 sin 200°36′10″ = −43.708

Bearing = 200°36′10″ ΔN= 124.209 cos 200°36′10″ = −116.265

∴ Coordinates of C are (456.292 m E; 533.735 m N)

(Checking with measurements from X; a discrepancy of 5 mm in the northings is acceptable.)

Bearing of YD = 225°00′00″ + 139°00′50″ − 360°00′00″ = 04°00′50″

YD Length = 157.674 ∴ ΔE = 157.674 sin 4°00′50″ = 11.037

Bearing = 4°00′50″ ΔN = 157.674 cos 4°00′50″ = 157.287

∴ Coordinates of D are (511.037 m E; 807.287 m E)

Level calculations

From X

RL of Y = 106.250 + 1.530 + 1.272 − 1.530 = 107.522 m AOD
RL of A = 106.250 + 1.530 + 0.145 − 1.530 = 106.395 m AOD
RL of B = 106.250 + 1.530 + 4.361 − 1.530 = 110.611 m AOD
RL of C = 106.250 + 1.530 − 6.723 − 1.530 = 99.527 m AOD

From Y

RL of C = 107.522 + 1.490 − 7.963 − 1.530 = 99.519 m AOD
RL of D = 107.522 + 1.490 + 5.235 − 1.530 = 112.717 m AOD
(The levels at C check within 8 mm.)

Note that the surveyor and chainman have neglected to set the pole height to equal that of the instrument at Y, but have remembered to note and book its height.

6.4 OPTICAL TACHEOMETRY

Optical tacheometry is a method of measuring distances indirectly using a level or a theodolite. At one end of the distance to be measured a staff or bar is set up; either the angle that a fixed length bar subtends is measured by the instrument at the other end, or the staff length subtending a fixed angle at the instrument is observed. There have been a number of systems devised; most have been made obsolete by the development of EDM, which is generally faster and more accurate. Nevertheless, perfectly adequate detailing (with contours) can be carried out using a theodolite and a levelling staff.

Fixed hair stadia system

The diaphragm glasses of levels and theodolites are marked with short horizontal hairs above and below the central horizontal cross-hair. These are the stadia hairs and are set so that a constant angle is subtended at the instrument by distant points with which they coincide (Fig. 6.9(a)).

Thus if a vertical levelling staff at one end of a line is observed through a theodolite at the other end, the staff readings against the stadia hairs, together with the zenith angle θ can be used to determine the horizontal distance. The difference between the upper and lower stadia hairs is called the stadia intercept S. For height determination the middle cross-hair staff value (m) is also taken (Fig. 6.9(b)).

As a check, m should be close (normally within 5 mm) to the mean of the two stadia readings.

Then $H = KS \sin^2 \theta$

and $V = KS \sin \theta \cos \theta$

Modern instruments are constructed so that K (the multiplying constant) = 100, and so

$$H = 100\,S \sin^2 \theta$$

$$V = 100\,S \sin \theta \cos \theta = 50\ S \sin 2\theta$$

and Rise A to B $= h + 50\,S\ \sin 2\theta - m$.

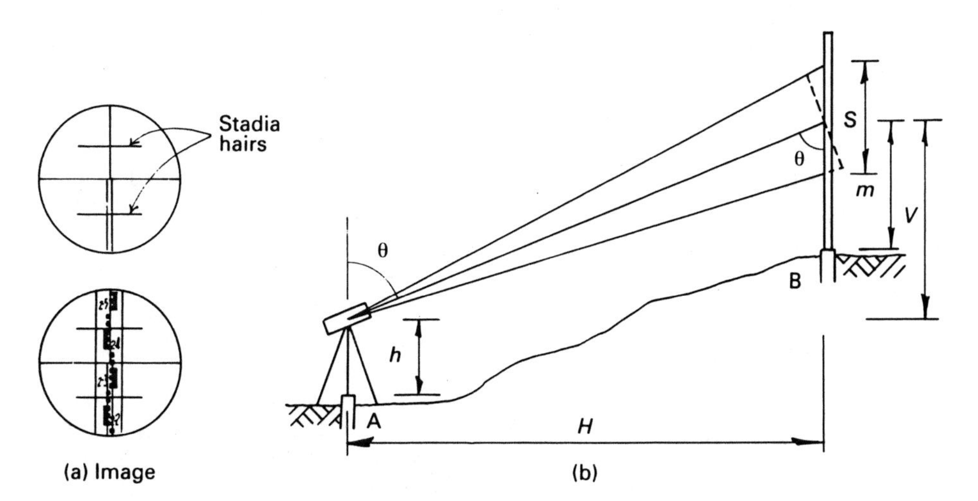

Fig. 6.9 Fixed hair stadia tacheometry.

Provided face left is used, the correct sign of V is automatically given. If the instrument reads vertical angles of elevation, $H = 100\,S\cos^2\theta$.

Field procedure

(1) Set up the theodolite over a control station. Measure the height of the trunnion axis above the station with a tape or staff. Sight a known reference object, setting the horizontal circle to zero or other required value.
Or
Set the theodolite at a convenient position. Take tacheometric readings at TWO known reference object stations. From these measurements, the coordinates of the instrument station can be calculated, and the trunnion axis level can be deduced.
(2) Direct the staffman to hold the staff vertically at the first detail point to be sighted. Turn the theodolite so that staff readings against the stadia and middle hairs can be taken. Calculations will be simplified if the telescope is tilted either to give an integral reading on the vertical circle or to arrange that the staff mid-reading equals the instrument height.
(3) Take and book the three staff readings, remembering that the middle one should equal the mean of upper and lower within 5 mm.
(4) Read the horizontal and vertical circles, noting the disposition of the vertical circle figuring. Readings to the nearest minute will be sufficiently precise.
(5) Make a clear sketch of the survey area and the detail points.
(6) Direct the staffman to a new point and continue tacheometry.

Note: It will usually be sensible to combine tacheometry with taped measurements, determining the position of features optically but measuring dimensions (width of road, size of buildings) directly.

Calculations and plotting

(1) Calculate H and V values.
(2) Calculate the reduced level at each staff station.
(3) Calculate the bearing of each detail point.
(4) Plot the points, either:

(a) Using a protractor (or adjustable set square) and scale. Or:
(b) By calculating rectangular coordinates.

(5) Write the reduced level by each point.
(6) Interpolate contours. (See Section 6.6.)
(7) The principles of plotting linear surveys should be followed.

Accuracy of stadia tacheometry

Sources of error:

(1) Non-vertical staff.
(2) Staff reading error.
(3) Vertical angle measurement.

There is also the multiplying constant of 100 used in calculating distances. Although relative precisions of 1 in 500 (horizontally) and 1 in 2000 (vertically) are possible, more realistic values are 1 in 250 and 1 in 1000. (Vertical relative precision is error divided by distance.)

Variable stadia systems

A variable stadia tacheometer is a theodolite with a moveable diaphragm on which curving stadia lines are engraved. A different part of the diaphragm is viewed at each elevation of the telescope, so that the fixed stadia hairs of a conventional theodolite are replaced by curving hairs giving an intercept varying as the telescope is tilted. The separation of the hairs is usually that for fixed hair systems multiplied by $\sin^2\theta$ for horizontal distance and by $\sin\theta\cos\theta$ (times a displayed factor, sometimes) for the vertical component. Measurement of vertical angles and the determination of trigonometric functions are therefore eliminated, but there is no improvement in accuracy.

6.5 GPS DETAILING

The Global Positioning System is increasing in use for detailing. A fixed receiver is set up at a station of known coordinates, while a roving (pole mounted) receiver is held at each detail point. After initialization to resolve ambiguities (see Chapter 5), the semi-kinematic ('stop and go') method can be used to locate detail.

During the survey, lock must be maintained on four or five satellites. Software is available to provide instant ('real time') display of coordinates with 10 mm accuracy in plan. Although GPS measurements are related to the WGS84 coordinates, transformation is carried out and local coordinates are displayed. Levels are accurate to about ±30 mm, but are related to the ellipsoid; correspondence between the ellipsoid and geoid for the area being surveyed is required.

The semi-kinematic method requires an open aspect for all survey points. Buildings, bridges and overhanging trees may interrupt the continuity of measurement causing lock on satellites to be lost. If this occurs a warning is displayed. Re-initialization can be carried out by reoccupying with the roving receiver a location whose position is known to within a few centimetres. Where there is a risk of lock being lost on several occasions it may be better to use the pseudo-kinematic method, whereby all points have to be reoccupied not less than one hour later. For a topographic detail survey this may not be practicable. GPS is not suitable for underground work.

If 10 mm accuracy is to be achieved, it requires considerable care in plumbing the receiver pole – a steadying raking ranging rod may be used. For any points to be located with greater accuracy it may be better to use the rapid-static method.

Plotting may be carried out directly from displayed (local) plan coordinates. Provided that it can be tied in to local levels, GPS can provide acceptable relative heighting. Programs are also available for land area calculations and the evaluation of earthwork volumes.

6.6 CONTOURING

A contour line is a line drawn on a plan connecting points of equal altitude. The irregularity of the ground and scale of the plan should indicate a suitable vertical interval, e.g. 0.5 m at 1:500, 1.0 m at 1:1000 and 1:1250, 2.0 m at 1:2500 and 5.0 m at 1:10 000 scales.

Contour lines are continuous and should form closed loops, although the closure may well occur outside the edge of a particular plan or map. Branching is not possible and lines will only run together at a vertical face (where they may be replaced by cliff or rock-face symbols), and cross at an overhang. The spacing of contour lines indicates the steepness of slopes, and the shape of grouped lines indicates recognizable features. Height values of contours are shown at intervals along the lines; when the value is viewed the correct way up, the slope

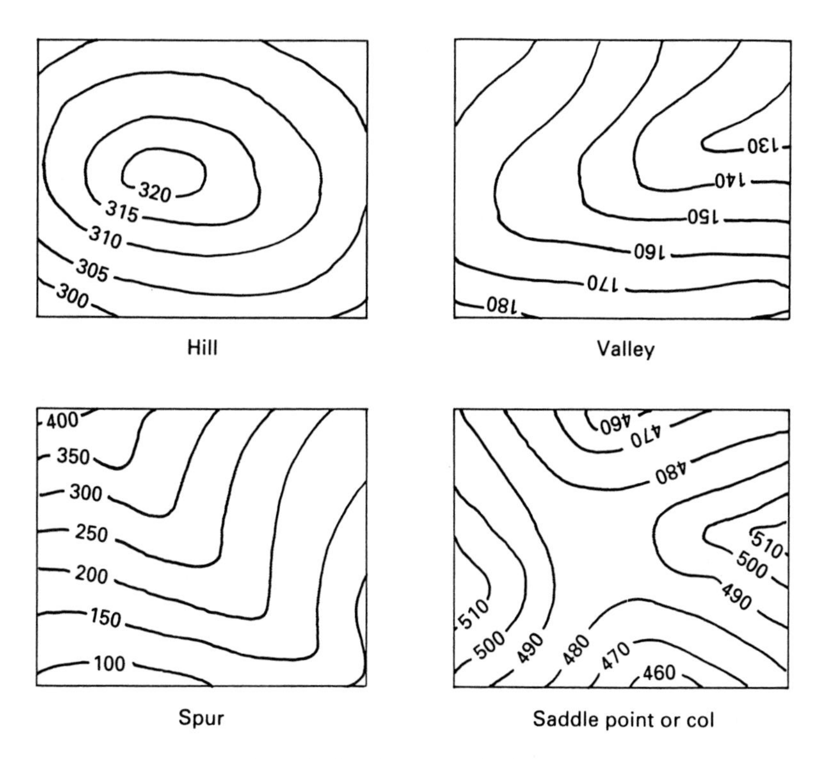

Fig. 6.10 Contoured features.

is rising 'up the paper', i.e. away from the viewer, Fig. 6.10 shows some typical features.

Contouring carried out from a ground survey will involve interpolation between points levelled. While ground levels are available at detail points surveyed by EDM, tacheometry or GPS, these will not normally be sufficient in number for linear interpolation to give an adequate representation of topography. Break-lines (sudden changes of gradient at streams, ridges or man-made earthworks, for example) must be properly surveyed and logged as consecutive points ('strings') so that plotted contours can change direction suddenly at the break. Contours may be interpolated from a grid of levels, in which case the grid interval must be chosen to give a satisfactory indication of topography, or from points forming a random pattern, which can be formed into a triangular network. Computer contour plotting programs involve such triangulation, the assumption being that the gradient along a triangle side is constant. Where a gradient is constant over an area, spot heights can be spread out, but in an area of surface irregularity they must be taken at much greater density. To enable the whole survey area to be contoured, boundaries must be surveyed, usually as strings.

A particular problem can arise at vertical surfaces, for example retaining walls. A boundary close to the bottom of the wall must be surveyed, with a further boundary established and picked up at the top.

Contour interpolation may be carried out by inspection, or by calculation, or by placing an interpolation graph on transparent material over the line to be divided by contours. Figure 6.11(a) shows the plot of part of a 10 m square level grid plotted with ground levels at adjacent points A and B of 47.63 and 49.28 m AOD.

$$48\,\text{m contour will cross AB at } \frac{48.00 - 47.63}{49.28 - 47.63} \times 10 = 2.2\,\text{m from A}$$

$$49\,\text{m contour will cross AB at } \frac{49.00 - 47.63}{49.28 - 47.63} \times 10 = 8.3\,\text{m from B}$$

In Fig. 6.11(b) an interpolation graph drawn on tracing paper is placed over a plotted level grid. The number '2.4' is lined up over 89.24 and '5.9' (on the next cycle of numerals) is lined up over 90.59. The zero line in the middle indicates a linear interpolation of the 90 m contour location.

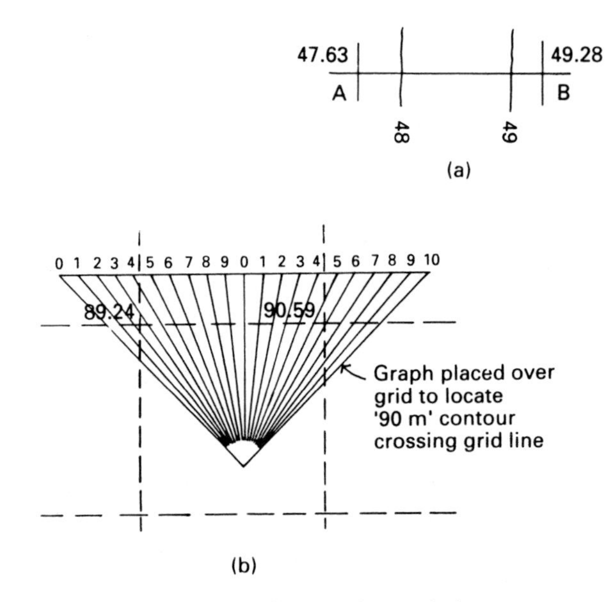

Fig. 6.11 Contour interpolation.

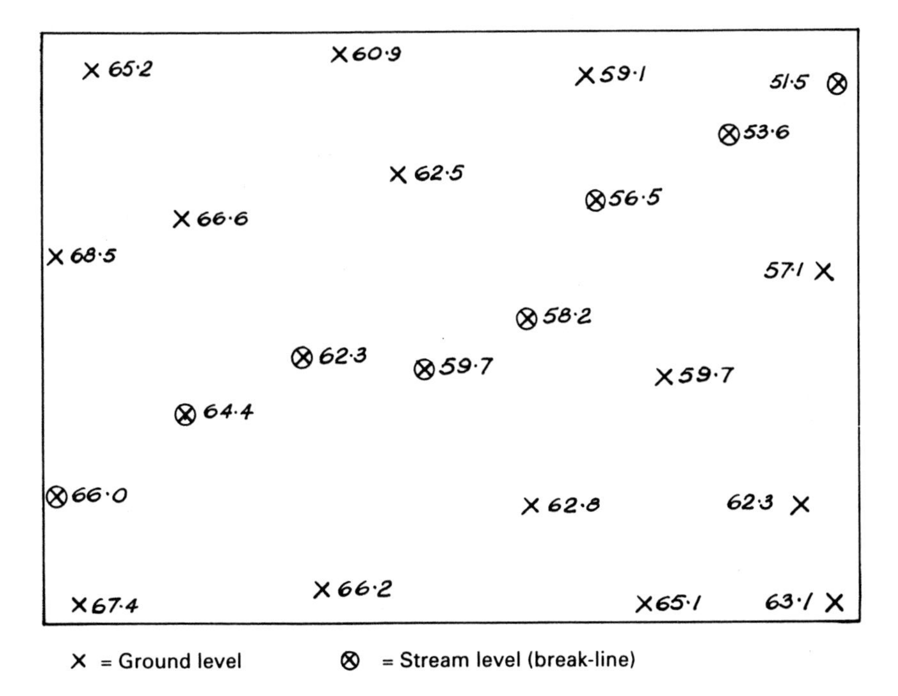

(a) Ground levels plotted

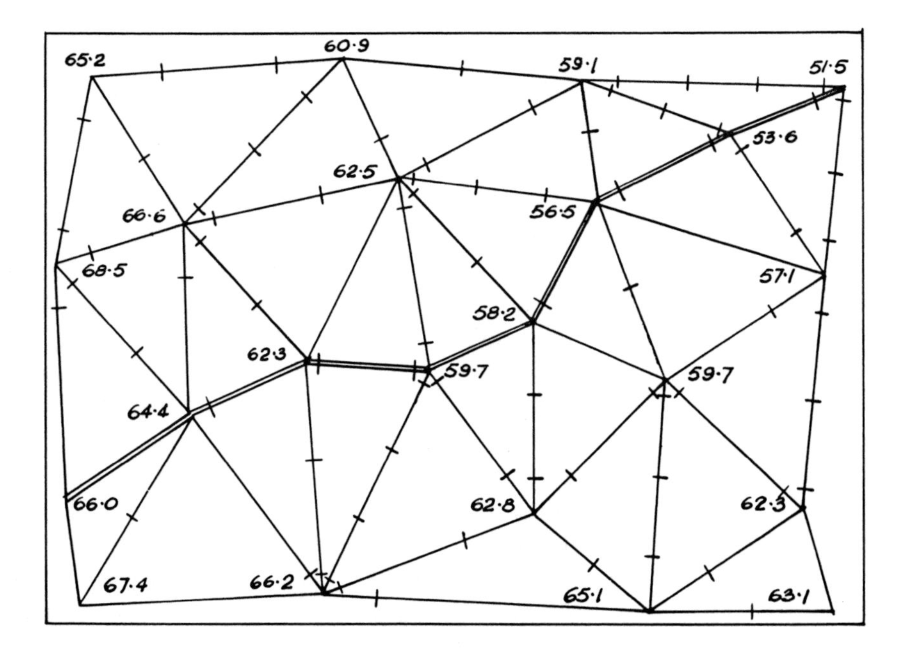

(b) Network drawn and 2 m integer values interpolated

Fig. 6.12 Contour plotting.

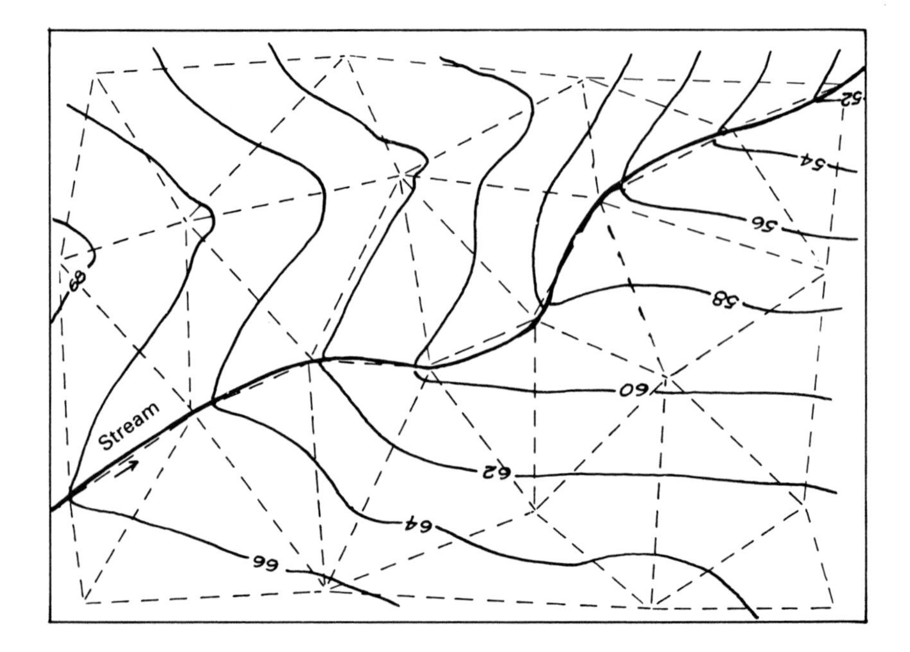

(c) Contours drawn

Fig. 6.12 contd.

Figure 6.12(a), (b) and (c) shows the three stages of a triangulated contour exercise. Spot levels have been taken along a stream (break-line) and on the ground either side. All spot levels are treated as nodes of triangles. The network is established by creating the smallest possible triangles and trying to keep them well conditioned. The crossing points of contours are interpolated along each triangle side and the contour lines are drawn in. Any contour line entering a triangle must leave it by the same or a different side.

Spot levels taken to suit the topography and converted to contours by the triangulation method should prove reliable. A level grid may be satisfactory for a limited area of regular topography – in an irregular area it is likely that some significant points will not be levelled, and that other points may be taken unnecessarily. Over a large area, the triangular method of interpolation will be very laborious to plot manually. It may be possible to estimate contours directly from spot levels, but there is a risk of impossible contour arrangements being produced. A check can be carried out by plotting a section through the dubious area. Figure 6.13 shows examples of impossibilities.

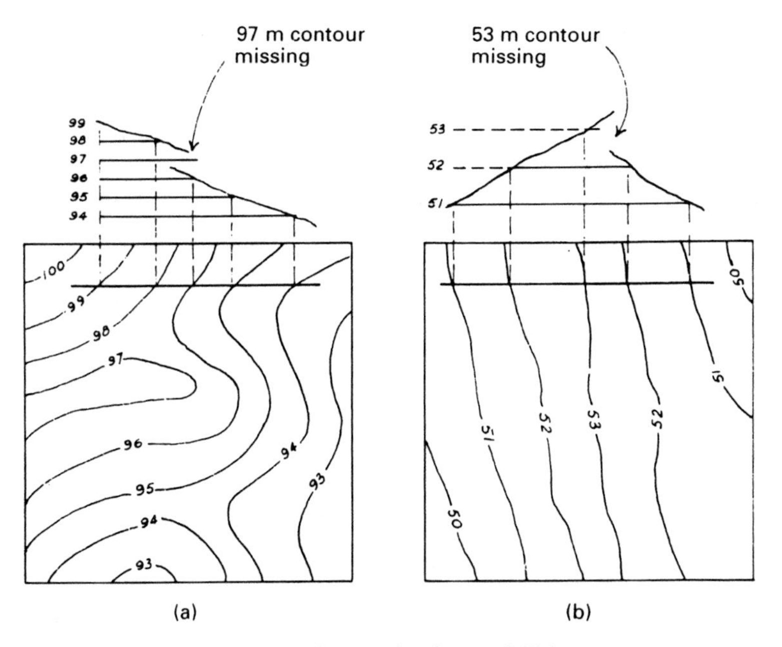

Fig. 6.13 Contouring impossibilities.

6.7 PHOTOGRAMMETRY

Detail surveys of large areas of land can be carried out by aerial photogrammetry. The survey area is photographed vertically from the air so that all parts appear on at least two photographs. Stereoscopic interpretation of pairs of overlapping prints can be carried out to produce (on a plotter) a plan and contours. A ground control survey is required to locate and orient the survey, to prevent cumulative overlap errors, scale (flying height variation) errors and tilt errors, and also to enable heighting to be tied to bench marks. The photography and interpretation is carried out by specialist organizations.

Terrestrial photogrammetry makes use of a pair of photo-theodolites, taking near horizontal overlapping photographs from each end of a baseline. The position and level of all points appearing on both prints can be determined.

While terrestrial photogrammetry can be used for detail work in a limited area, it is more often employed in surveys of vertical faces and the monitoring of movement and deformation.

More information on photogrammetry can be found in 'Further reading'.

6.8 DATA LOGGING AND PROCESSING

All field data, particularly if surveyed electromagnetically, can be stored by a computer to form a digital ground (or terrain) model, DGM or DTM. Coordinates in x, y and z directions are recorded, as is descriptive information for each point surveyed. Although the necessary programming for logging, transfer, retrieval, calculations, storage and plotting can be carried out by a surveyor with sound computing ability, it will generally be more convenient for the purchaser of data recording equipment to buy the necessary software. Most suppliers offer such a service, which they will tailor to suit a customer's requirements.

The components of a comprehensive system are as follows.

Field data logger

A conventional data logger, in appearance, is a cross between a calculator and a laptop computer. According to manufacture, it may be instrument mounted, tripod fixed or hand held. A cable connects the logger to the theodolite/EDM. Power is provided by a rechargeable nickel-cadmium battery, and typically there is sufficient memory space for a day's work to be stored. Some loggers have interchangeable memory cards or modules. In the office, loggers can be connected by a special cable to microcomputers.

Dedicated loggers are produced by surveying equipment manufacturers to provide automatic recording when interfaced with the appropriate field instruments (electronic theodolites and EDM). They are ready programmed for recording and transfer (Fig. 6.14).

A general purpose logger requires programming for manual and automatic recording, in the latter instance to suit the instrument with which it is to be used. It will also require programming so that data can be transferred to an office microcomputer, or printed or plotted directly (Fig. 6.15).

While a logging system is clearly most useful in a survey process, it can also bring considerable benefits to setting out. If data for points to be set out can be entered in the logger, it can interact in the field with observations and display information to assist correct setting out.

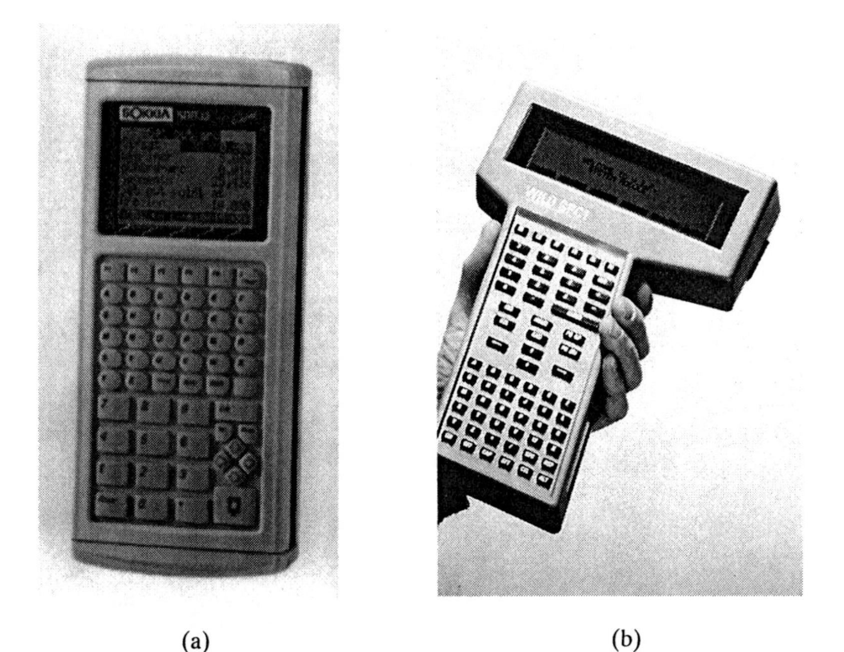

(a) (b)

Fig. 6.14 Data loggers: (a) Sokkia SDR33; (b) Leica Wild GPC1.

Data cards

A development taken up by several manufacturers is the data card. About the size of a credit card, but 5 mm thick, one of these memory cards can be inserted in a special slot on certain total stations (Fig. 6.16(a)). Field observations are automatically logged, and controlled (and annotated) on the instrument keyboard. In the office the card is accepted by a specially designed card reader which is coupled to a computer for data handling, storage and plotting (Fig. 6.16(b)). A fresh card can be put into the total station allowing fieldwork to be continued while the first batch of data is processed.

In general, the card merely records data and does not provide some of the manipulative features of a separate logger. One development, in the form of the Nikon DTM700 series is to provide a program card as well as a data card. This provides interaction in the field between collected observations and prepared data.

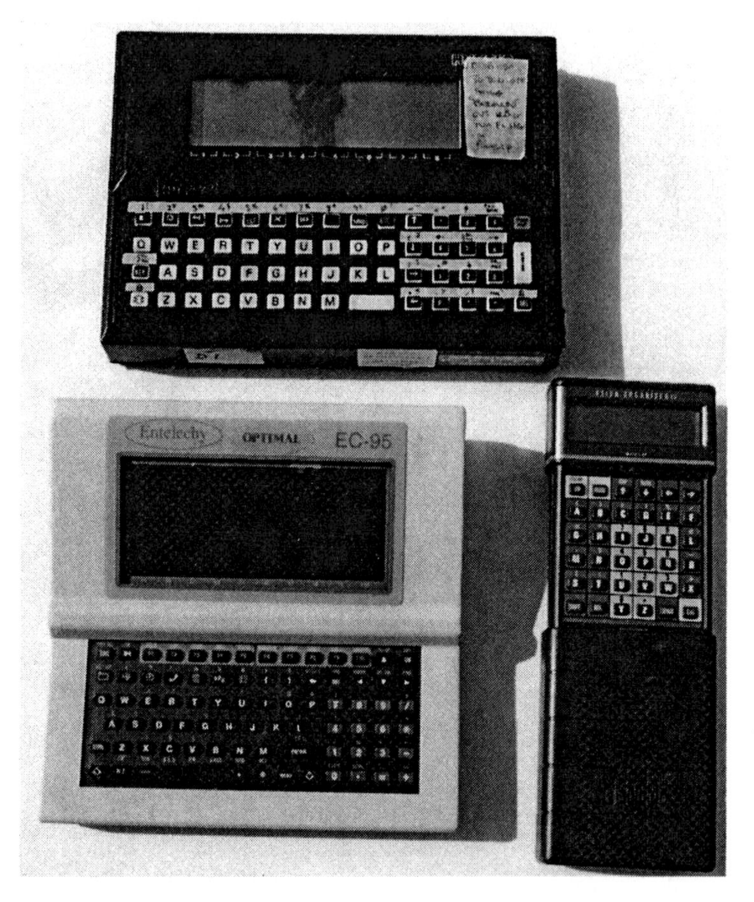

Fig. 6.15 Data loggers (non-dedicated).

Keyboard loggers

Another way of achieving flexible logging is to utilize a detachable keyboard-cum-logger, as provided with a Geodimeter 600 series. The keyboard/logger can be taken to the office for data plotting while a second keyboard is attached to the total station for continuation of measurement (Fig. 6.17).

GPS logging

All GPS outfits incorporate loggers. Where post processing has to be carried out, the received information must be stored at the field

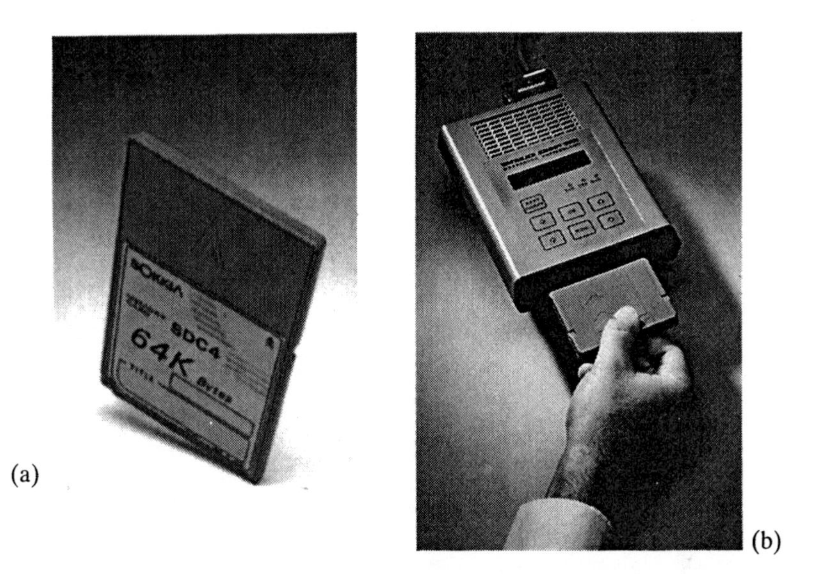

Fig. 6.16 (a) Sokkia memory card; (b) Leica GIF10 card reader.

Fig. 6.17 Geodimeter 600 detachable keyboard/logger.

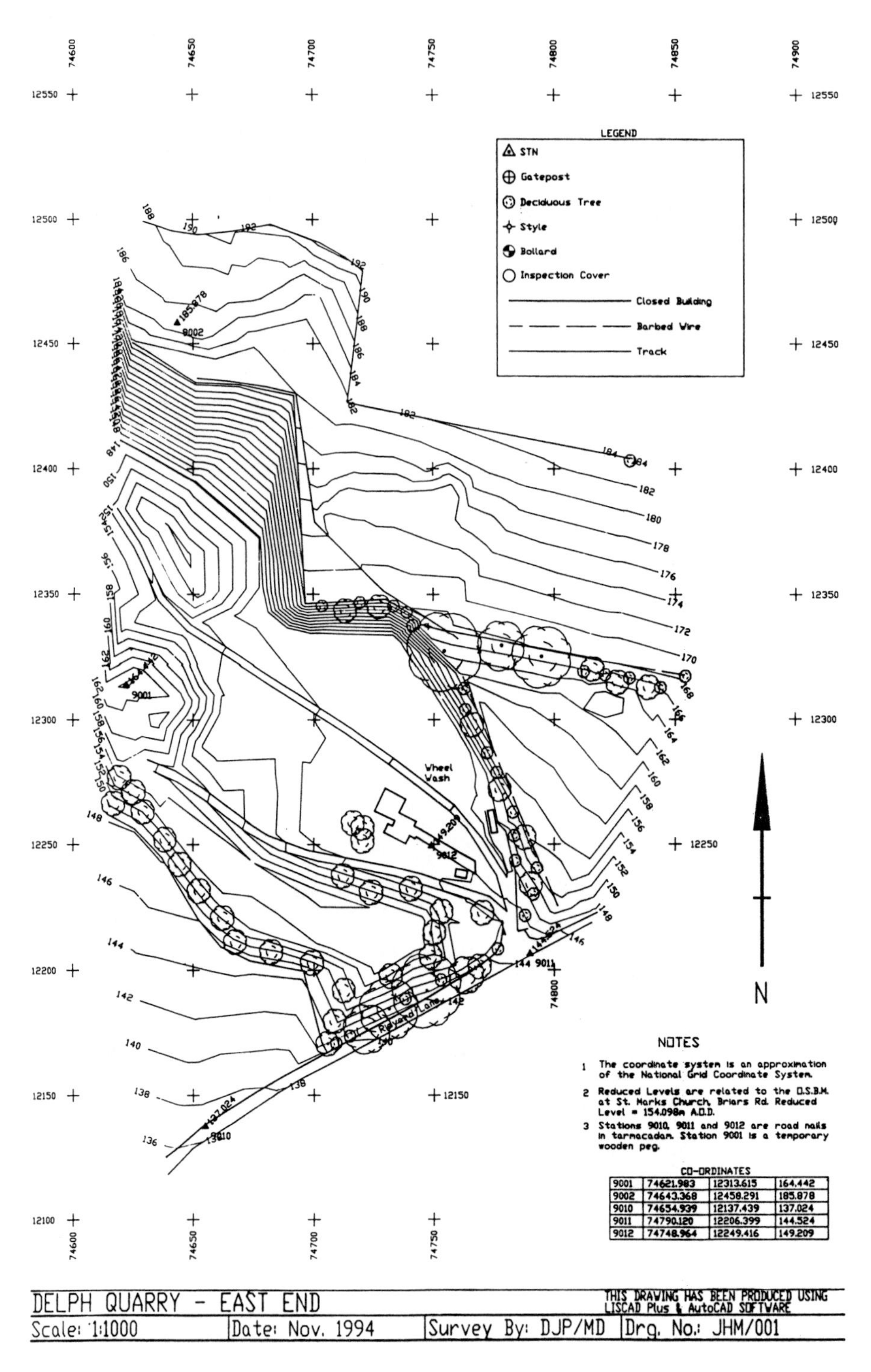

CO-ORDINATES			
9001	74621.983	12313.615	164.442
9002	74643.368	12458.291	185.878
9010	74654.939	12137.439	137.024
9011	74790.120	12206.399	144.524
9012	74748.964	12249.416	149.209

Fig. 6.18 Computer produced plan.

station, and handled subsequently by an office computer. For real time kinematic surveys, data has to be handled, displayed and stored for plotting. Hand-held data loggers and memory card systems are available.

Office equipment

A computer capable of handling surveying data must be available. Surveying equipment manufacturers and specialized companies produce programs for logging, storage and handling. Figure 6.18 shows a computer produced plan from a survey using a total station with a logging card.

In addition to the storage of information in rectangular coordinate format, programs can produce (via a plotter) plans with levels or contours, sections of the ground, and isometric views giving a three-dimensional impression.

Further operations are possible, for example the calculation of earthwork volumes, and engineering design, such as highway alignment, can be superimposed.

6.9 FURTHER READING

Atkinson, K.B. (ed.) (1980) *Developments in Close Range Photogrammetry*. London: Applied Science Publishers.

Burnside, C.D. (1985) *Mapping from Aerial Photographs*. 2nd edn. Oxford: Blackwell Science.

Karara, H.M. (ed.) (1989) *Non-topographic Photogrammetry*. Washington (USA): American Society for Photogrammetry and Remote Sensing.

Kilford, W.K. (1979) *Elementary Air Survey*. 4th edn. London: Pitman.

6.10 EXERCISES

Exercise 6.1

Figure 6.19 is a plan of part of an area surveyed by linear methods.

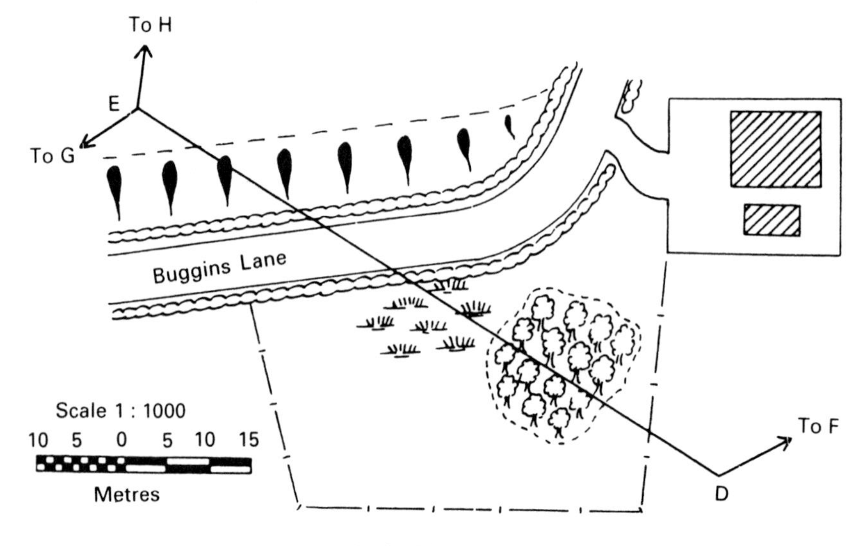

Fig. 6.19 Linear survey.

Draw up field book entries for line DE. Dimensions may be scaled or estimated.

Comment critically on the selection of the stations.

Exercise 6.2

A total station was used to survey an area close to the London–Glasgow railway line. The instrument was set up at X with coordinates of (1173.500 m N; 1629.750 m E) and the horizontal circle was set to zero on Y, with a bearing of 135°00′00″ from X. Observations were made to seven detail points at positions marked 1–6 on the sketch plan Fig. 6.20.

Points 1 and 2 were on the same rail where the line runs straight and at a constant grade on an embankment. Point 3 was on the road where it passes under the railway bridge. A remote object elevation sight was taken to the underside of the bridge deck immediately above point 3 on the road. Point 4 was at the toe of the embankment, close to point 1, and points 5 and 6 were the covers of drain manholes, the respective depths to inverts being 3.650 m and 2.135 m. The instrument axis level was 236.51 m AOD. All sights were made to a pole-mounted prism with a height of 1.450 m except for the remote object elevation sight which was directly to a point on the underside of the bridge.

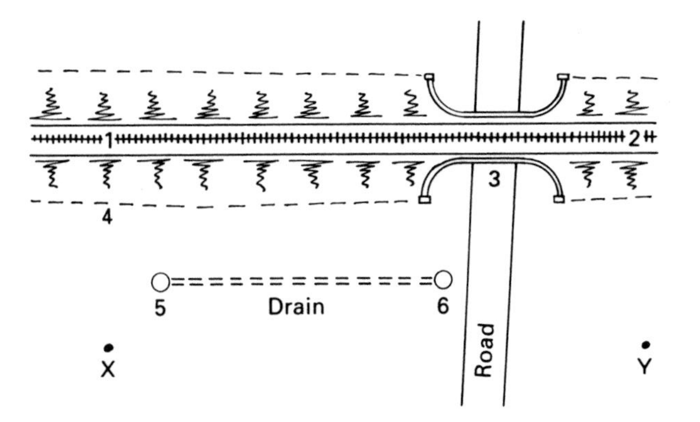

Fig. 6.20 EDM detail survey.

Table 6.3

Location	Horizontal reading	Zenith angle	Slope distance (m)	Notes
1	270°00′00″	84°54′11″	83.294	Rail
2	328°30′00″	87°57′12″	159.607	Rail
3	320°14′00″	90°49′24″	126.985	Road
3	320°14′00″	88°32′42″	–	U/S bridge
4	270°00′00″	90°00′29″	70.000	Ground
5	281°00′00″	90°14′53″	48.500	M/H cover
6	331°17′12″	91°34′30″	104.051	M/H cover

From the data in Table 6.3, calculate the plan distance from X and the reduced level of each point surveyed. Hence determine:

(a) The railway gradient.
(b) The clearance under the bridge.
(c) The manhole coordinates and the drain gradient.
(d) The approximate height of the embankment at point 1.

Exercise 6.3

A trainee surveyor, having forgotten the accepted method of booking staff readings, recorded the following notes while levelling points forming a 20 m square grid. The staff readings have been noted in the order of observation, the first sight from each station being a back sight, the final sight being a fore sight, as in Table 6.4.

Table 6.4

Instrument station	Staff station	Staff reading
P	TBM 'X'	2.92
	A1	3.70
	A2	1.93
	A3	0.35
Q	A3	2.40
	A4	1.42
	A5	0.46
	B5	1.29
	B4	2.17
	B3	3.54
R	B3	0.24
	B2	2.12
	B1	3.87
S	B1	2.88
	C1	3.25
	C2	2.01
	C3	1.10
	C4	0.62
	C5	0.09

The reduced level of the TBM was 62.15 m AOD. Book the readings in the accepted manner and calculate levels by rise and fall or by collimation method.

Grid lines A, B and C ran north–south, A being the most westerly; lines 1 to 5 ran east–west, line 1 being the most southerly. Draw the grid at 1 to 500 scale, plot the levels and interpolate contours at 1 m (integral) values.

Exercise 6.4

Figure 6.21 is a plan to the scale shown of points levelled for contouring. Plot the points at 1 to 250 scale, draw a suitable triangular network and interpolate contours at 2.00 m vertical intervals from the ground levels given in Table 6.5.

Table 6.5

Point	Ground level (m AOD)
A	92.60
B	93.20
C	94.80
D	91.80
E	99.00
F	95.00
G	100.20
H	91.80
I	94.60
J	100.30

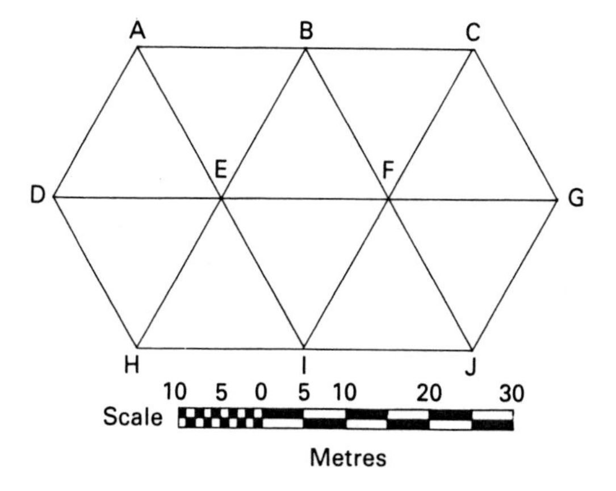

Fig. 6.21 Levelling network.

7 Errors and Adjustments

7.1 TYPES OF ERROR

Surveying fieldwork is liable to three types of error: mistakes, systematic errors and random errors.

Mistakes

Mistakes occur when the surveyor blunders through lack of experience, lack of care or human error. The effects are unpredictable and often large. Frequent causes of mistakes by the inexperienced surveyor are selection of unsuitable procedures, selection of unsuitable equipment and unskilled handling of equipment. The practised surveyor is less likely to err in basic field technique, though faulty calculations and incorrect reading of drawings can cause spectacular blunders!

Mistakes should be identified and eliminated. Wherever possible the work should be self-checking; alternatively, repetition by the same, or preferably a different, method should be carried out. Where a mistake has occurred the calculations should first be checked; if this check does not reveal the source, it is likely that some or all of the fieldwork will have to be repeated. (Calculation processes should also have checks built into them.)

Fieldwork should be free from mistakes before calculations and plotting are carried out, although computation may be used to detect and isolate a lurking mistake.

Systematic errors

When predictable variations in equipment performance occur, or when conditions result in the measurement made being different from the measurement required, systematic errors are produced. Though small, their effects are usually significant and often cumulative. Corrections must be applied to the field measurements to remove such errors. Measurements made with instruments which are out of adjustment (calibration error) can be corrected provided that the magnitude and sign of the maladjustment can be found. Slope measurements must be reduced to give plan lengths, and operating conditions (overtensioned tape, non-standard atmospheric conditions for EDM measurements) should be taken into account.

Thus the sources of error must be known, and any further measurements necessary to enable corrections to be applied must be taken (slope angle for a plan length).

Error analysis can be performed to investigate the effects of systematic errors on derived quantities for calculations and plotting.

Random errors

Mistakes too small to be identified and systematic errors caused by imperceptible variations in conditions and equipment form random errors. They are small and unpredictable.

Where related measurements display inconsistencies through random errors, adjustments can be applied to achieve consistency.

Example 7.1

A closing error of 1.005 m is found after returning levelling to the bench mark from which it started. A mistake of 1.000 m has probably occurred.

Example 7.2

In Fig. 7.1(a) the plan length is required. A measurement on the slope has been made with EDM and an atmospheric correction of +5 parts per million is to be applied. The difference between the slope and plan lengths is a systematic error, as is the error caused by atmospheric variation from standard conditions.

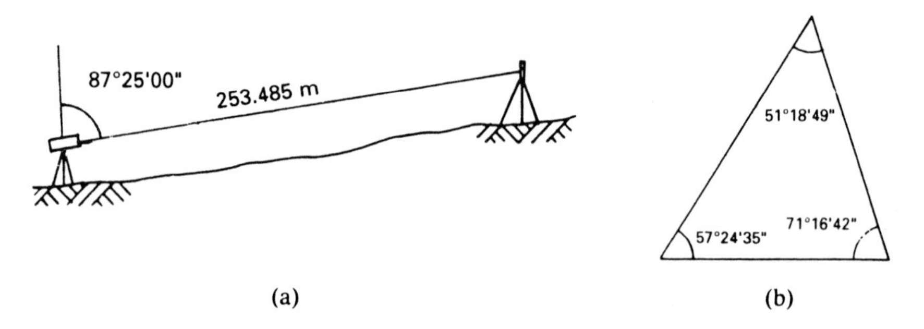

Fig. 7.1 Systematic and random errors.

$$\text{Plan length} = \left(253.485 + \frac{0.005 \times 253.485}{1000}\right) \sin 87°25'00''$$
$$= 253.229\,\text{m}$$

Example 7.3

Figure 7.1(b) shows a plane triangle where the angles sum to 180°00′06″. The error, likely to be random, is 6″. If these angles are to be used in any calculations or plotting, they must be adjusted; the adjustment of each by −2″ is often acceptable.

7.2 ANALYSIS OF ERRORS

In carrying out his or her duties, a surveyor is trying to establish reliable values of quantities. Absolute accuracy is impossible to achieve, and an examination of the likely errors enables the surveyor to select methods and equipment which should produce errors smaller than the permitted tolerances. An error analysis performed after taking measurements then allows the surveyor to see whether the desired accuracy has probably been achieved.

From experience the practised surveyor should have a good idea of suitable equipment for a task; on site, simple closing checks can often indicate whether acceptable accuracy has been achieved. Nevertheless, there will be times when a mathematical analysis of errors must be carried out, particularly when very close tolerances are set.

The need for adjustment

Fieldwork is usually organized to provide checks. A series of measurements with an acceptably small closing error will almost certainly contain inconsistencies. Control surveying by triangulation will produce closing errors of angles while angular and linear errors will be present in traverses. The inconsistencies must be removed so that calculations can be carried out based on a single consistent set of results.

On construction sites, surveying or setting out frequently does not require adjustment. Where the closing error is smaller than the specified tolerance, the individual measurements will *probably* be within tolerance and can be accepted.

Characteristics of adjustment

(1) Observed measurements are adjusted to give consistency. (The term adjustment is used in preference to 'correction' employed by a number of authors; correction here is reserved for the treatment of systematic errors.)
(2) Adjustments should not be applied to rectify mistakes or to correct systematic errors.
(3) After adjustment the measurements should be mathematically consistent and *as a whole* will *probably* be 'more correct'. However, some individual measurements may possibly be less correct than the original observations!
(4) The difference between an arbitrarily adjusted value and a rigorously adjusted one is often so small that the extra effort in applying a rigorous adjustment may appear wasted. All adjustment processes should be unbiased, but rigorous methods involving the theory of least squares give the 'most probable' results.
(5) The development of modern computing facilities has greatly assisted rigorous adjustment of complex systems, and the adoption of a particular adjustment method may depend on the type and precision of the field survey, the surveyor's judgement and the computer and program availability. Time is precious and the good engineering surveyor has to judge how much time must be spent to achieve adequate precision.

7.3 RELIABILITY OF A SINGLE QUANTITY

There is a likelihood that the accuracy of a measurement of a quantity will be improved if a mean is taken from several observations. The term accuracy refers to the nearness to the truth of the observations; as the true value cannot be determined, the precision of the measurements is taken as an indicator of the accuracy. Precision refers to the degree of refinement in the measuring technique and the repeatability of the result. (A group of precise readings will not give an accurate result if a systematic error is present.)

It is assumed that a group of measurements (free from systematic errors) of a quantity will conform to a normal distribution. Figures 7.2(a) and (b) show a histogram with grouped values and the normal distribution curve where the class intervals have been decreased to produce a smooth curve.

The distribution is used to portray the variation of individuals within a group, for example height variation of adults, diameter variation within a batch of steel rods. A narrow hump indicates small overall variation; a wide hump indicates a greater spread of measurements. A mean value occurs at the top of the curve which is symmetrical; from the data the standard deviation (a measure of the spread) can be calculated. Properties of the distribution are:

(1) Positive and negative variations from the mean are equally distributed.
(2) Small variations from the mean are common.
(3) Large variations from the mean are rare.

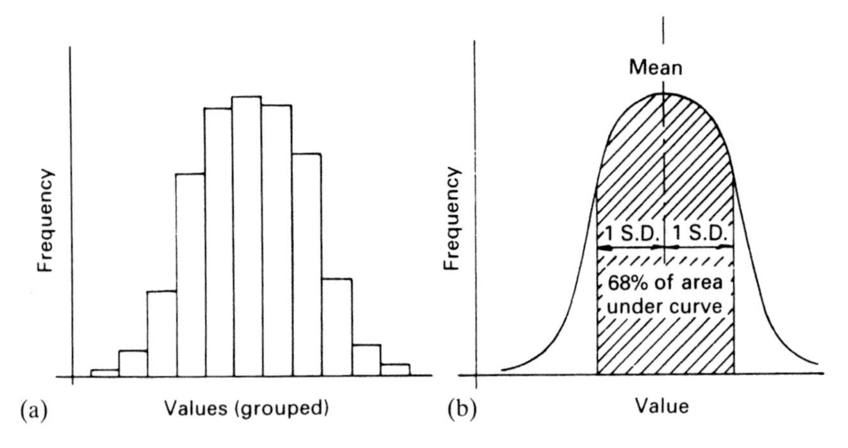

Fig. 7.2 The normal distribution.

In surveying, this distribution represents the variation of values for a quantity measured several times. The properties listed accord well with characteristics of random errors. The mean becomes the most probable value of the quantity, while the standard deviation becomes the standard error indicating the precision of the measurements.

1 standard deviation either side of the mean covers 68% of the 'population'
2 standard deviations cover 95%
3 standard deviations cover 99.75%
3.29 standard deviations cover 99.90%.

For a population (infinitely large number of observations), the mean is μ, and:

Standard deviation, of a measurement $\sigma = \pm\left(\frac{\Sigma(x - \mu)^2}{n}\right)^{1/2}$

where x represents each of the n observations.

For a sample (limited number of observations):

$$\text{Mean } (\bar{x}) = \frac{\Sigma x}{n}$$

and $\bar{x}$ provides an estimate of μ. $\Sigma(x - \bar{x})^2$ has a minimum value, and thus is less than $\Sigma(x - \mu)^2$. To compensate, n is replaced by $(n - 1)$ and standard deviation, σ, is replaced by standard error, s.

$$\text{Then } s = \pm\left(\frac{\Sigma(x - \bar{x})^2}{n - 1}\right)^{1/2}$$

Means and standard errors can be evaluated on a scientific calculator.

The surveyor should ascertain whether the denominator is n or $n - 1$; it is usually the latter, although some instruments can calculate for both.

The standard error given is that for a single measurement within the distribution; it must be divided by $n^{1/2}$ to give the standard error of the mean, hence:

$$s_{\bar{x}} = \pm\left(\frac{\Sigma(x - \bar{x})^2}{n(n - 1)}\right)^{1/2}$$

The standard error can have several uses:

(1) To give an indication of the precision of the measurement of a

quantity; a most probable value can be quoted as the mean with the standard error indicating a range. The variance (square of the standard error) is also useful, particularly in the weighting and analysis of related measurements.

(2) To give an estimate of the number of observations of a quantity to produce a specified standard error; the error for a single observation must be initially available.
(3) To provide rejection criteria for observations not representative of the quantity.
(4) To combine with the standard errors for other quantities where measured quantities are to be combined to provide a derived quantity.

Example 7.4

Find the most probable value and standard error of an angle for which the following five readings have been taken: 58°29′40″, 58°29′33″, 58°29′31″, 58°29′29″, 58°29′37″.

Solution

$$\text{Mean, } \bar{x} = 58°29' + \frac{40'' + 33'' + 31'' + 29'' + 37''}{5} = 58°29'34''$$

$$\text{Standard error, } s = \pm\left(\frac{(6)^2 + (-1)^2 + (-3)^2 + (-5)^2 + (3)^2}{(5-1)}\right)^{1/2}$$

$$= \pm 4.47'' \rightarrow \pm 4.5''$$

$$\text{Standard error of the mean, } s_{\bar{x}} = \pm\frac{4.47}{(5)^{1/2}} = \pm 2.0''$$

$$\text{Most probable value, (MPV)} = 58°29'34'' \pm 2''\text{(SE)}$$

Example 7.5

Determine the number of readings to be taken of the angle in example 7.4 to give a standard error (of the mean) of ±1″.

Solution

Standard error = ±4.47″; assume that this figure will remain constant for any number of observations.

$$s_{\bar{x}} = \frac{4.47}{(n)^{1/2}} = 1; \quad n = (4.47)^2 = 19.98$$

Take 20 readings.

Rejection criteria

In a set of observations, it may be found that some lie a considerable distance from the mean. For a large number of observations, the mean will not be affected by an isolated rogue result. Where the number of observations is small, usually the case in surveying measurements, readings lying outside certain limits must be investigated. For a group of twenty or fewer observations, any readings lying more than 2.5 standard errors from the mean should be rejected if supporting evidence (field book notes, surveyor's recollection) can be found. If such evidence is not available, it may be better to retain the doubtful values; subsequent calculations may confirm or deny the wisdom of this. After outlying readings have been removed, the mean and standard error should be recalculated.

7.4 WEIGHTING

Frequently, observations are not of equal precision. The precision should reflect the type and condition of equipment, the weather conditions, the skill of the surveyor and the number of readings of the quantity. Where groups of readings of differing precision have to be combined, account must be taken of the weights.

The weights are purely comparative measures. For a population, the weight of the mean is inversely proportional to the square of the standard error (or variance), thus for a sample:

$$\text{Weight} \propto \frac{1}{(\text{standard error})^2}$$

Consider an angle measured several times by two observers giving results of 30°29′57″ ±3″ and 30°29′54″ ±2″ (mean ± standard error). The weights are proportional to:

$$\frac{1}{3^2} \text{ and } \frac{1}{2^2}, \quad \text{i.e.} \quad \frac{1}{9} \text{ and } \frac{1}{4}$$

and may be given as:

4:9 or 1:2.25

However, the assignment of weights should be carried out with some care; samples are often small and quite unrepresentative standard errors often occur. The surveyor should be prepared to replace calculated values with more sensible ones according to his or her judgement. Manufacturers' specifications and values quoted by experienced surveyors for typical operations using particular equipment may be referred to.

Mean of observations of different weights

For observations x of different weights w:

$$\text{The Mean, } \bar{x} = \frac{\Sigma(wx)}{\Sigma w}$$

Standard error of an observation of weight w_0

$$= \pm\left(\frac{\Sigma(w(x - \bar{x})^2)}{w_0(n - 1)}\right)^{1/2}$$

Standard error of the weighted mean

$$s_{\bar{x}} = \pm\left(\frac{\Sigma(w(x - \bar{x})^2)}{(n - 1)\Sigma w}\right)^{1/2}$$

Two or more sets of observations of one quantity

Where two (or more) sets of readings each giving a mean and a standard error for the same quantity are available, they can give a combined MPV. The combined mean is calculated as a weighted mean. To calculate the combined standard error, the weight of each set of observations is considered.

Combined weight = Sum of weight of each set

For two sets A and B giving a combined value C, $w_C = w_A + w_B$, and $(w_A + w_B)s_C{}^2 = w_A s_A{}^2 = w_B s_B{}^2$. Hence:

$$s_C = s_A\left(\frac{w_A}{w_A + w_B}\right)^{1/2} = s_B\left(\frac{w_B}{w_A + w_B}\right)^{1/2}$$

Example 7.6

Two sets of observations for a baseline have been determined as 104.050 ± 0.007 m (standard error) and 104.045 ± 0.011 m. Calculate the MPV and SE for both sets combined.

Solution

$$\text{Weights are } \frac{1}{(0.007)^2} : \frac{1}{(0.011)^2} = \frac{(0.011)^2}{(0.007)^2} : \frac{(0.011)^2}{(0.011)^2} = 2.47 : 1$$

$$\text{Combined mean} = \frac{(104.050)(2.47) + (104.045)(1)}{(2.47 + 1)} = 104.049\,\text{m}$$

$$\text{Combined standard error} = \pm 0.007\left(\frac{2.47}{3.47}\right)^{1/2} = \pm 0.006\,\text{m}$$

$$\text{Combined MPV} = 104.049\,\text{m} \pm 0.006\,\text{m (SE)}$$

7.5 QUANTITIES DERIVED FROM TWO OR MORE OBSERVED QUANTITIES

Where F is a function of several independent measurements, i.e. $F = f(a, b, c, \ldots)$:

$$s_F = \pm\left[s_a^2\left(\frac{\partial F}{\partial a}\right)^2 + s_b^2\left(\frac{\partial F}{\partial b}\right)^2 + s_c^2\left(\frac{\partial F}{\partial c}\right)^2 + \ldots\right]^{1/2}$$

This formula can be simplified for functions that are sums, differences and, in some cases, products.

Addition

$$F = a + b + c; \frac{\partial F}{\partial a} = \frac{\partial F}{\partial b} = \frac{\partial F}{\partial c} = 1$$

and therefore $s_F = \pm(s_a^2 + s_b^2 + s_c^2)^{1/2}$.

Subtraction

$$F = a - b; \frac{\partial F}{\partial a} = 1; \frac{\partial F}{\partial b} = -1$$

and therefore $s_F = \pm(s_a^2 + s_b^2)^{1/2}$.

Multiplication and division

Where F is a product of *simple* variables ($= a \times b \times c$) so that

$$\frac{\partial F}{\partial a} = bc, \frac{\partial F}{\partial b} = ac \text{ (i.e. not powers or trigonometric quantities)}$$

$$s_F = \pm abc\left[\left(\frac{s_a}{a}\right)^2 + \left(\frac{s_b}{b}\right)^2 + \left(\frac{s_c}{c}\right)^2\right]^{1/2}$$

where $F = \frac{a}{b}$, (simple variables) $s_F = \pm\frac{a}{b}\left[\left(\frac{s_a}{a}\right)^2 + \left(\frac{s_b}{b}\right)^2\right]^{1/2}$

Example 7.7

Find the length (MPV and SE) of a baseline measured in three bays as 99.365 ± 0.005 m, 87.260 ± 0.010 m and 98.665 ± 0.002 m (standard errors).

Solution

$$\text{Length} = 285.290 \pm [(0.005)^2 + (0.010)^2 + (0.002)^2]^{1/2}$$

$$\text{MPV} = 285.290\,\text{m} \pm 0.011\,\text{m (SE)}$$

Example 7.8

Find the area of a rectangle with sides 27.444 ± 0.007 m and 63.235 ± 0.005 m (standard errors).

Solution

$$\text{Area} = 27.444 \times 63.235$$

$$\pm (27.444 \times 63.235) \times \left[\left(\frac{0.007}{27.444}\right)^2\left(\frac{0.005}{63.235}\right)^2\right]^{1/2}$$

$$\text{MPV} = 1735.42 \pm 0.47\,\text{m}^2 \text{ (SE)}$$

Example 7.9

A new station, P, has been surveyed using a total station set up at an existing station A. The coordinates of A are (200.000 ± 0.010 m E; 100.00 ± 0.010 m N) and its reduced level is 100.000 ± 0.010 m AOD. A reference object X with a bearing from A of 60°00′00″ ± 20″ has been sighted. (Standard errors quoted.)

From the following observations (and standard errors), calculate the probable three-dimensional coordinates (with standard errors) of P.

Horizontal angle PÂX	= 54°32′00″ ± 10″
Zenith angle (to P)	= 86°30′00″ ± 20″
Slope distance AP	= 236.630 ± 0.005 m
Height of instrument	= 1.530 ± 0.005 m
Height of prism at P	= 1.530 ± 0.005 m

Solution

Referring to Fig. 7.3

$$E_P \text{ (Easting of P)} = E_A \text{ (Easting of A)} + D \sin\theta \sin(\text{N}\hat{\text{A}}\text{X} - \text{P}\hat{\text{A}}\text{X}) = 222.501\,\text{m}$$

$$\frac{\partial E_P}{\partial E_A} = 1; \ \frac{\partial E_P}{\partial D} = \sin\theta \sin(\text{N}\hat{\text{A}}\text{X} - \text{P}\hat{\text{A}}\text{X}) = 0.095089$$

$$\frac{\partial E_P}{\partial \theta} = D\cos\theta \sin(\text{N}\hat{\text{A}}\text{X} - \text{P}\hat{\text{A}}\text{X}) = 1.37621\,\text{m}$$

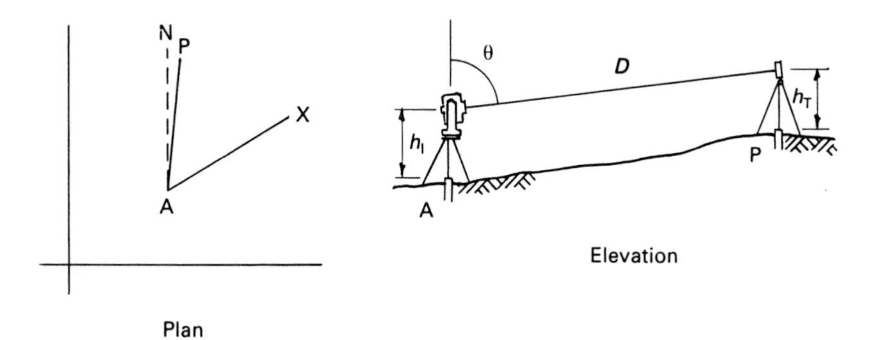

Fig. 7.3 EDM positioning.

$$\frac{\partial E_P}{\partial \hat{NAX}} = D \sin\theta \cos(\hat{NAX} - \hat{PAX}) = 235.1144\,\text{m}$$

$$\frac{\partial E_P}{\partial \hat{PAX}} = -D \sin\theta \cos(\hat{NAX} - \hat{PAX}) = -235.1144\,\text{m}$$

$$s_E = \pm\left[(0.010 \times 1)^2 + (0.005 \times 0.0951)^2 + \left(\frac{20}{206\,265} \times 1.37621\right)^2 + \left(\frac{20}{206\,265} \times 235.1144\right)^2 + \left(\frac{10}{206\,265} \times -235.1144\right)^2\right]^{1/2}\text{m}$$

(1/206 265 factor is to convert seconds to radians)

$$= \pm(0.0001 + 0.0000002 + 0.000000017 + 0.0005197 + 0.0001299)^{1/2}$$

↓ E_A error ↓ D error ↓ θ error ↓ NÂX error ↓ PÂX error

$$= \pm 0.027\,\text{m}$$

Easting of P is 222.501 ± 0.027 m (SE).

Principal source of error is in measurement of angle NÂX; slope distance and zenith angle errors are insignificant.

$$N_P(\text{Northing of P}) = N_A(\text{Northing of A}) + D \sin\theta \cos(\hat{NAX} - \hat{PAX}) = 335.114\,\text{m}$$

$$\frac{\partial N_P}{\partial N_A} = 1;\ \frac{\partial N_P}{\partial D} = \sin\theta \cos(\hat{NAX} - \hat{PAX}) = 0.993595$$

$$\frac{\partial N_P}{\partial \theta} = D \cos\theta \cos(\hat{NAX} - \hat{PAX}) = 14.38021\,\text{m}$$

$$\frac{\partial N_P}{\partial \hat{NAX}} = -D \sin\theta \sin(\hat{NAX} - \hat{PAX}) = -22.5009\,\text{m}$$

$$\frac{\partial N_P}{\partial \hat{PAX}} = D \sin\theta \sin(\hat{NAX} - \hat{PAX}) = 22.5009\,\text{m}$$

$$s_N = \pm\left[(0.010 \times 1)^2 + (0.005 \times 0.993595)^2 + \left(\frac{20}{206\,265} \times 14.38021\right)^2 + \left(\frac{20}{206\,265} \times -22.5009\right)^2 + \left(\frac{10}{206\,265} \times 22.5009\right)^2\right]^{1/2}\text{m}$$

$$= \pm(0.0001 + 0.0000247 + 0.0000019 + 0.0000048 + 0.0000012)^{1/2}$$

$\downarrow$ N_A error, $\downarrow$ D error, $\downarrow$ θ error, $\downarrow$ NÂX error, $\downarrow$ PÂX error

$$= \pm 0.012\,\text{m}$$

Northing of P is 335.114 ± 0.012 m (SE)

Principal source of error is in the given northing of A; the slope distance error has a slight effect.

$$L_P \text{ (Level of P)} = L_A \text{ (Level of A)} + h_I + D\cos\theta - h_T$$
$$= 114.446\,\text{m AOD}$$

$$\frac{\partial L_P}{\partial L_A} = 1; \frac{\partial L_P}{\partial h_I} = 1; \frac{\partial L_P}{\partial D} = \cos\theta = 0.061049$$

$$\frac{\partial L_P}{\partial \theta} = -D\sin\theta = -236.18864\,\text{m}; \frac{\partial L_P}{\partial h_I} = -1$$

$$s_L = \pm\left[(0.010 \times 1)^2 + (0.005 \times 1)^2 + (0.005 \times 0.061049)^2 + \left(\frac{20}{206\,265} \times -236.18864\right)^2 + (0.005 \times -1)^2\right]^{1/2} \text{m}$$

$$= \pm(0.0001 + 0.000025 + 0.00000009 + 0.0005245 + 0.000025)^{1/2}$$

$\downarrow$ A level error, $\downarrow$ h_I error, $\downarrow$ D error, $\downarrow$ θ error, $\downarrow$ h_T error

$$= \pm 0.026\,\text{m}$$

Reduced level of P = 114.446 m AOD ± 0.026 m

Principal source of error is in zenith angle measurement; the errors in the given level of A and the instrument and prism heights also have some effect.

Powers and roots

The standard errors for powers and roots must be calculated from the basic formula. Note that the SE for a power is *not* the same as for a product of equal amounts; there is only one source of error having increasing effect as the quantity is raised to a power.

7.6 ADJUSTMENT OF RELATED QUANTITIES

Where a relationship exists between several quantities independently measured, it is likely that a closing error will be found. Provided that mistakes and systematic errors have been eliminated, and that the closing error is small, the measured quantities may be adjusted to satisfy the relationship.

Consider a level baseline measured in three bays and as a whole (Fig. 7.4):

AB = 33.265 m	Weight 1
BC = 37.374 m	Weight 1
CD = 24.253 m	Weight 2
AD = 94.914 m	Weight 4

A single relationship exists, AB + BC + CD − AD = zero. If the error is small, adjustments can be made; these should be in inverse proportion to the weights. (This satisfies the theory of least squares, described in the next section.)

$$\text{AB} + \text{BC} + \text{CD} - \text{AD} = -0.022\,\text{m} = \text{error}$$

Distribute this in proportions:

$$\frac{1}{1}, \frac{1}{1}, \frac{1}{2}, \frac{1}{4} \text{ (or 1, 1, 0.5, 0.25)}$$

Dividing each of these amounts by their sum, the proportions may be rewritten:

$$\frac{1}{2.75}, \frac{1}{2.75}, \frac{0.5}{2.75}, \frac{0.25}{2.75} \text{ (i.e. 0.36, 0.36, 0.18, 0.09)}$$

and the adjustments to AB will be $0.36 \times 0.022 = 0.008\,\text{m}$; the remaining adjustments will be 0.008, 0.004 and 0.002 m.

The first three adjustments must be added to the measured values while the final one must be subtracted, giving adjusted lengths of:

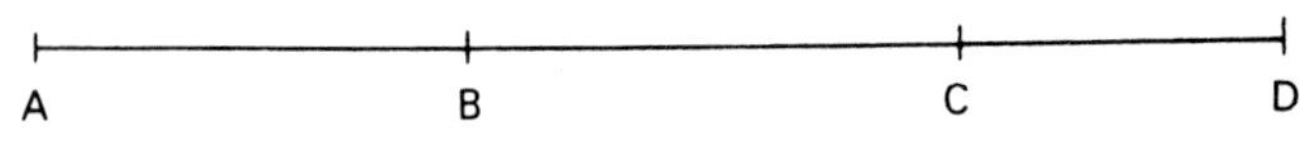

Fig. 7.4 Baseline measurement.

$$\left.\begin{aligned} AB &= 33.273 \\ BC &= 37.382 \\ CD &= 24.257 \end{aligned}\right\} \text{sum} = 94.912$$

AD = 94.912 which checks

If, additionally, lengths AC and BD had been measured, three closing errors would become apparent and a simultaneous proportional adjustment could not be applied. A sequential adjustment might be possible and a technique has been described in Chapter 5 for traverse adjustment (Bowditch's method). Sequential methods are not mathematically rigorous, requiring some values to be kept fixed while others are adjusted. Although established procedures will produce adjusted results, the sequential adjustment of a survey network may produce errors which are chased around ad infinitum. There is also a likelihood that adjustments applied latterly in such a process will be disproportionately large.

The theory of least squares

The difference between the MPV of a quantity and an individual observation is a residual. The theory of least squares states that 'the sum of the weighted squares of the residuals will be a minimum for the most probable values', or:

$\Sigma w(\bar{x} - x)^2$ is a minimum

For the MPV of a single quantity, the (weighted) mean of the constituent observations satisfies this theory. In other words we have already been applying least squares theory in earlier sections of this chapter!

To find values to satisfy the least squares condition, the expression must be differentiated with respect to each variable and put equal to zero. The function can only have a minimum value; no maximum can exist as there is no limit to the size of errors that could, in theory, occur.

There are two well-chronicled methods of applying least squares theory to related quantities: the variation method and the condition method. There is usually little to choose between them for manual adjustment of fairly simple sets of data; the variation method is preferable for adjusting control surveys combining linear and angular measurements. It is also used in programming computer solutions, and only this method will be described.

Variation method (observations equation method)

The method works by setting up a consistent model and finding variations which satisfy the least squares theory. Thus, consistent probable values are derived. If the model uses values which are close to the probable ones, variations should be small, making the arithmetic easier to handle.

Procedure

(1) Determine the number of independent quantities. This may be done by building up the model (on paper) using observed quantities one at a time (without incorporating redundant measures) until the model is complete.
(2) Identify suitable independent quantities and give them provisional values; each will require modification by a variation as yet unknown.
(3) Form observation equations, one for each measured quantity, expressing each residual as the MPV (provisional value plus variation) minus the observed value.
(4) By applying the least squares theory, form normal equations which can be simultaneously solved to calculate the variations. Weights for all observed quantities are required.
(5) The probable values (each provisional plus variation) will be consistent and will now satisfy the theory of least squares.

The method is best demonstrated by considering the example previously studied, with the addition of two further measurements, AC and BD:

AB = 33.265 m Weight 1
BC = 37.374 m Weight 1
CD = 24.253 m Weight 2
AD = 94.914 m Weight 4
AC = 70.634 m Weight 2
BD = 61.635 m Weight 2

It is necessary to determine the number of independent quantities; in this example there are six observed quantities. Three obviously independent ones are AB, BC and CD which, after adjustment, allow AC, BD and AD to be deduced. There are thus three independent and three redundant quantities. It would be acceptable to select instead AB, BC and AD (allowing CD, AC and BD to be deduced) but not AB, BC and AC where CD cannot be found.

Take the measured values of AB, BC and CD as the provisional ones and let variations of l_1, l_2 and l_3, be applied to them respectively. Then the MPVs for all the observed lengths are:

$AB = 33.265 + l_1$	Weight 1
$BC = 37.374 + l_2$	Weight 1
$CD = 24.253 + l_3$	Weight 2
$AD = 94.892 + l_1 + l_2 + l_3$	Weight 4 (AB + BC + CD)
$AC = 70.639 + l_1 + l_2$	Weight 2 (AB + BC)
$BD = 61.627 + l_2 + l_3$	Weight 2 (BC + CD)

Letting the residuals be $v_1 \ldots v_6$
and writing MPV − Observation = Residual:

$$
\begin{aligned}
&33.265 + l_1 && -33.265 = v_1 \\
&37.374 + l_2 && -37.374 = v_2 \\
&24.253 + l_3 && -24.253 = v_3 \\
&94.892 + l_1 + l_2 + l_3 && -94.914 = v_4 \\
&70.639 + l_1 + l_2 && -70.634 = v_5 \\
&61.627 + l_2 + l_3 && -61.635 = v_6
\end{aligned}
$$

These are the observation equations.

Simplifying, and working in millimetres:

$$
\begin{aligned}
&+l_1 && = v_1 \text{ weight } 1 \\
&+ l_2 && = v_2 \text{ weight } 1 \\
&+ l_3 && = v_3 \text{ weight } 2 \\
&+l_1 + l_2 + l_3 - 22 && = v_4 \text{ weight } 4 \\
&+l_1 + l_2 + 5 && = v_5 \text{ weight } 2 \\
&+ l_2 + l_3 - 8 && = v_6 \text{ weight } 2
\end{aligned}
$$

$$\text{Let } E = w_1v_1^2 + w_2v_2^2 + w_3v_3^2 + w_4v_4^2 + w_5v_5^2 + w_6v_6^2$$
$$= l_1^2 + l_2^2 + 2l_3^2 + 4(l_1 + l_2 + l_3 - 22)^2 + 2(l_1 + l_2 + 5)^2 + 2(l_2 + l_3 - 8)^2$$

substituting for v_1 to v_6 and putting in the weights.

For a minimum value of E, differentiate partially with respect to l_1, l_2 and l_3 in turn and equate each expression to zero.

$$\frac{\partial E}{\partial l_1} = 2l_1 + 8(l_1 + l_2 + l_3 - 22) + 4(l_1 + l_2 + 5) = 0$$
$$\frac{\partial E}{\partial l_2} = 2l_2 + 8(l_1 + l_2 + l_3 - 22) + 4(l_1 + l_2 + 5) + 4(l_2 + l_3 - 8) = 0$$
$$\frac{\partial E}{\partial l_3} = 4l_3 + 8(l_1 + l_2 + l_3 - 22) + 4(l_2 + l_3 - 8) = 0$$

These are the normal equations.

Collecting terms:

$$14l_1 + 12l_2 + 8l_3 = 156$$
$$12l_1 + 18l_2 + 12l_3 = 188$$
$$8l_1 + 12l_2 + 16l_3 = 208$$

The coefficients of the l terms form a matrix which is skew-symmetrical, i.e. about a diagonal from top left to bottom right. If this symmetry is not present, an arithmetic mistake has occurred.

Solving the equations:

$l_1 = +5$
$l_2 = 0$
$l_3 = +10$ to the nearest millimetre.

The MPVs are:

AB = 33.265 + 0.005 = 33.270 m
BC = 37.374 + 0.000 = 37.374 m
CD = 24.253 + 0.010 = 24.263 m

and the length of the baseline, AD, is 94.907 m.

Two points may be noted:

(1) The unknown residuals $v_1 \ldots v_6$ are used for identification; they are not employed in calculation of the adjusted quantities, but should be subsequently evaluated to enable a comparison to be made with the standard errors of the observations. They are

also required in the determination of the standard errors of the adjusted quantities.

(2) The normal equations can be formed more quickly as follows:
  (a) For $\partial E/\partial l_1$, take the left side of each observation equation in which l_1, appears, multiply it by its l_1 coefficient and by its weight; sum to zero.
  (b) Repeat for $\partial E/\partial l_2$ etc.; the normal equations as found previously will be formed but with all coefficients and constants halved.

Example 7.10

Figure 7.5 shows a plan of a levelling network between four stations, A, B, C and D. Station A is a bench mark with a reduced level of 95.100 m AOD. From the given data, calculate the most probable reduced levels of the stations.

Level run	Level difference (m)	Distance (m)
A to B	+4.185	600
B to C	+1.814	600
C to D	−2.552	600
D to A	−3.425	200
A to C	+5.991	300
B to D	−0.752	200

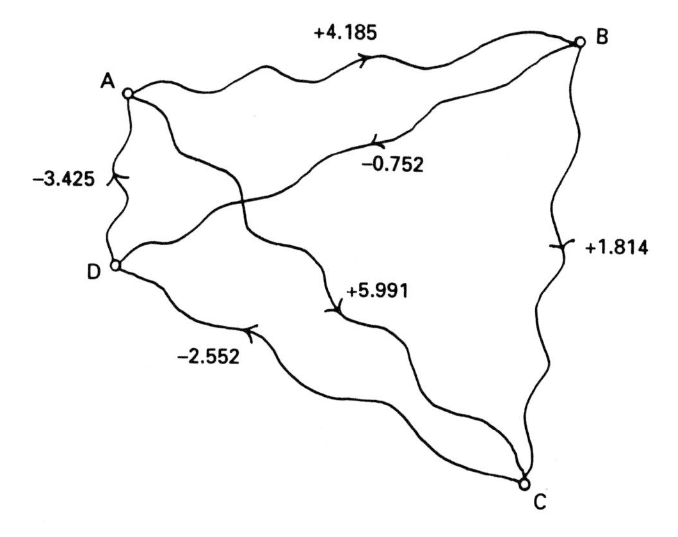

Fig. 7.5 Levelling network.

Solution

If provisional reduced levels are assigned to stations B, C and D, it can be seen that out of the six level runs, three are redundant. Take the level runs from A to B, B to C and C to D to give the provisional levels with variations of h_1, h_2 and h_3 respectively.

Provisional RL:

of B = 95.100 + 4.185 = 99.285 m AOD variation h_1

of C = 99.285 + 1.814 = 101.099 variation h_2

of D = 101.099 − 2.552 = 98.547 variation h_3

The error over a levelling leg is usually taken as proportional to the square root of the distance; thus weights may be taken as inversely proportional to distances, i.e.:

$$\frac{1}{6}:\frac{1}{6}:\frac{1}{6}:\frac{1}{2}:\frac{1}{3}:\frac{1}{2} \text{ or } 1:1:1:3:2:3$$

Let the residuals be $v_1 \ldots v_6$.

Forming the observation equations:

$$\begin{aligned}
\text{for AB: } & (99.285 + h_1) - 95.100 - 4.185 &= v_1\\
\text{BC: } & (101.099 + h_2) - (99.285 + h_1) - 1.814 &= v_2\\
\text{CD: } & (98.547 + h_3) - (101.099 + h_2) - (-2.552) &= v_3\\
\text{DA: } & 95.100 - (98.547 + h_3) - (-3.425) &= v_4\\
\text{AC: } & (101.099 + h_2) - 95.100 - 5.991 &= v_5\\
\text{BD: } & (98.547 + h_3) - (99.285 + h_1) - (-0.752) &= v_6
\end{aligned}$$

Collecting terms and working in millimetres:

$$\begin{aligned}
+h_1 & = v_1 \text{ weight 1}\\
-h_1 + h_2 & = v_2 \text{ weight 1}\\
-h_2 + h_3 & = v_3 \text{ weight 1}\\
-h_3 - 22 & = v_4 \text{ weight 3}\\
+h_2 + 8 & = v_5 \text{ weight 2}\\
-h_1 + h_3 + 14 & = v_6 \text{ weight 3}
\end{aligned}$$

Normal equations:

For h_1: $+h_1 + (-1)(-h_1 + h_2) + (-1)(3)(-h_1 + h_3 + 14) \quad = 0$

(each residual containing h_1 multiplied by the h_1 coefficient and by the weight)

For h_2: $(-h_1 + h_2) + (-1)(-h_2 + h_3) + (2)(+h_2 + 8) \quad = 0$

For h_3: $(-h_2 + h_3) + (-1)(3)(-h_3 - 22) + (3)(-h_1 + h_3 + 14) = 0$

Collecting terms:

$$\begin{aligned} 5h_1 - h_2 - 3h_3 &= 42 \\ -h_1 + 4h_2 - h_3 &= -16 \\ -3h_1 - h_2 + 7h_3 &= -108 \end{aligned}$$

Solving:

$$\begin{aligned} h_1 &= -5\,\text{mm} \\ h_2 &= -10\,\text{mm} \\ h_3 &= -19\,\text{mm} \end{aligned}$$

Thus the MPVs of the stations are:

B: 99.285 − 0.005 = 99.280 m AOD

C: 101.099 − 0.010 = 101.089

D: 98.547 − 0.019 = 98.528

Matrix methods may be used to form and solve normal equations. An explanation is given in Section 7.8.

Example 7.11

Four stations, A, B, C and D form a small plan triangulation survey as shown in Fig. 7.6.

Horizontal angles have been measured as follows:

Angle	Mean value	No of observations
$B\hat{A}D$	28°31′58″	1
$A\hat{D}B$	31°21′28″	1
$B\hat{D}C$	44°18′30″	2
$A\hat{D}C$	75°40′07″	4
$D\hat{C}B$	41°53′30″	2
$C\hat{B}D$	93°48′18″	2
$D\hat{B}A$	120°06′40″	1
$A\hat{B}C$	146°04′56″	2

Find the most probable values of the angles by a method of least squares.

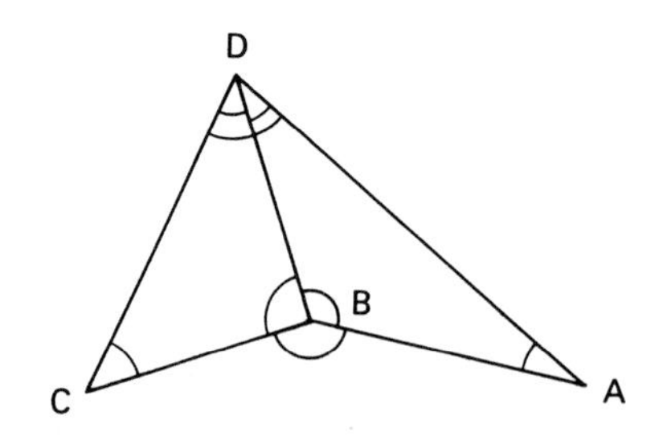

Fig. 7.6 Triangulation survey.

Solution

There are eight observed angles. To define both of the small triangles, two (adjusted) angles in each are required; thus there are four independent angles and four redundant ones.

Take the observed values of $B\hat{A}D$, $A\hat{D}B$, $B\hat{D}C$ and $D\hat{C}B$ as provisional values of the four independents and let variations of θ_1, θ_2, θ_3 and θ_4 be respectively applied to them.

Values in terms of the provisional quantities and these variations can then be found for angles $A\hat{D}C$, $C\hat{B}D$, $D\hat{B}A$ and $A\hat{B}C$ as follows:

$$A\hat{D}C = 75°39'58'' + \theta_2 + \theta_3$$
$$C\hat{B}D = 93°48'00'' - \theta_3 - \theta_4$$
$$D\hat{B}A = 120°06'34'' - \theta_1 - \theta_2$$
$$A\hat{B}C = 146°05'26'' + \theta_1 + \theta_2 + \theta_3 + \theta_4$$

Forming observation equations (with residuals $v_1 \ldots v_8$)

$$\text{for } B\hat{A}D\text{: } (28°31'58'' + \theta_1) - 28°31'58'' = v_1$$
$$\text{for } A\hat{D}B\text{: } (31°21'28'' + \theta_2) - 31°21'28'' = v_2$$
$$\text{for } B\hat{D}C\text{: } (44°18'30'' + \theta_3) - 44°18'30'' = v_3$$
$$\text{for } A\hat{D}C\text{: } (75°39'58'' + \theta_2 + \theta_3) - 75°40'07'' = v_4$$
$$\text{for } D\hat{C}B\text{: } (41°53'30'' + \theta_4) - 41°53'30'' = v_5$$
$$\text{for } C\hat{B}D\text{: } (93°48'00'' - \theta_3 - \theta_4) - 93°48'18'' = v_6$$
$$\text{for } D\hat{B}A\text{: } (120°06'34'' - \theta_1 - \theta_2) - 120°06'40'' = v_7$$
$$\text{for } A\hat{B}C\text{: } (146°05'26'' + \theta_1 + \theta_2 + \theta_3 + \theta_4) - 146°04'56'' = v_8$$

The weights may be taken as equal to the numbers of observations.

Simplifying and working in seconds:

$$
\begin{aligned}
+\theta_1 &= v_1 \text{ weight } 1\\
+\theta_2 &= v_2 \text{ weight } 1\\
+\theta_3 &= v_3 \text{ weight } 2\\
+\theta_2 + \theta_3 - 9 &= v_4 \text{ weight } 4\\
+\theta_4 &= v_5 \text{ weight } 2\\
-\theta_3 - \theta_4 - 18 &= v_6 \text{ weight } 2\\
-\theta_1 - \theta_2 - 6 &= v_7 \text{ weight } 1\\
+\theta_1 + \theta_2 + \theta_3 + \theta_4 + 30 &= v_8 \text{ weight } 2
\end{aligned}
$$

Normal equations:

for θ_1:

$$(1)(\theta_1) + (-1)(1)(-\theta_1 - \theta_2 - 6) + (2)(\theta_1 + \theta_2 + \theta_3 + \theta_4 + 30) = 0$$

for θ_2:

$$(1)(\theta_2) + (4)(\theta_2 + \theta_3 - 9) + (-1)(1)(-\theta_1 - \theta_2 - 6) + (2)(\theta_1 + \theta_2 + \theta_3 + \theta_4 + 30) = 0$$

for θ_3:

$$(2)(\theta_3) + (4)(\theta_2 + \theta_3 - 9) + (-1)(2)(-\theta_3 - \theta_4 - 18) + (2)(\theta_1 + \theta_2 + \theta_3 + \theta_4 + 30) = 0$$

for θ_4:

$$(2)(\theta_4) + (-1)(2)(-\theta_3 - \theta_4 - 18) + (2)(\theta_1 + \theta_2 + \theta_3 + \theta_4 + 30) = 0$$

Collecting terms:

$$
\begin{aligned}
4\theta_1 + 3\theta_2 + 2\theta_3 + 2\theta_4 &= -66 \qquad \text{Solving: } \theta_1 = -14''\\
3\theta_1 + 8\theta_2 + 6\theta_3 + 2\theta_4 &= -30 \qquad \theta_2 = +6''\\
2\theta_1 + 6\theta_2 + 10\theta_3 + 4\theta_4 &= -60 \qquad \theta_3 = -2''\\
2\theta_1 + 2\theta_2 + 4\theta_3 + 6\theta_4 &= -96 \qquad \theta_4 = -12''
\end{aligned}
$$

(Note skew-symmetry of coefficients)

Most probable values are:

$B\hat{A}D = 28°31'44''$ $\quad A\hat{D}B = 31°21'34''$ $\quad B\hat{D}C = 44°18'28''$

$A\hat{D}C = 75°40'02''$ $\quad D\hat{C}B = 41°53'18''$ $\quad C\hat{B}D = 93°48'14''$

$D\hat{B}A = 120°06'42''$ $\quad A\hat{B}C = 146°05'04''$

Example 7.12 Prism constant estimation

Tests were carried out to determine the constant of a prism used in EDM measurements. Four collinear stations A, B, C and D were established on level ground and horizontal distances were measured, with equal weight, as follows:

AB = 95.178 m		AC = 289.378 m
BC = 194.240 m		BD = 397.510 m
CD = 203.306 m		AD = 492.664 m

Use a method of least squares to establish the probable value of the prism constant.

Solution

There are six measured quantities of which three are independent. For provisional quantities, take the measured values of AB, BC and CD modified by an unknown prism constant, e, in millimetres (to be subtracted) and apply variations of l_1, l_2 and l_3 (also in millimetres).

Forming the observation equations (in millimetres),

$$\text{for AB: } (95\,178 - e + l_1) - (95\,178 - e) = v_1$$
$$\text{BC: } (194\,240 - e + l_2) - (194\,240 - e) = v_2$$
$$\text{CD: } (203\,306 - e + l_3) - (203\,306 - e) = v_3$$
$$\text{AC: } (95\,178 - e + l_1) + (194\,240 - e + l_2) - (289\,378 - e) = v_4$$
$$\text{BD: } (194\,240 - e + l_2) + (203\,306 - e + l_3) - (397\,510 - e) = v_5$$
$$\text{AD: } (95\,178 - e + l_1) + (194\,240 - e + l_2) + (203\,306 - e + l_3) - (492\,664 - e) = v_6$$

Simplifying:

$$\begin{aligned}
+l_1 \quad\quad\quad\quad\quad\quad &= v_1 \\
+ l_2 \quad\quad\quad\quad &= v_2 \\
+ l_3 \quad\quad &= v_3 \\
+l_1 + l_2 \quad - e + 40 &= v_4 \quad \text{all equal weight of 1} \\
+ l_2 + l_3 - e + 36 &= v_5 \\
+l_1 + l_2 + l_3 - 2e + 60 &= v_6
\end{aligned}$$

Normal equations:

For l_1:

$$l_1 + (l_1 + l_2 - e + 40) + (l_1 + l_2 + l_3 - 2e + 60) = 0$$

For l_2:

$$l_2 + (l_1 + l_2 - e + 40) + (l_2 + l_3 - e + 36) + (l_1 + l_2 + l_3 - 2e + 60) = 0$$

For l_3:

$$l_3 + (l_2 + l_3 - e + 36) + (l_1 + l_2 + l_3 - 2e + 60) = 0$$

For e:

$$(-1)(l_1 + l_2 - e + 40) + (-1)(l_2 + l_3 - e + 36) + (-2)(l_1 + l_2 + l_3 - 2e + 60) = 0$$

Collecting terms:

$$\begin{aligned}
+3l_1 + 2l_2 + l_3 - 3e &= -100\\
+2l_1 + 4l_2 + 2l_3 - 4e &= -136\\
+l_1 + 2l_2 + 3l_3 - 3e &= -96\\
-3l_1 - 4l_2 - 3l_3 + 6e &= 196
\end{aligned}$$

When the method of developing the normal equations is understood, it is possible to form them directly from the simplified observation equations.

Solving, $l_1 = -1.0\,\text{mm}$, $l_2 = -4.0\,\text{mm}$, $l_3 = 1.0\,\text{mm}$, $e = 30.0\,\text{mm}$

The prism constant is therefore 30 mm (to be subtracted).

Example 7.13 EDM calibration

An EDM outfit suspected of having a zero error and a proportional error was checked against six horizontal baselines with accepted lengths of 100.000 m, 150.000 m, 200.000 m, 250.000 m, 300.000 m and 400.000 m.

The horizontal distances displayed were 100.008 m, 149.998 m, 200.004 m, 250.005 m, 299.998 m and 400.003 m all deemed to be of equal weight. Calculate, using the principle of least squares, the probable values of the zero error (in millimetres) and the proportional error (in mm per km).

Solution

This problem is different from others so far encountered; usually related surveying measurements displaying mathematical inconsistency require adjustment to provide that consistency. Here we have six fixed quantities with which field measurements have to be compared to enable corrections to be estimated, such corrections then being applicable to future measurements.

The unknowns, to be found, are the zero error and the proportional error; let these be $+a$ mm and $+b$ mm per km. The observation equations are formed by subtracting each observed value from its fixed value modified by the corrections to give the residual.

Forming the observation equations (in millimetres):

$$\left.\begin{aligned}
100\,000 + a + 0.10b - 100\,008 &= v_1\\
150\,000 + a + 0.15b - 149\,998 &= v_2\\
200\,000 + a + 0.20b - 200\,004 &= v_3\\
250\,000 + a + 0.25b - 250\,005 &= v_4\\
300\,000 + a + 0.30b - 299\,998 &= v_5\\
400\,000 + a + 0.40b - 400\,003 &= v_6
\end{aligned}\right\}\text{ all of equal weight}$$

Simplifying:

$$\begin{aligned}
+ a + 0.10b - 8 &= v_1\\
+ a + 0.15b + 2 &= v_2\\
+ a + 0.20b - 4 &= v_3\\
+ a + 0.25b - 5 &= v_4\\
+ a + 0.30b + 2 &= v_5\\
+ a + 0.40b - 3 &= v_6
\end{aligned}$$

Normal equations:

For a:

$$(a + 0.10b - 8) + (a + 0.15b + 2) + (a + 0.20b - 4) + (a + 0.25b - 5) + (a + 0.30b + 2) + (a + 0.4b - 3) = 0$$

For b:

$$(0.10)(a + 0.10b - 8) + (0.15)(a + 0.15b + 2) + (0.20)(a + 0.20b - 4) + (0.25)(a + 0.25b - 5) + (0.30)(a + 0.30b + 2) + (0.40)(a + 0.40b - 3) = 0$$

Collecting terms:

$$\begin{aligned}
+6a + 1.40b &= +16\\
+1.40a + 0.3850b &= +3.15
\end{aligned}$$

Solving: $a = +5.0\,\text{mm}$, $b = -10.0\,\text{mm}$ per km.

These are then the probable values of the zero and proportional errors respectively. If the instrument cannot be adjusted, these values (with the signs changed) should be applied to displayed measurements.

It may be noted that this process is in fact one of linear regression.

7.7 VARIATION OF COORDINATES

In control surveys involving any combination of distances and angles, the method of variation of coordinates may be used for adjustment. Provisional coordinates are assigned to the stations and variations to satisfy the least squares condition are calculated. The method operates in the same manner as the variation adjustments of a baseline and a level run, as described. The process is likely to be longer for several reasons:

(1) From the provisional coordinates, provisional lengths and angles must be determined to form observation equations.
(2) The large number of measurements requires the formation of many observation equations and, for each station to be located, there will be two unknowns and hence two normal equations.
(3) The process is an iterative one; this is because coefficients are trigonometric functions based on provisional coordinates (a non-linear relationship). If the coordinate variations are not small, the coefficients will need amending and the process should be repeated using the revised coordinates. If the provisional coordinates are based on field observations, the adjustments (variations) should be small and a single iteration will often suffice. Where a computer program is available, further iterations should be carried out.

Figure 7.7 shows a plan of a tied traverse between fixed stations A and E. Reference objects at these two stations have been sighted and angles have been measured at A, B, C, D and E. Lengths AB, BC, CD and DE have been measured. There are therefore nine observed quantities to be adjusted to give angular and linear consistency. If provisional coordinates are assigned to B, C and D, it can be seen that there are six independent quantities: the easting and northing of each station. Six normal equations must be formed and solved to determine the three pairs of variations.

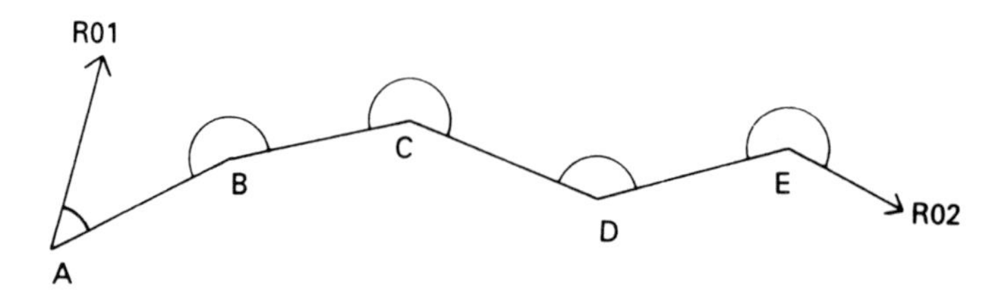

Fig. 7.7 Tied traverse.

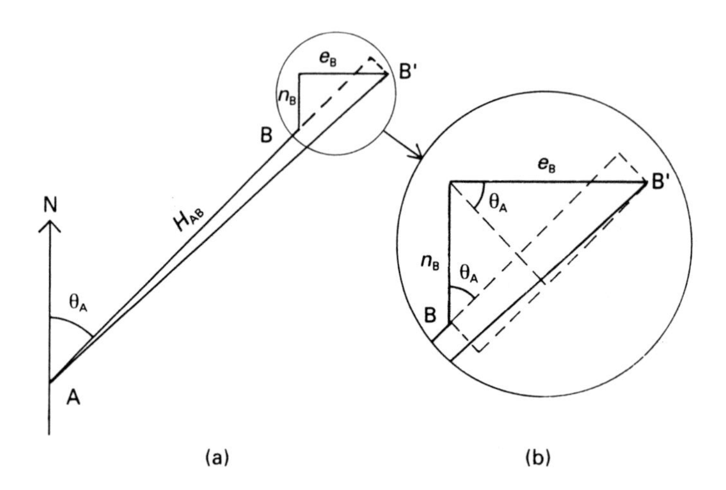

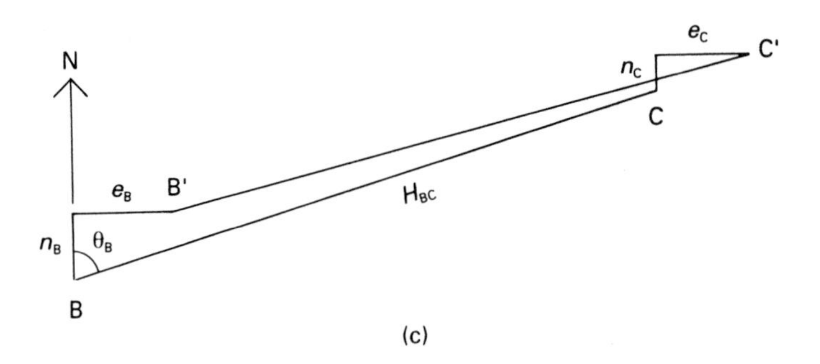

Fig. 7.8 Traverse legs.

It will usually be convenient to determine the provisional coordinates from *some* of the observed quantities.

In Fig. 7.8(a) AB is a line in a plan control survey with a provisional length of H_{AB} and a provisional bearing of θ_A. (The length and bearing may be measured values used to set the provisional coordinates, or they may be derived from the provisional coordinates determined from other measurements.) Suppose that A is kept fixed, while B is to be adjusted by unknown variations of coordinates e_B and n_B to move it to B′ (Fig. 7.8(b)).

$$\text{Increase in length AB} = e_B \sin\theta_A + n_B \cos\theta_A$$

$$\text{Increase in bearing of AB} = \frac{e_B \cos\theta_A - n_B \sin\theta_A}{H_{AB}} \text{ radians}$$

with negligible error if the adjustments are small.

If both ends of the line are adjusted, the coordinate variation of both ends must be considered. For line BC (Fig. 7.8(c)):

$$\text{Increase in line length} = (e_C - e_B)\sin\theta_B + (n_C - n_B)\cos\theta_B$$

Call this δH_{BC}

Increase in line bearing

$$= \frac{(e_C - e_B) \times 206\,265 \cos\theta_B - (n_C - n_B) \times 206\,265 \sin\theta_B}{H_{BC}} \text{ seconds}$$

Similarly the increase in the bearing of CD

$$= \frac{(e_D - e_C) \times 206\,265 \cos\theta_C - (n_D - n_C) \times 206\,265 \sin\theta_C}{H_{CD}} \text{ seconds}$$

Therefore the increase in angle $B\hat{C}D$:

$$= 206\,265\left[\frac{(e_D - e_C)\cos\theta_C - (n_D - n_C)\sin\theta_C}{H_{CD}} - \frac{(e_C - e_B)\cos\theta_B - (n_C - n_B)\sin\theta_B}{H_{BC}}\right] \text{ seconds}$$

Call this $\delta\theta_C$

Unknowns are e_D, n_D, e_C, n_C, e_B, n_B while θ_B, θ_C, H_{CD}, H_{BC} are the provisional quantities.

Observation equations are formed, e.g.:

$$H_{CD} + \delta H_{CD} - \text{measured CD} = v_{CD}$$

$$\text{Provisional } B\hat{C}D + \delta\theta_C - \text{measured } B\hat{C}D = v_{\theta_C}$$

Normal equations are formed and solved. If the variations are small and a further iteration is carried out using the new coordinates, the observation and normal equations will have virtually the same coefficients, and further variations will be negligible. For this reason, provisional coordinates should be based on selected observed quantities.

In *trilateration* observation equations are formed for lengths only.

In *triangulation*, *intersection* and *resection*, observation equations are formed for angles only.

In *closed traverses* and *networks* observation equations are formed for lengths and angles.

Weights

If weights are taken as the reciprocals of the squares of the standard errors of observations, the least squares expression (and the normal equations) will be dimensionless. The standard errors must be in the same units as the observation equations; for a manual solution it is convenient to work in millimetres and seconds, thus field measurements in metres and degrees, minutes and seconds must be converted when forming the observation equations. The variations, when evaluated, will be in millimetres. (A computer solution will usually be in metres and degrees or radians.)

If weights are assigned by other methods, care must be taken to achieve consistency of units. In a manual solution, the weights may be rounded off to convenient numbers to ease calculations. A small change in weight values has negligible effect on the adjustment; this is fortunate as a certain amount of estimation and judgement may be required in deriving weights.

Example 7.14

Plan control station P has been observed from two fixed stations X and Y with local coordinates

X (294.635 m E; 96.105 m N); Y (443.249 m E; 251.738 m N)

Use two of the following observations to calculate provisional coordinates of P and apply the method of variation of coordinates to determine its most probable position:

$$XP = 369.247 \pm 0.010\,\text{m (horizontal distance)}$$
$$YP = 330.193 \pm 0.010\,\text{m (horizontal distance)}$$
$$P\hat{X}Y = 62°24'20'' \pm 15'' \quad \text{(horizontal angle)}$$

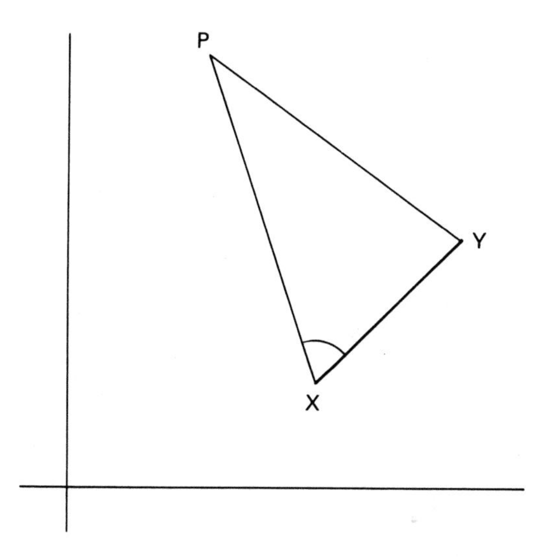

Fig. 7.9 Point location.

Solution

Figure 7.9 is a plan. Take the measured values of length XP and angle $\hat{PXY}$ to give provisional coordinates of P.

$$\text{Bearing of XY} = \tan^{-1}\frac{443.249 - 294.635}{251.738 - 96.105} = 43°40'42.3''$$

$$\therefore \text{Bearing of XP} = 43°40'42.3'' + 360°00'00'' - 62°24'20''$$
$$= 341°16'22.3''$$

For XP,
$\Delta E = -118.5510\,\text{m}$
$\Delta N = 349.6984\,\text{m}$ (polar to rectangular conversion)

Add to X coordinates: provisionally P is at (176.0840 m E; 445.8034 m N)

Let the variations in coordinates of P be e and n.
The provisional length and bearing of YP must be found.

For YP
$\Delta E = 176.0840 - 443.249 = -267.1650\,\text{m}$
$\Delta N = 445.8034 - 251.738 = 194.0654\,\text{m}$
Length = 330.2098 m; Bearing = −54°00′20.9″; add 360° to give 305°59′39.1″.

Then for each line

$$\delta H = e \sin\theta + n \cos\theta$$

The observation equations are formed:

For XP:
$$369.247 + (e \sin 341°16'22.3'' + n \cos 341°16'22.3'') - 369.247 = v_1$$
For YP:
$$330.2098 + (e \sin 305°59'39.1'' + n \cos 305°59'39.1'') - 330.193 = v_2$$

Form the observation equation for $P\hat{X}Y$; the increase in angle will be equal in magnitude but opposite in sign to the increase in bearing of XP because XY is fixed.

$$\delta\theta = -\frac{e \cos\theta - n \sin\theta}{H} \text{radians}$$

$$\therefore\ 62°24'20'' - \frac{e \cos 341°16'22.3'' - n \sin 341°16'22.3''}{369.247 \times 1000} \times 206\,265 - 62°24'20'' = v_3$$

(working in millimetres and seconds).

Weights are taken as the reciprocals of the squares of the standard errors, i.e.

$$\frac{1}{100} : \frac{1}{100} : \frac{1}{225}$$

Evaluating the trig functions and collecting terms we obtain:

$$-0.32106e + 0.94706n = v_1 \quad \text{weight } 1/100$$
$$-0.80908e + 0.58770n + 16.8 = v_2 \quad 1/100$$
$$-0.52904e - 0.17935n = v_3 \quad 1/225$$

(working in millimetres and seconds).

Forming the normal equations:

For e: $(-0.32106)(1/100)(-0.32106e + 0.94706n)$
$+ (-0.80908)(1/100)(-0.80908e + 0.58770n + 16.8)$
$+ (-0.52904)(1/225)(-0.52904e - 0.17935n) = 0$

For n: $(0.94706)(1/100)(-0.32106e + 0.94706n)$
$+ (0.58770)(1/100)(-0.80908e + 0.58770n + 16.8)$
$+ (-0.17935)(1/225)(-0.52904e - 0.17935n) = 0$

Collecting terms: $0.0088208e - 0.0073739n = 0.135925$
$-0.0073739e + 0.0125661n = -0.098734$

Solving: $e = +17.35\,\text{mm}$, $n = +2.33\,\text{mm}$

The coordinates of P are thus (176.101 m E; 445.806 m N).

A further iteration gives zero variations and these are therefore the probable coordinates.

7.8 MATRIX SOLUTION

The observation equations can be written:

$$\begin{aligned}
a_{11}x_1 + a_{12}x_2 + a_{13}x_3 + \ldots + a_{1n}x_n - k_1 &= v_1 \\
a_{21}x_1 + a_{22}x_2 + a_{23}x_3 + \ldots + a_{2n}x_n - k_2 &= v_2 \\
a_{31}x_1 + a_{32}x_2 + a_{33}x_3 + \ldots + a_{3n}x_n - k_3 &= v_3 \\
\ldots\ldots\ldots\ldots\ldots\ldots\ldots\ldots\ldots\ldots\ldots \\
\ldots\ldots\ldots\ldots\ldots\ldots\ldots\ldots\ldots\ldots\ldots \\
a_{m1}x_1 + a_{m2}x_2 + a_{m3}x_3 + \ldots + a_{mn}x_n - k_m &= v_m
\end{aligned}$$

For m observed quantities of which n are independent.

where $x_1 \ldots x_n$ are the (unknown) adjustments, $a_{11} \ldots a_{mn}$ are the coefficients, $k_1 \ldots k_m$ are the differences between observed and provisional quantities and $v_1 \ldots v_m$ are the residuals.

In matrix notation:

$$\mathbf{Ax} - \mathbf{k} = \mathbf{v}$$

The weight matrix (**W**) is:

$$\begin{pmatrix}
w_1 & 0 & 0 & \ldots & 0 \\
0 & w_2 & 0 & \ldots & 0 \\
0 & 0 & w_3 & \ldots & 0 \\
\cdot & \cdot & \cdot & \cdot\cdot\cdot & \cdot \\
\cdot & \cdot & \cdot & \cdot\cdot\cdot & \cdot \\
\cdot & \cdot & \cdot & \cdot\cdot\cdot & \cdot \\
0 & 0 & 0 & \ldots & w_m
\end{pmatrix}$$

The least squares condition states that Σwv^2 is a minimum, i.e.:

$$\mathbf{v}^{\mathrm{T}}\mathbf{W}\mathbf{v} = (\mathbf{Ax} - \mathbf{k})^{\mathrm{T}}\mathbf{W}(\mathbf{Ax} - \mathbf{k}) \text{ is a minimum}$$

Differentiating w.r.t. **x** and equating to zero,

$$2\mathbf{A}^{\mathrm{T}}\mathbf{WAx} - \mathbf{A}^{\mathrm{T}}\mathbf{Wk} - \mathbf{A}^{\mathrm{T}}\mathbf{Wk} = 0 \quad \therefore (\mathbf{A}^{\mathrm{T}}\mathbf{WA})\mathbf{x} = \mathbf{A}^{\mathrm{T}}\mathbf{Wk}$$

Hence $\mathbf{x} = (\mathbf{A}^{\mathrm{T}}\mathbf{WA})^{-1}\mathbf{A}^{\mathrm{T}}\mathbf{Wk}$.

This provides a solution, although the matrix inversion may be difficult to carry out (even by computer). Solution of $(\mathbf{A}^T\mathbf{W}\mathbf{A})\mathbf{x} = \mathbf{A}^T\mathbf{W}\mathbf{k}$ may be simpler and advantage may be taken of the coefficient symmetry (e.g. by using Choleski's Method).

Assessment of precision

The adjustment of surveying measurements by a least squares process produces results that are mathematically consistent, providing compensation for random errors. Closing checks, carried out before adjustment, should allow the surveyor to assess the acceptability of fieldwork, i.e. whether mistakes have been eliminated, systematic errors corrected and what the cumulative effect of random errors may be.

Even after adjustment, surveying measurements may not be acceptable. Bad geometry or poor practice during measurement (poorly conditioned triangles, for example) can give a weak determination of a probable value, and poorly chosen standard errors (for weighting) can produce a distorted adjusted network.

By calculating the residuals v_1, v_2 etc., the surveyor can form an assessment of the validity of the adjusted results: the residuals should approximate to the standard errors.

The strength of the survey network (dependent on geometry and sound practice) can be assessed before the fieldwork is undertaken, provided that the method and geometry are known. This will usually require an approximate survey to be carried out to enable the coefficients a_{11} etc. to be evaluated.

The variance – covariance matrix of the unknowns is given by:

$$\boldsymbol{\sigma}_{xx} = \sigma_0^2(\mathbf{A}^T\mathbf{W}\mathbf{A})^{-1} = \begin{pmatrix} \boldsymbol{\sigma}_{x_1}^{\ 2} & \boldsymbol{\sigma}_{x_1x_2} & \boldsymbol{\sigma}_{x_1x_3} & \cdot\ \cdot\ \cdot & \boldsymbol{\sigma}_{x_1x_n} \\ \boldsymbol{\sigma}_{x_2x_1} & \boldsymbol{\sigma}_{x_2}^{\ 2} & \boldsymbol{\sigma}_{x_2x_3} & \cdot\ \cdot\ \cdot & \boldsymbol{\sigma}_{x_2x_n} \\ \cdot & \cdot & \cdot & \cdot\ \cdot\ \cdot & \cdot \\ \cdot & \cdot & \cdot & \cdot\ \cdot\ \cdot & \cdot \\ \boldsymbol{\sigma}_{x_nx_1} & \boldsymbol{\sigma}_{x_nx_2} & \boldsymbol{\sigma}_{x_nx_3} & \cdot\ \cdot\ \cdot & \boldsymbol{\sigma}_{x_n}^{\ 2} \end{pmatrix}$$

where $\boldsymbol{\sigma}_{x_1}^{\ 2}$, etc. are variances and $\boldsymbol{\sigma}_{x_1x_2}$ are covariances.

$$\sigma_0^2 \simeq s_0^2 \text{ and } s_0^2 = \frac{\Sigma wv^2}{m - n} = \frac{\mathbf{v}^T\mathbf{W}\mathbf{v}}{m - n}$$

where m is the number of observation equations, and n is the number of unknowns.

σ_0^2 is the unit variance and s_0^2 should approximate to unity; this will be the case if the weights, taken as the reciprocals of the standard errors, have been reliably estimated. If the geometry of a survey is fixed before the fieldwork is carried out, the coefficients a_{11}, etc. will be predetermined and, by taking $\sigma_0^2 = 1$, the variance–covariance matrix may be evaluated. The variances ($\sigma_{x_1}^2$, $\sigma_{x_2}^2$, etc.) can be compared with the expected values so that an *a priori* assessment can be made of the weights and of the geometry; changes can be implemented before measurements are taken. The covariances refer to the correlation between measurements.

The same evaluation can be carried out after the survey (*a posteriori*); this time $\Sigma wv^2/(m - n)$ can be determined. It should approximate to unity if each weight has been taken as the reciprocal of its variance. (This will not be the case if the weights have been multiplied by a factor; the factor will be cancelled out when multiplication by $(\mathbf{A}^T\mathbf{W}\mathbf{A})^{-1}$ is carried out so that the variance–covariance matrix will not be affected.) Large unexplained deviations of s_0^2 from unity (less than 0.5 or more than 2.0) should be investigated, particularly where weighting of dissimilar quantities has occurred. This check is not really reliable where $m - n$ is small, for example in a simple traverse where it equals 3.

Once the geometry and the weights appear satisfactory, standard errors for the calculated adjusted quantities can be quoted.

7.9 ERROR ELLIPSES

For a plan survey, the variance – covariance matrix will be of the form

$$\sigma_{\mathrm{E,N}} = \begin{pmatrix} \sigma_{\mathrm{E_A}}^2 & \sigma_{\mathrm{E_AN_A}} & \sigma_{\mathrm{E_AE_B}} & \sigma_{\mathrm{E_AN_B}} & \cdot & \cdot \\ \sigma_{\mathrm{N_AE_A}} & \sigma_{\mathrm{N_A}}^2 & \sigma_{\mathrm{N_AE_B}} & \sigma_{\mathrm{N_AN_B}} & \cdot & \cdot \\ \sigma_{\mathrm{E_BE_A}} & \sigma_{\mathrm{E_BN_A}} & \sigma_{\mathrm{E_B}}^2 & \sigma_{\mathrm{E_BN_B}} & \cdot & \cdot \\ \sigma_{\mathrm{N_BE_A}} & \sigma_{\mathrm{N_BN_A}} & \sigma_{\mathrm{N_BE_B}} & \sigma_{\mathrm{N_B}}^2 & \cdot & \cdot \\ \cdot & \cdot & \cdot & \cdot & \cdot & \cdot \\ \cdot & \cdot & \cdot & \cdot & \cdot & \cdot \end{pmatrix}$$

For point A, the relevant submatrix is

$$\begin{pmatrix} \sigma_{\mathrm{E_A}}^2 & \sigma_{\mathrm{E_AN_A}} \\ \sigma_{\mathrm{N_AE_A}} & \sigma_{\mathrm{N_A}}^2 \end{pmatrix} \quad \text{where } \sigma_{\mathrm{E_AN_A}} = \sigma_{\mathrm{N_AE_A}} = \sigma_{\mathrm{EN_A}}$$

$\sigma_{\mathrm{E_A}}^2$ and $\sigma_{\mathrm{N_A}}^2$ are the variances in east and north directions and $\sigma_{\mathrm{EN_A}}$ is the covariance. The maximum and minimum values of the standard error of a point will not usually lie along the coordinate axes; in fact the locus of the standard error in all directions will be a ‘figure-of-eight' shape which is the pedal curve of an ellipse. The semi-major and semi-minor axes are the maximum and minimum values of the standard error.

Figure 7.10 shows the ellipse and pedal curve with the major axis having a bearing of ϕ. It can be shown that, for the maximum and minimum values of the standard error:

$$2\phi_M = \tan^{-1}\frac{2\sigma_{EN}}{\sigma_N^2 - \sigma_E^2}$$

σ_{EN}	$\sigma_N^2 - \sigma_E^2$	$2\phi_M$
+	+	0°– 90°
+	–	90°–180°
–	–	180°–270°
–	+	270°–360°

Two values of ϕ_M are given; the signs of σ_{EN} and $(\sigma_N^2 - \sigma_E^2)$ enable the correct value of $2\phi_M$ (and hence ϕ_M) for the major axis to be found.

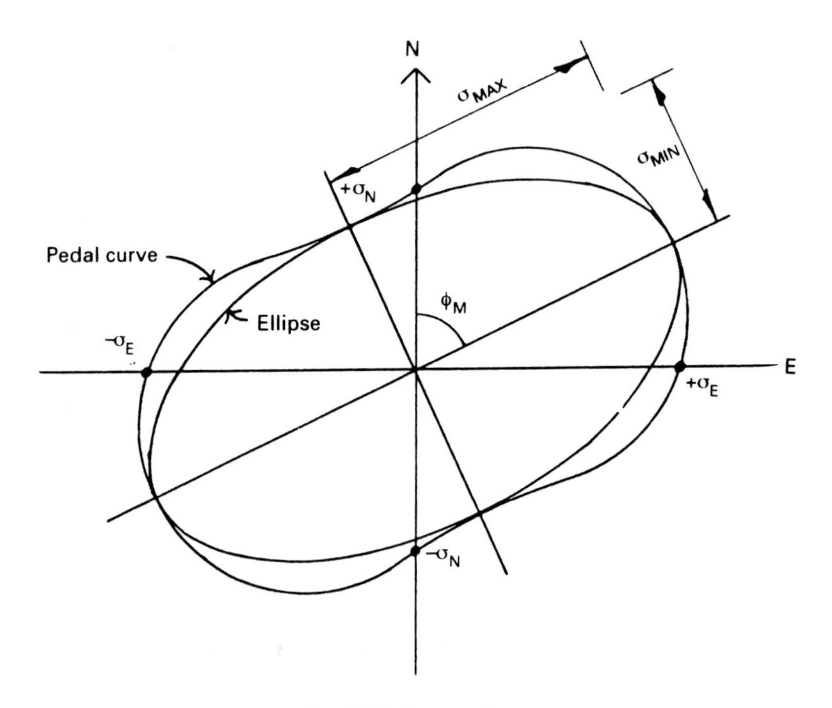

Fig. 7.10 Error ellipse.

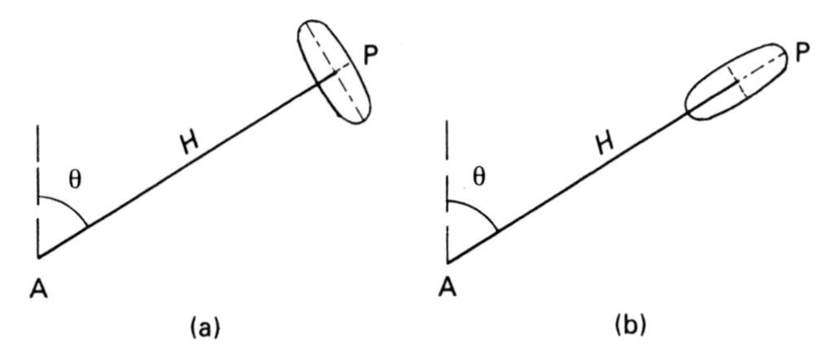

Fig. 7.11 Polar fix of P.

The maximum and minimum values are:

$$\sigma_{MAX}^2 = \frac{1}{2}\left\{\sigma_E^2 + \sigma_N^2 + [(\sigma_E^2 - \sigma_N^2)^2 + 4\sigma_{EN}^2]^{1/2}\right\}$$

$$\sigma_{MIN}^2 = \frac{1}{2}\left\{\sigma_E^2 + \sigma_N^2 - [(\sigma_E^2 - \sigma_N^2)^2 + 4\sigma_{EN}^2]^{1/2}\right\}$$

The error ellipses indicate the precision of location of a point. They can also indicate whether field measurements are satisfactory or not. Consider a point B located in plan by a distance and a bearing from a fixed station A. Figure 7.11(a) indicates that the distance is far more precise than the bearing measurement, while Fig. 7.11(b) indicates the opposite. (The ellipse axes for such a measurement are clearly parallel to and perpendicular to the bearing.)

The probability of a point lying within the error ellipse is 0.394. This is the probability for a joint event for one standard deviation corresponding to 0.68 for a single function. The following table shows the probabilities, expressed as percentage confidence limits, for various numbers of standard error for a two-dimensional fix.

Confidence limit %	39.4	50.0	90.0	95.0	97.5	99.0
Standard error	1.00	1.18	2.15	2.45	2.76	3.03

Frequently 95% confidence limit is taken, and σ_{MAX} and σ_{MIN} are multiplied by 2.45 before being quoted.

Example 7.15

For Example 7.14 previously worked, Σwv^2 can be calculated. The coordinate differences and hence the polar coordinates for XP and YP are first required.

From the adjusted coordinates of P:

For XP: $\Delta E = -118.5337\,\text{m}$, $\Delta N = 349.7007\,\text{m}$. Polar coordinates are: $369.2436\,\text{m}$ (distance), $62°24'10.5''$ (bearing)

For YP: $\Delta E = -267.1477\,\text{m}$, $\Delta N = 194.9677\,\text{m}$. Polar coordinates are: $330.1972\,\text{m}$ (distance), $305°59'46.6''$ (bearing)

Residuals are therefore $v_1 = -3.4\,\text{mm}$, $v_2 = +4.2\,\text{mm}$, $v_3 = -9.5''$

Residual	Weight	wv^2
$v_1 = -3.4\,\text{mm}$	$w_1 = 1/100$	0.1156
$v_2 = +4.2\,\text{mm}$	$w_2 = 1/100$	0.1764
$v_3 = -9.5''$	$w_3 = 1/225$	0.4011
		$\Sigma = 0.693$

$$m = 3, n = 2 \quad \sigma_0^2 \triangleq s_0^2 = \frac{0.693}{3-2} = 0.693$$

This value lies between the recommended limits of 0.5 and 2.0, although no great confidence can be attached to it for a survey of only 3 observed quantities.

$$\therefore \mathbf{A}^\mathrm{T}\mathbf{W}\mathbf{A} = \begin{pmatrix} +0.0088208 & -0.0073739 \\ -0.0073739 & +0.0125661 \end{pmatrix} \text{ (Coefficients in normal equations)}$$

$$\therefore (\mathbf{A}^\mathrm{T}\mathbf{W}\mathbf{A})^{-1}$$

$$= \frac{1}{(0.0088208)(0.0125661) - (-0.0073739)^2} \begin{pmatrix} +0.0125661 & +0.0073739 \\ +0.0073739 & +0.0088208 \end{pmatrix}$$

$$\therefore \sigma_{xx} = 0.693 \times 17708.94 \begin{pmatrix} +0.0125661 & +0.0073739 \\ +0.0073739 & +0.0088208 \end{pmatrix}$$

$$= \begin{pmatrix} +154.21 & +90.49 \\ +90.49 & +108.25 \end{pmatrix}$$

$$\therefore \quad \sigma_E^2 = 154.21 \qquad \sigma_E = \pm 12.4\,\text{mm}$$
$$\sigma_N^2 = 108.25 \qquad \sigma_N = \pm 10.4\,\text{mm}$$
$$\sigma_{EN} = +90.49 \qquad 2\phi_M = -75°45'03.3'' = 104°14'56.7''$$
$$\sigma_{MAX} = 15.0\,\text{mm}$$
$$\sigma_{MIN} = 6.2\,\text{mm} \qquad \phi_M = 52°07'28''$$

For 95% confidence limits, σ_{MAX} and σ_{MIN} are multiplied by 2.45 to give semi-major and semi-minor axes of 37 mm and 15 mm.

Standard errors for the adjusted values of the measured quantities can also be calculated by combining the appropriate coefficients from the observation equations with the associated elements from the variance–covariance matrix.

For XP:

$$\sigma_{XP}{}^2 = [-0.32106 + 0.94706] \begin{bmatrix} +154.21 & +\ 90.49 \\ +\ 90.49 & +108.25 \end{bmatrix} \begin{bmatrix} -0.32106 \\ +0.94706 \end{bmatrix}$$

$$= 57.96$$

$$\therefore \sigma_{XP} = \pm 7.6\,\text{mm}$$

For YP:

$$\sigma_{YP}{}^2 = [-0.80908 + 0.58770] \begin{bmatrix} +154.21 & +\ 90.49 \\ +\ 90.49 & +108.25 \end{bmatrix} \begin{bmatrix} -0.80908 \\ +0.58770 \end{bmatrix}$$

$$= 52.28$$

$$\therefore \sigma_{YP} = \pm 7.2\,\text{mm}$$

For $P\hat{X}Y$:

$$\sigma_{P\hat{X}Y}{}^2 = [-0.52904 \ -0.17935] \begin{bmatrix} +154.21 & +\ 90.49 \\ +\ 90.49 & +108.25 \end{bmatrix} \begin{bmatrix} -0.52904 \\ -0.17935 \end{bmatrix}$$

$$= 63.81$$

$$\therefore \sigma_{P\hat{X}Y} = \pm 8.0''$$

7.10 TIED TRAVERSE EXAMPLE

Example 7.16

A tied traverse has been conducted between station A (316.750 m E; 205.412 m N) and station D (864.600 m E; 247.900 m N) via stations B and C. Reference object X with a bearing of 28°43′00″ from A and RO Y with a bearing of 124°30′30″ from D were sighted. Angles were measured at A, B, C and D with a standard error of 15″ and distances AB, BC and CD were measured with a standard error of 10 mm.

Using the observations given below, find the probable coordinates of B and C using a method of least squares.

Included angle	Value	Leg	Horizontal length (m)
XÂB	27°38′20″		
		AB	216.350
AB̂C	239°21′36″		
		BC	230.419
BĈD	146°16′24″		
		CD	161.755
CD̂Y	222°31′54″		

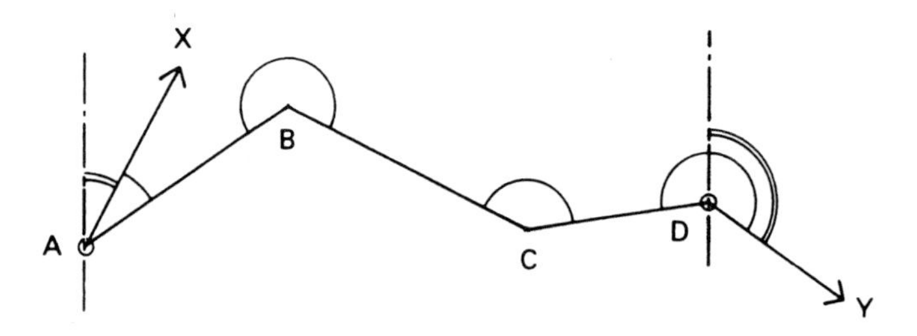

Fig. 7.12 Tied traverse example.

Solution

Figure 7.12 shows the traverse. There are seven observed quantities; if provisional coordinates are given to B and C it follows that there are three redundant quantities. Calculations for an open traverse are worked; these enable the provisional coordinates of B and C to be assigned and the angular and linear closing errors to be evaluated (Table 7.2).

The closing errors are therefore 44″, 0.0341 m(E) and 0.0587 m(N). If these indicate field results of acceptable precision, the traverse may be adjusted. The initial consistent model traverse will be similar to that shown above, but with the fixed coordinates of D inserted and the consequent provisional values of the angles at C and D and the distance CD replacing the observed quantities (Table 7.3).

Let variations (in millimetres) of e_B, n_B, e_C and n_C be applied to stations B and C. The corresponding variations to the observations (in seconds and millimetres) will be:

Table 7.2 Open traverse.

Leg and plan length (m)	Included angle	Bearing	Coord. diffs ΔE	ΔN	Station coords E	N	Stn
AX		28°43′00″					
	27°38′20″				316.750	205.412	A
AB 216.350		56°21′20″	180.1096	119.8660			
	239°21′36″				496.8596	325.2780	B
BC 230.419		115°42′56″	207.5981	−99.9797			
	146°16′24″				704.4577	225.2983	C
CD 161.755		81°59′20″	160.1764	22.5430			
	222°31′54″				864.6341	247.8413	D
DY		124°31′14″					

Table 7.3 Initial model traverse.

Leg and plan length (m)	Included angle	Bearing	Coord. diffs ΔE	ΔN	Station coords E	N	Stn
AX		28°43′00″					
	27°38′20″				316.750	205.412	A
AB 216.350		56°21′20″	180.1096	119.8660			
	239°21′36″				496.8596	325.2780	B
BC 230.419		115°42′56″	207.5981	−99.9797			
	146°15′04″				704.4577	225.2983	C
CD 161.7294		81°58′00″	160.1423	22.6017			
	222°32′30″				864.6000	247.9000	D
DY		124°30′30″					

Angle A $\dfrac{e_B \cos 56°21'20'' - n_B \sin 56°21'20''}{216.305 \times 1000} \times 206\,265$

Angle B $\dfrac{(e_C - e_B)\cos 115°42'56'' - (n_C - n_B)\sin 115°42'56''}{230.419 \times 1000} \times 206\,265$

$- \dfrac{e_B \cos 56°21'20'' - n_B \sin 56°21'20''}{216.350 \times 1000} \times 206\,265$

Angle C $\dfrac{-e_C \cos 81°58'00'' + n_C \sin 81°58'00''}{161.7294 \times 1000} \times 206\,265$

$- \dfrac{(e_C - e_B)\cos 115°42'56'' - (n_C - n_B)\sin 115°42'56''}{230.419 \times 1000} \times 206\,265$

Angle D $\dfrac{e_C \cos 81°58'00'' - n_C \sin 81°58'00''}{161.7294 \times 1000} \times 206\,265$

AB $e_B \sin 56°21'20'' + n_B \cos 56°21'20''$
BC $(e_C - e_B)\sin 115°42'56'' + (n_C - n_B)\cos 115°42'56''$
CD $-e_C \sin 81°58'00'' - n_C \cos 81°58'00''$

The angular: linear weights will be $\dfrac{1}{15^2}:\dfrac{1}{10^2}$ i.e. $\dfrac{1}{225}:\dfrac{1}{100}$

Simplifying the variations and forming the observation equations:

For A:

$$27°38'40'' + (0.52821e_B - 0.79369n_B) - 27°38'40'' = v_1 \text{ weight } 1/225$$

For B:

$$239°21'36'' + (-0.38842e_C - 0.80651n_C - 0.13979e_B + 1.60020n_B) - 239°21'36'' = v_2 \text{ wt } 1/225$$

For C:

$$146°15'04'' + (0.21019e_C + 2.06937n_C - 0.38842e_B - 0.80651n_B) - 146°16'24'' = v_3 \text{ wt } 1/225$$

For D:

$$222°32'30'' + (0.17823e_C - 1.26286n_C) - 222°31'54'' = v_4 \text{ wt } 1/225$$

For AB:

$$216.350 + (0.83249e_B + 0.55404n_B) - 216.350 = v_5 \text{ wt } 1/100$$

For BC:

$$230.419 + (0.90096e_C - 0.43390n_C - 0.90096e_B + 0.43390n_B) - 230.419 = v_6 \text{ wt } 1/100$$

For CD:

$$161.7294 + (-0.99019e_C - 0.13975n_C) - 161.755 = v_7 \text{ wt } 1/100$$

Simplifying:

$$\begin{array}{lr}
0.52821e_B - 0.79369n_B & = v_1 \text{ wt } 1/225 \\
-0.13979e_B + 1.60020n_B - 0.38842e_C - 0.80651n_C & = v_2 \text{ wt } 1/225 \\
-0.38842e_B - 0.80651n_B + 0.21019e_C + 2.06937n_C - 80'' & = v_3 \text{ wt } 1/225 \\
0.17823e_C - 1.26286n_C + 36'' & = v_4 \text{ wt } 1/225 \\
0.83249e_B + 0.55404n_B & = v_5 \text{ wt } 1/100 \\
-0.90096e_B + 0.43390n_B + 0.90096e_C - 0.43390n_C & = v_6 \text{ wt } 1/100 \\
- 0.99019e_C - 0.13975n_C - 25.6 & = v_7 \text{ wt } 1/100
\end{array}$$

Forming the normal equation for e_B

$$\begin{aligned}
&(0.52821)(1/225)(0.52821e_B - 0.79369n_B) \\
&\quad + (-0.13979)(1/225)(-0.13979e_B + 1.60020n_B - 0.38842e_C \\
&\quad - 0.80651n_C) + (-0.38842)(1/225)(-0.38842e_B - 0.80651n_B \\
&\quad + 0.21019e_C + 2.06937n_C - 80) + (+0.83249)(1/100)(0.83249e_B \\
&\quad + 0.55404n_B) + (-0.90096)(1/100)(-0.90096e_B + 0.43390n_B \\
&\quad + 0.90096e_C - 0.43390n_C) = 0
\end{aligned}$$

Collecting terms and forming the other normal equations similarly, we get:

$$\begin{array}{l}
+0.0170451e_B - 0.0007621n_B - 0.0082388e_C + 0.0008380n_C = -0.138105 \\
-0.0007621e_B + 0.0220236n_B + 0.0003934e_C - 0.0150362n_C = -0.286759 \\
-0.0082388e_B + 0.0003934n_B + 0.0189301e_C - 0.0002004n_C = -0.207271 \\
+0.0008380e_B - 0.0150362n_B - 0.0002004e_C + 0.0310894n_C = +0.902058
\end{array}$$

Solving:

$e_B = -18.4$, $n_B = 10.1$ A further iteration gives no change in
$e_C = -18.8$, $n_C = 34.3$ these values.

The adjusted coordinates are:

B (496.841 m E; 325.288 m N)
C (704.439 m E; 225.333 m N)

Note: Corresponding Bowditch coordinates are:

B (496.841 m E; 325.288 m N)
C (704.436 m E; 225.331 m N)

Error ellipse information is now calculated.

$\dfrac{\Sigma wv^2}{m - n}$ can be evaluated as follows:

Residual	Weight	wv^2
$v_1 = -17.7''$	$w_1 = 1/225$	1.3924
$v_2 = -\ 1.7''$	$w_2 = 1/225$	0.0128
$v_3 = -14.1''$	$w_3 = 1/225$	0.8836
$v_4 = -10.5''$	$w_4 = 1/225$	0.4900
$v_5 = -\ 9.7$ mm	$w_5 = 1/100$	0.9409
$v_6 = -10.9$ mm	$w_6 = 1/100$	1.1881
$v_7 = -11.8$ mm	$w_7 = 1/100$	1.3924
		$\Sigma = 6.3002$

$$m = 7, n = 4 \quad \sigma_0^2 = s_0^2 = \frac{6.3002}{7-4} = 2.1001$$

This value is only marginally outside the suggested limit and may be considered acceptable.

By inverting the coefficient matrix for the normal equations,

$$(\mathbf{A}^{\mathrm{T}}\mathbf{W}\mathbf{A})^{-1} = \begin{pmatrix} +74.42 & +1.15 & +32.35 & -1.24 \\ +1.15 & +67.83 & -0.56 & +32.77 \\ +32.35 & -0.56 & +66.91 & -0.71 \\ -1.24 & +32.77 & -0.71 & +48.04 \end{pmatrix}$$

Multiplying by 2.1001 gives the variance – covariance matrix:

$$\sigma_{xx} = \begin{pmatrix} +156.29 & +2.42 & +67.94 & -2.60 \\ +2.42 & +142.45 & -1.18 & +68.82 \\ +67.94 & -1.18 & +140.52 & -1.49 \\ -2.60 & +68.82 & -1.49 & +100.89 \end{pmatrix}$$

$$\begin{aligned} \sigma_{E_B} &= \pm 12.50\,\text{mm} \\ \sigma_{N_B} &= \pm 11.94\,\text{mm} \quad \sigma_{EN_B} = +2.42\,\text{mm}^2 \\ \sigma_{E_C} &= \pm 11.85\,\text{mm} \\ \sigma_{N_C} &= \pm 10.04\,\text{mm} \quad \sigma_{EN_C} = -1.49\,\text{mm}^2 \end{aligned}$$

And for the error ellipses:

$$\begin{aligned} \sigma_{B_{MAX}} &= 12.52\,\text{mm} \quad \sigma_{C_{MAX}} = 11.86\,\text{mm} \\ \sigma_{B_{MIN}} &= 11.92\,\text{mm} \quad \sigma_{C_{MIN}} = 10.04\,\text{mm} \\ \phi_{B_{MAX}} &= 80°22' \quad\quad \phi_{C_{MAX}} = 92°09' \end{aligned}$$

For 95% confidence limits, σ_{MAX} and σ_{MIN} are multiplied by 2.45 to give semi-major and semi-minor axes:

For B: $\sigma_{MAX} = 31$ mm For C: $\sigma_{MAX} = 29$ mm
$\sigma_{MIN} = 29$ mm $\sigma_{MIN} = 25$ mm

7.11 CONTROL SURVEY ADJUSTMENT

In the simple traverse worked, there is little difference between the adjustments according to least squares theory and the Bowditch method.

In a manual solution the extra work in least squares is hardly worthwhile. For more extensive control surveys, for surveys which have to be tied in to a number of fixed points and for surveys with extra measurements (a cross run in a traverse), a least squares process is the only satisfactory method of analysis and adjustment. With the use of EDM, the different methods of control survey can be combined. Field considerations are the strength of fix achieved and the provision of a number of redundant measures. If a preliminary survey of an area has been carried out, it may be possible to plan the control work and to test the suitability with an a priori analysis; after the survey a further assessment, in particular of the weights, can be carried out. A computer program is virtually essential for this type of work.

In a variation of coordinates computer program any combinations of measured distances and angles can be entered. Azimuth observations (angles related to stations of fixed bearing from the stations occupied) can also be entered. Standard errors for all input observations are also required. Fixed station coordinates and any fixed bearings must be entered. Depending on the sequence of entering data, and the degree of refinement of the program, it may or may not be necessary to calculate provisional coordinates for the stations to be located. Programs have been known to 'crash' for an apparent want (according to error messages) of a plethora of data; often the addition of provisional coordinates of a few stations has restored program operation.

Consistent and correct numbering or lettering of stations is important, and angles must be described according to the 'from-at-to' or 'at-from-to' format of the program.

A program should be designed to carry out a sufficient number of iterations to produce final adjustments of negligible amounts. The solution should converge in only a few iterations, even when very approximate initial provisional coordinates have been entered. A warning is usually given for a solution not quickly converging, or for a unit variance outside the recommended tolerance.

The program will normally display

station coordinates
adjusted observations
residuals
standard errors of coordinates
standard errors of adjusted observations
error ellipse parameters.

Error ellipses may be plotted (on screen or on print out) at an enlarged scale over a plan of the control network.

It should then be possible to assess the survey in terms of:

the assignment of standard errors to observations
the geometry of the network
the precision of critical measurements
the provision of strengthening measurement and redundants.

Such assessment may well be helpful in planning the work for future surveys.

7.12 FURTHER READING

Cooper, M.A.R. (1987) *Control Surveys in Civil Engineering*. Oxford: Blackwell Science.

Cooper, M.A.R. (1974) *Fundamentals of Survey Measurement and Analysis*. Oxford: Blackwell Science.

Schofield, W. (1993) *Engineering Surveying*, 4th edn. London: Butterworth-Heinemann.

Shepherd, F.A. (1981) *Advanced Engineering Surveying*. London: Arnold.

7.13 EXERCISES

Exercise 7.1

(a) Find the horizontal length and standard error of a baseline measured in four bays from the following information:

Section	Mean length (m)	Standard error (m)	Zenith angle	Standard error
AB	50.304	+0.005	Level	
BC	37.365	+0.004	Level	
CD	123.120	+0.013	84°17′15″	±20″
DE	68.731	+0.008	94°27′54″	±20″

(b) Dertermine the probable coordinates (with standard errors) of the station Y from the following information:

Station X
  Eastings 1435.670 m ± 0.003 m (standard error)
  Northings 1736.119 m ± 0.003 m
Slope length XY 223.653 m ± 0.006 m
Zenith angle to Y 83°22′15″ ± 10″
Bearing of XY 71°25′37″ ± 5″

Exercise 7.2

K, L, M and N are points extending a triangulation scheme. Angles have been measured (clockwise) as follows:

$K\hat{L}M = 101°47'25''$ $\quad M\hat{N}K = 50°45'04''$
$L\hat{M}K = 38°30'06''$ $\quad N\hat{K}M = 50°17'34''$
$K\hat{M}N = 78°57'23''$ $\quad M\hat{K}L = 39°42'33''$
$L\hat{M}N = 117°27'22''$

Adjust the angles, all of which have equal weight, by a method of least squares so that angle $N\hat{K}L$ is kept fixed at 90°00′00″.

Exercise 7.3

Levelling has been carried out from three fixed benchmarks, P, Q and R to establish values of two new stations, X and Y. From the following information, calculate the most probable levels of X and Y:

P Level	112.37 m AOD
Q	137.24
R	121.33

Level run	Rise (m)	Horizontal distance (m)
P to X	+12.05	126.5
X to Q	+12.74	200.0
Q to Y	− 7.38	151.2
Y to R	− 8.15	126.5
X to Y	+ 5.37	178.9
P to Y	+17.46	200.0

Exercise 7.4

From control stations A, B and C, electromagnetic distance measurements have been taken to a point P whose coordinates are approximately (518.400 m E; 412.400 m N). Coordinates of the control stations are:

A (187.370 m E: 770.680 m N)
B (312.765 m E; 130.230 m N)
C (971.415 m E; 633.455 m N)

Horizontal measurements were:

AP 487.822 ± 0.003 m (standard error)
BP 349.096 ± 0.005 m
CP 504.100 ± 0.007 m

Use the variation of coordinates method to determine the probable coordinates of P.

Exercise 7.5

To locate a control station P, a horizontal distance has been measured with EDM from fixed station A. Horizontal angles $P\hat{B}C$ and $B\hat{C}P$ have been measured at fixed stations B and C. Use two of these measurements to derive provisional coordinates and apply a variation of coordinates adjustment to determine the most probable coordinates of P.

Fixed station coordinates	Observations
A (306.053 m E; 259.763 m N)	AP 198.145 m ± 10 mm (standard error)
B (353.892 m E; 507.345 m N)	P$\hat{B}$C 21°22′36″ ± 15″
C (451.370 m E; 227.467 m N)	B$\hat{C}$P 25°24′06″ ± 15″

8 Area and Volume Measurement

8.1 INTRODUCTION

The tasks of measuring areas and volumes relevant to construction works are usually assigned to the site engineer. Measurements will be needed:

(1) At the design stage for planning works and estimating quantities.
(2) During construction so that the correct quantities of materials can be ordered and placed.
(3) After construction so that the appropriate payment can be made.

It is worth mentioning that whereas the contractor will generally be paid for the quantities described in the contract documents, the supplier of the materials will expect payment for all materials delivered to site. The site engineer will have to explain any discrepancies between the two sets of figures.

Areas

Examples of area measurement are:

(1) The plan area of parcels of land at the planning stage of a construction project.
(2) The plan area of paving, flooring and roofing and the elevation area of walls for design, ordering and payment purposes.
(3) Plan or cross-section areas for volume calculations.

Volumes

Examples of volume measurement are:

(1) Earthwork measurement for design and payment purposes.
(2) Construction materials measurement for design, ordering and payment purposes.
(3) Water measurement for reservoir calculations.

Accuracy

While the field measurements need not be carried out with the same precision as in site surveys and setting out (the irregular nature of undisturbed ground would make a nonsense of precise work), care must be taken that unacceptably large errors are not introduced when measurements are multiplied.

Major roadworks usually involve large quantities of earthmoving. The cost of such works may be a significant proportion of the contract price, and accumulative errors may make the difference between success and failure in winning a contract, or between profit and loss on the contract.

8.2 REGULAR AREAS

In many instances, regular shapes can be used in the determination of areas. Quantities of construction materials (number of bricks in a wall, amount of roofing felt required) can easily be calculated where they are to occupy a regular area. Many irregular shapes can be approximated to regular ones to facilitate calculations without undue loss of accuracy. Figure 8.1 shows regular shapes and their areas.

Area within a loop traverse

For an n-sided traverse, numbered anti-clockwise:

$$\text{Area} = \frac{1}{2}[E_1(N_2 - N_n) + E_2(N_3 - N_1) + E_3(N_4 - N_2) + \ldots + E_n(N_1 - N_{n-1})]$$

where E_n, N_n are the eastings and northings of the nth station. For a clockwise traverse, the area will be given as negative and the sign must be changed.

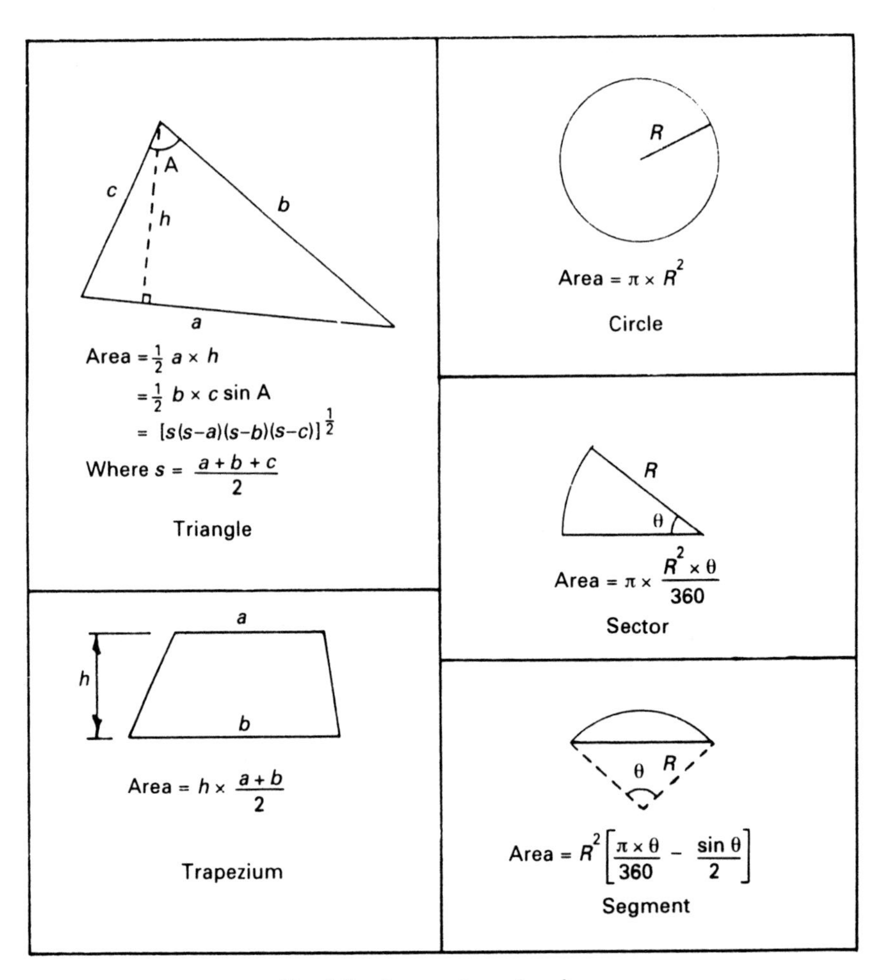

Fig. 8.1 Areas of regular shapes.

The formula is also useful where land inside a loop traverse has to be divided in a given ratio by a line passing through one of the stations. By assigning unknown coordinates of E and N to the intersection point of the far end of the line and the traverse, two simultaneous equations can be developed and solved.

8.3 IRREGULAR AREAS

Areas involving land measurements often have irregular perimeters, typically cross-sections, areas within contours and plan areas within

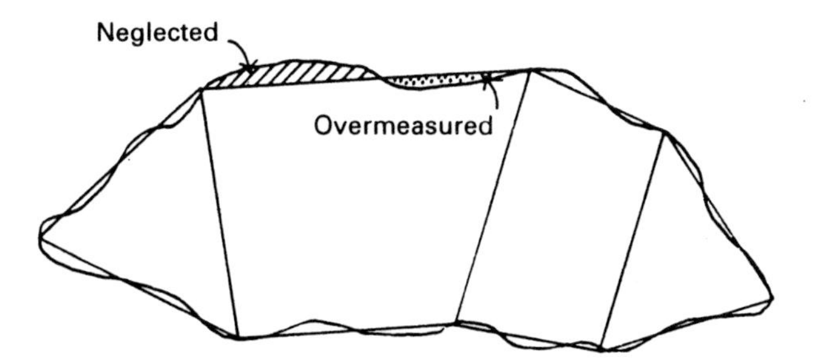

Fig. 8.2 Give and take lines.

irregular boundaries. There are two approaches in determining such areas: one is to take measurements on site to facilitate the direct calculation of the area; the other is to compute the area from a scale plot of the land following a survey.

Direct calculation: approximation to regular figures

In Fig. 8.2, the plan area has been approximated to two trapezia and two triangles of the same total area. This has been achieved by constructing 'give and take' lines along irregular boundaries, so that, within the limits of accuracy, the amount neglected equals the amount overmeasured. The regular figures are measured in the field by marking out the apices of the polygons with ranging rods, and taping between them.

The method is simple and, on a plan, regular shapes can easily be superimposed. It is more difficult in the field, requiring a reconnaissance to determine a suitable arrangement of regular shapes.

Direct calculation: trapezoidal method

In this method, the area concerned is divided into a series of strips of equal width which are taken to be trapezia. The parallel edges of these trapezia ('ordinates') are then measured (Fig. 8.3).

By summing the areas of the trapezia:

$$\text{Total area} = \frac{d}{2}[l_1 + 2(l_2 + l_3 + \ldots + l_{n-1}) + l_n]$$

Clearly, the smaller the strip width d, the more accurate will be the

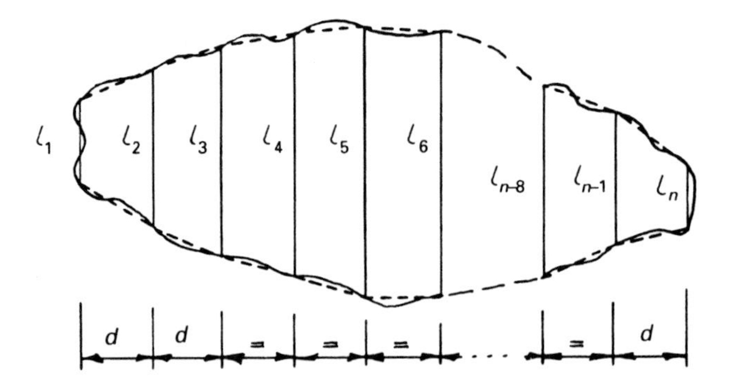

Fig. 8.3 Trapezoidal rule.

area determination. For plan areas, the fieldwork may be accomplished by measuring perpendicular offsets at constant intervals along a 'chain' line laid along the length of the land. Vertical sections can be measured by taking ground levels at regular intervals and calculating depths of cut or fill from design levels to give the ordinates.

Direct calculations: Simpson's method

The fieldwork for this method is the same as for the trapezoidal method, although it is desirable to arrange for an even number of strips to be measured. The irregular boundary of a neighbouring pair of strips is considered to be parabolic.

The area enclosed by a chord and the arc of a parabola is 2/3 that of the surrounding parallelogram; it can be shown that the area of the first two strips

$$= \frac{d}{3}(l_1 + 4l_2 + l_3)$$

Hence for an *even* number of strips (*odd* number of ordinates):

$$\text{Total area} = \frac{d}{3}(l_1 + 4l_2 + 2l_3 + 4l_4 + \ldots + 2l_{n-2} + 4l_{n-1} + l_n)$$

$$\text{i.e. area} = \frac{d}{3}\begin{bmatrix} \text{first and last ordinates.........singly} \\ \text{even ordinates...................quadrupled} \\ \text{odd ordinates ....double} \end{bmatrix}$$

Note that the penultimate ordinate should be quadrupled.

If Simpson's method is applied where there are an even number of ordinates, it must be used for one less ordinate, the final strip being calculated as a trapezium and added on.

Notes: (1) Zero ordinates should be written down to prevent factors from becoming displaced.
(2) Use a convenient strip width; for a length not a multiple of a strip width, calculate the final strip area as a trapezium of unique width.

Examples 8.1

Figure 8.4 shows a sketch plan of an irregularly bounded area where a linear survey has been carried out taking offsets at 5.00 m intervals. The three stations are located on the perimeter with horizontal distances of AB = 30.00 m, BC = 35.00 m and CA = 25.00 m. Determine the area within the boundary.

Solution

Area within triangle

$$s = (30.00 + 35.00 + 25.00)/2 = 45.00$$

$$\text{Area} = [45.00(45.00 - 30.00)(45.00 - 35.00)(45.00 - 25.00)]^{1/2}$$
$$= 367.42\,\text{m}^2$$

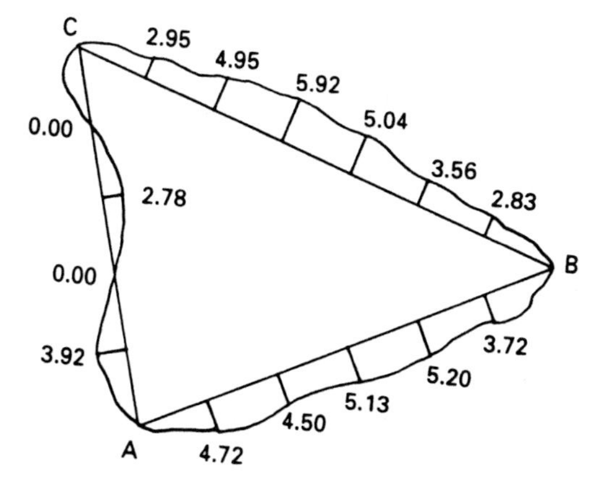

Fig. 8.4 Example.

Line AB

$$\text{Trapezoidal area} = \frac{5.00}{2}\left[0.00 + 2(4.72 + 4.50 + 5.13 + 5.20 + 3.72) + 0.00\right]$$
$$= 116.35\,\text{m}^2$$

$$\text{Simpson's area} = \frac{5.00}{3}\left[0.00 + 4(4.72 + 5.13 + 3.72) + 2(4.50 + 5.20) + 0.00\right]$$
$$= 122.80\,\text{m}^2$$

Line BC

$$\text{Trapezoidal area} = \frac{5.00}{2}\left[0.00 + 2(2.83 + 3.56 + 5.04 + 5.92 + 4.95 + 2.95) + 0.00\right]$$
$$= 126.25\,\text{m}^2$$

$$\text{Simpson's area} = \frac{5.00}{3}\left[0.00 + 4(2.83 + 5.04 + 4.95) + 2(3.56 + 5.92) + 2.95\right]$$
$$+ \frac{5.00}{2}(2.95 + 0.00)$$
$$= 129.36\,\text{m}^2$$

Line CA

$$\text{Trapezoidal area} = \frac{5.00}{2}\left[0.00 + 2(0.00 - 2.78 + 0.00 + 3.92) + 0.00\right]$$
$$= 5.70\,\text{m}^2$$

$$\text{Simpson's area} = \frac{5.00}{3}\left[0.00 + 4(0.00 + 0.00) + 2(-2.78) + 3.92\right]$$
$$+ \frac{5.00}{2}(3.92 + 0.00)$$
$$= 7.07\,\text{m}^2$$

(Note the negative ordinate where the chain line is outside the perimeter.)

$$\text{Total area (trapezoidal)} = 615.72 = 616\,\text{m}^2$$
$$\text{(Simpson's)} = 626.65 = 627\,\text{m}^2$$

Area from a plotted plan

An area from a plot can be determined by:

(1) Applying any of the methods so far described.
(2) Using a planimeter.
(3) Counting the squares when the area has been plotted on squared paper (a method for the patient!).

The planimeter

The traditional planimeter consists of two hinged arms: a radius arm with a pole block at its outer end, and a tracing arm with a tracing point at its outer end. Close to the hinge on the tracing arm is a small wheel which is in contact with the paper and rotates when the point is traced around a perimeter. A dial provides a reading (Fig. 8.5(a)).

The pole should, if possible, be placed outside the perimeter. An initial reading should be taken, the point traced in a complete loop around the perimeter and a final reading taken. The difference between the two readings, usually multiplied by a scale factor, will give the area.

Unless some care is taken, large errors can occur, and the process should be repeated several times until three or more consistent readings are obtained.

If the pole has to be placed inside the perimeter, a 'zero circle' constant should be added or subtracted. Details of this and of the scale factors are provided by the manufacturers.

Electronic planimeters are now produced which are easier to operate and to read, though care still needs to be exercised to achieve satisfactory results (Fig. 8.5 (b)).

8.4 ROAD CROSS-SECTIONS

Road cross-sections are bounded by the formation and the side slopes which are straight, and the original ground which is usually approximated to one or two straight lines. The simplest section is a trapezium, where both original ground and formation are level.

More commonly, original ground will not be level, and it will be necessary, from levels recorded on site, to determine the offset

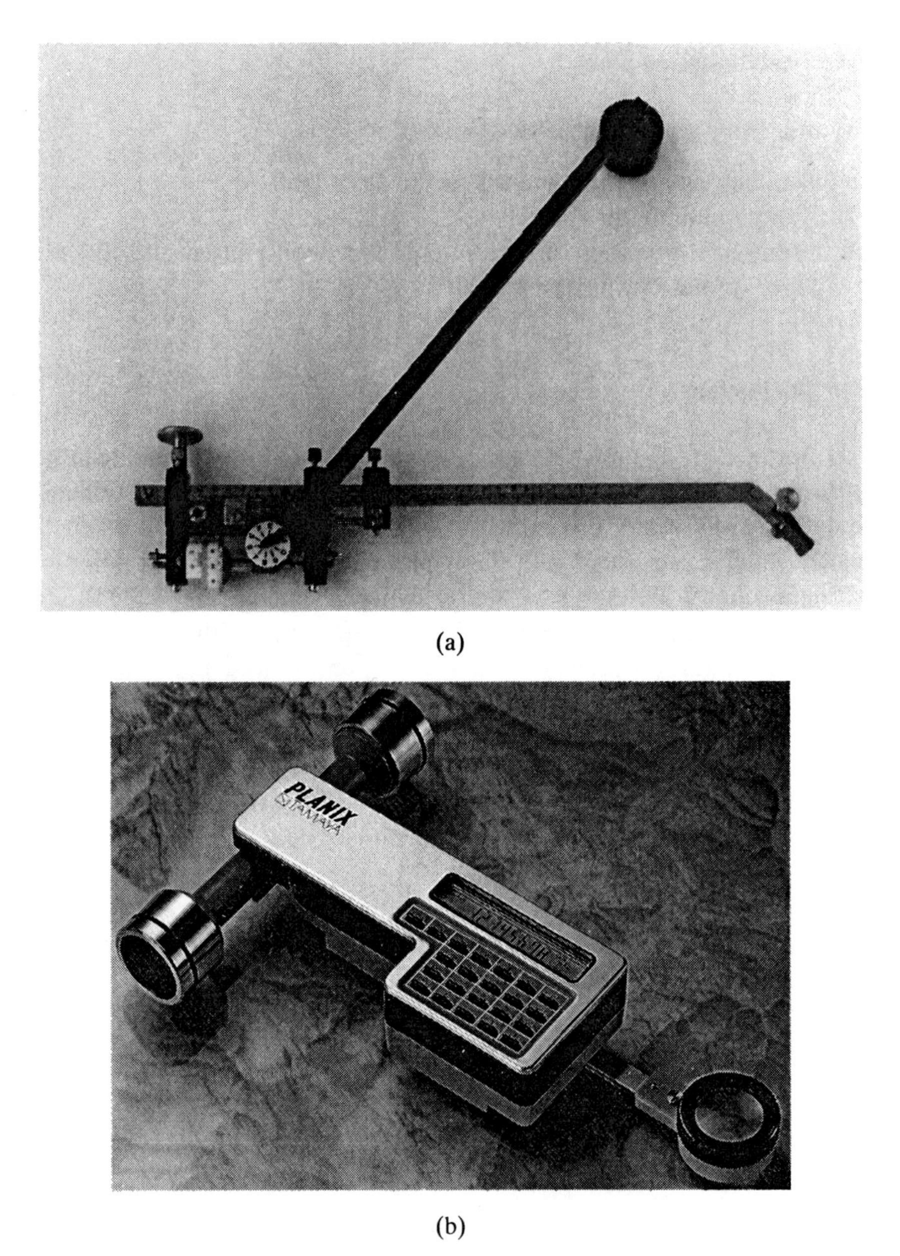

(a)

(b)

Fig. 8.5 Planimeters: (a) mechanical; (b) electronic (Planix).

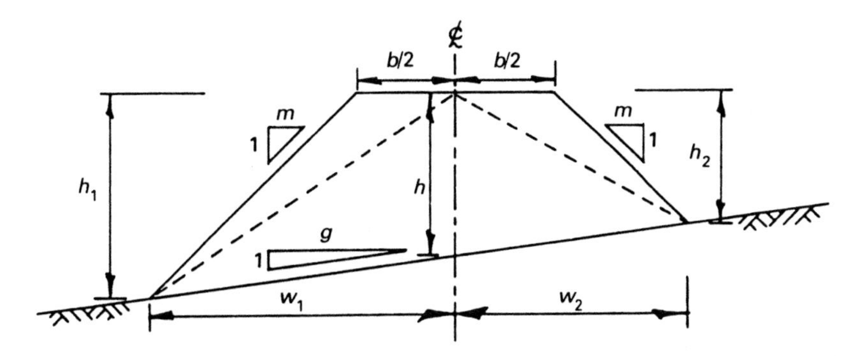

Fig. 8.6 Embankment section.

distance of the top of cutting or toe of embankment. The formation depth below or height above these points is also required.

Consider a section through an embankment as shown in Fig. 8.6. By examining the height of the formation above the left toe:

$$h + \frac{w_1}{g} = \frac{w_1 - b/2}{m} \quad (=h_1)$$

$$\therefore\ w_1\left(\frac{1}{m} - \frac{1}{g}\right) = \frac{b}{2m} + h$$

$$\therefore\ w_1 = \frac{g(b/2 + mh)}{g - m} \quad \text{and} \quad h_1 = \frac{w_1 - b/2}{m} = \frac{gh + b/2}{g - m}$$

Similarly:

$$w_2 = \frac{g(b/2 + mh)}{g + m}, \qquad h_2 = \frac{w_2 - b/2}{m} = \frac{gh - b/2}{g + m}$$

By splitting the section into three triangles:

$$\text{Area} = (w_1 + w_2)\frac{h}{2} + (w - b/2)\frac{b}{4m} + (w_2 - b/2)\frac{b}{4m}$$

$$= (w_1 + w_2)[h + b/(2m)]/2 - b^2/(4m)$$

$$= \frac{g^2}{m(g^2 - m^2)}(b/2 + mh)^2 - \frac{b^2}{4m}$$

The same reasoning can be applied to sections in cut, sections on sidelong ground and 'three level' sections. ('Two level' sections are so called because two ground levels define the profile of the original ground: a single crossfall. A double crossfall, with the gradient changing at the centreline, requires the ground to be levelled at three

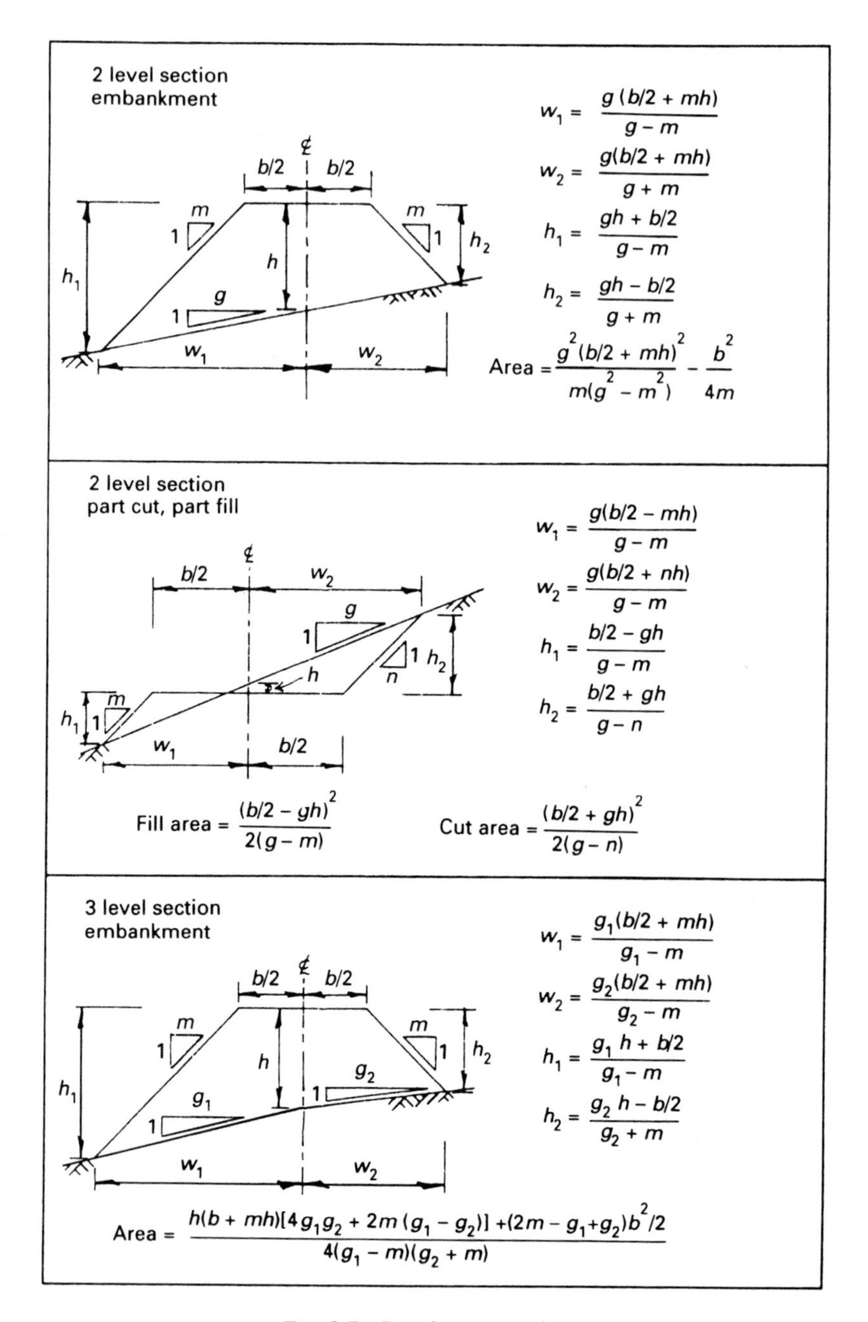

Fig. 8.7 Road cross-sections.

positions, hence 'three level'.) Figure 8.7 shows areas for various cross-sections.

For more complex sections the formula for determining the area within a loop traverse may be applied. Where there are many sections, a computer solution may be advisable; the program can combine the areas to give a volume.

8.5 VOLUMES OF REGULAR SOLIDS

The formulae for regular solids can nearly always be used for calculating volumes of concrete and other materials, and frequently earthworks can be approximated to one or more regular solids. Where a volume is a product of three dimensions, they must be in directions mutually perpendicular.

Useful solids

Many complex solids can be divided into tetrahedra, pyramids, cones and wedges. Figure 8.8 shows their volumes.

The prismoid

The prismoid (Fig. 8.9) is the solid between two plane parallel faces connected directly by plane sides. If the areas of these faces are A_1 and A_2, their distance apart is D and the area of mid-section is A_m,

$$\text{Volume} = \frac{A_1 + 4A_m + A_2}{6} \times D$$

Note that generally A_m will not be the mean of A_1 and A_2. The formulae for tetrahedra, pyramids, cones, wedges and even spheres can be derived from the prismoid.

8.6 VOLUMES OF IRREGULAR SOLIDS

As with irregular areas, it is impossible to calculate true quantities for the volumes of irregular solids. It will be the engineer's task to select a method of approximation which will produce acceptable accuracy.

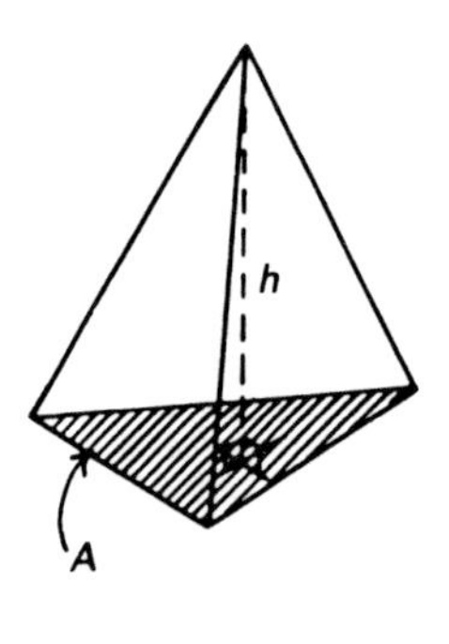

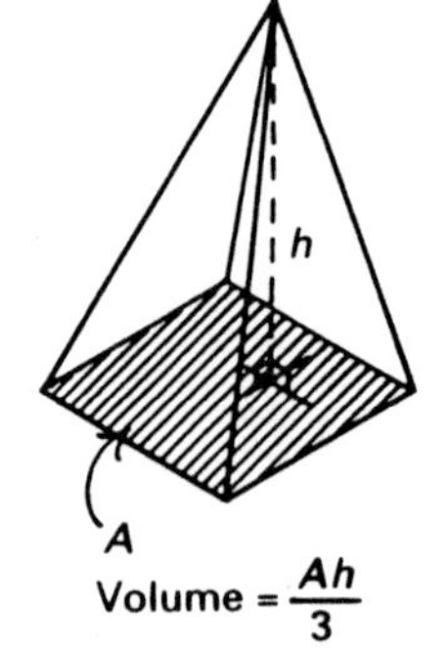

h

A

Volume = $\frac{Ah}{3}$

Tetrahedron Pyramid Cone

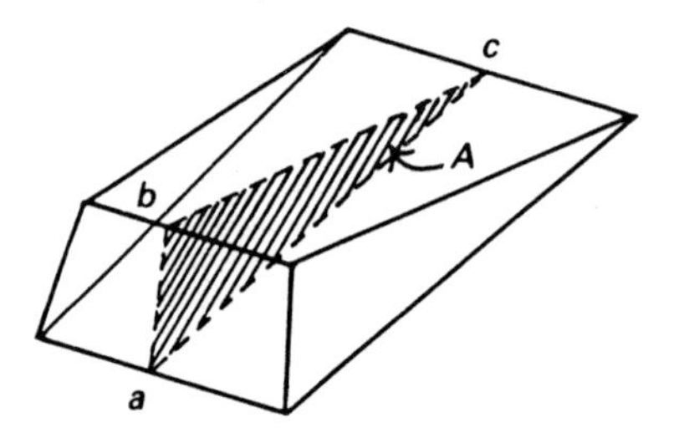

Three edges, *a,b,c,* perpendicular to cross-section *A*

Volume = $A \times \frac{a+b+c}{3}$

Wedge

Fig. 8.8 Regular solids.

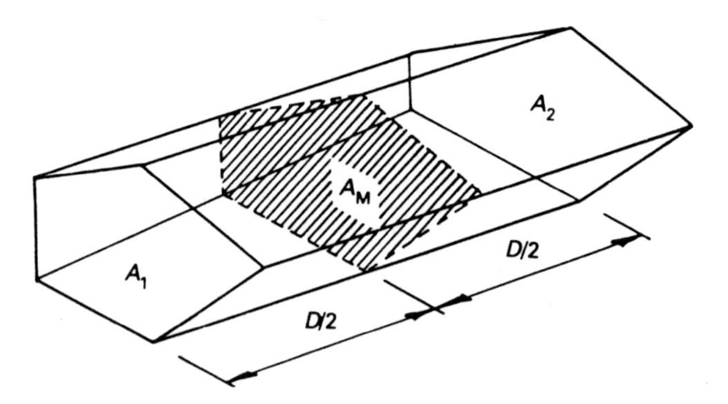

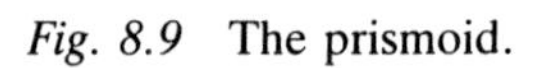

Fig. 8.9 The prismoid.

Approximation to regular solids

Many volume calculations can be performed by this method, the solid often being approximated to one or more prismoids.

Volume from section areas at regular intervals

Where cross-sections at regular chainages or plan areas at regular depths have been determined, the volume can be found by two methods.

Trapezoidal method for volumes (end area method)

In Fig. 8.10, by taking the volume between two parallel end areas as the product of the mean of the areas and their distance apart and applying this method to a volume of many sections, we see that:

$$\text{Total volume} = \frac{d}{2}[A_1 + 2(A_2 + A_3 + \ldots + A_{n-1}) + A_n]$$

Prismoidal method (Simpson's volume method)

Taking Fig. 8.10 again and considering the volume between sections 1 and 3 to be a prismoid, we see that:

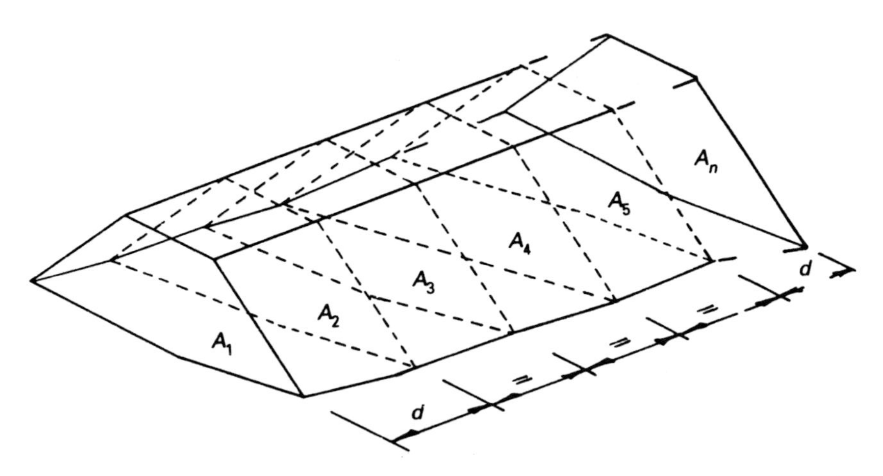

Fig. 8.10 Trapezoidal rule.

$$\text{Volume (sections 1 to 3)} = \frac{d}{3}(A_1 + 4A_2 + A_3)$$

and, similarly:

$$\text{Volume (sections 3 to 5)} = \frac{d}{3}(A_3 + 4A_4 + A_5)$$

Thus:

$$\text{Total volume} = \frac{d}{3}(A_1 + 4A_2 + 2A_3 + 4A_4 + \ldots + 2A_{n-2} + 4A_{n-1} + A_n)$$

where n must be *odd*.

$$\text{i.e. Volume} = \frac{d}{3}\begin{bmatrix} \text{first and last areas .... singly} \\ \text{even areas .... quadrupled} \\ \text{odd areas.................... doubled} \end{bmatrix}$$

If the prismoidal rule is applied where n is even, it must be applied to $n - 1$ sections, the volume between the last two sections being calculated by the trapezoidal method.

In both methods, closer sections will yield better accuracy. The prismoidal method is generally thought to be more accurate.

Applications

(1) Earthwork volumes, particularly for roads calculated from cross-sections.
(2) Volume of deep excavations or topsoil heaps calculated from plan areas at regular depths or heights.
(3) Water volume in an impounding reservoir calculated from areas within ground (and face of dam) contours. The relationship between capacity and water level can be investigated by using the trapezoidal method incrementally.

Example 8.2

A road is to run on an embankment between chainages 230.00 m and 307.00 m, and cross-section areas have been determined every 10.00 m as follows:

Chainage (m)	Area (m^2)
230.00	0.0
240.00	20.5
250.00	45.7
260.00	96.6
270.00	127.3
280.00	125.9
290.00	88.9
300.00	45.2
307.00	0.0

Calculate the volume of fill required.

Solution: trapezoidal

The final portion being 7.00 m long must be calculated separately.

$$\text{Volume} = \frac{10.00}{2}\left[0.0 + 2(20.5 + 45.7 + 96.6 + 127.3 + 125.9 + 88.9) + 45.2\right] + \frac{7.00}{2}(45.2 + 0.0)$$

$$= 5433.2 = 5430\,\text{m}^3$$

Solution: prismoidal

Calculate for 7 sections, and add the last two portions using the trapezoidal method.

$$\text{Volume} = \frac{10.00}{3}\left[0.0 + 88.9 + 4(20.5 + 96.6 + 125.9) + 2(45.7 + 127.3)\right]$$

$$+ \frac{10.00}{2}(88.9 + 45.2) + \frac{7.00}{2}(45.2 + 0.0)$$

$$= 5518.4 = 5520\,\text{m}^3$$

Earthwork volume as fill changes to cut

Where a length of road changes from running on an embankment to running in a cutting, both cut and fill volumes will have to be calculated between the sections where the changeover takes place.

Figure 8.11 is a perspective view of such a length of road. Portion A lies between a normal embankment section and the final 'all fill' section. Portion B lies between the final 'all fill' section and first all cut' section. Portion C lies between the final 'all cut' section and a normal cut section. Portion A will involve fill only; portion B will incorporate fill and cut while portion C will be cut only. The chainages of the final all fill and first all cut sections should first be estimated. Interpolation from existing sections will be required. The volumes for portions A and C can be determined by the trapezoidal method.

Portion B may have one or more sections of part cut/part fill plotted and evaluated for areas. If not, a staight line is drawn diagonally across the formation between the points where the embankment toe becomes the cutting top. This will produce two pyramids, one for fill, one for cut, and the respective volumes can be calculated accordingly (1/3 × section area × distance along road). Where intermediate part cut/part fill sections have been produced, a series of straight lines should be drawn connecting points on adjacent sections at formation

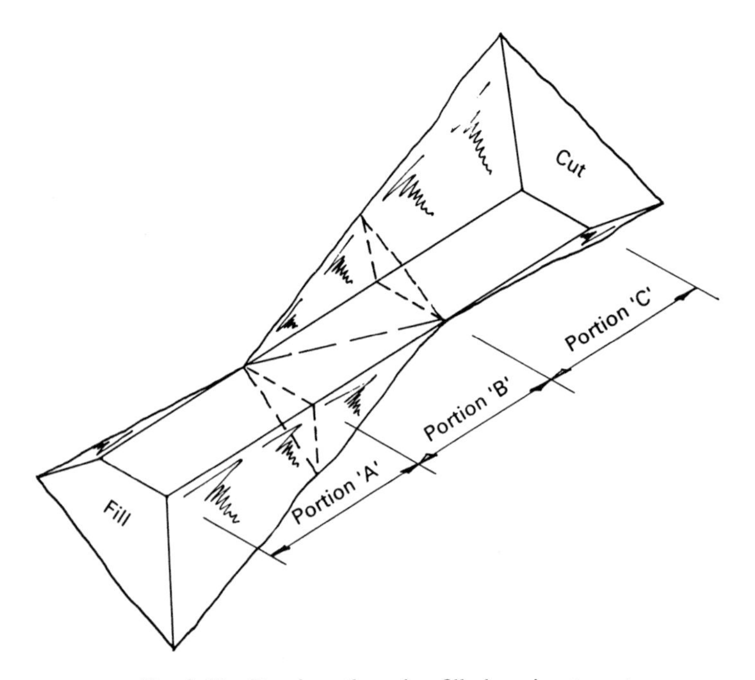

Fig. 8.11 Road earthworks: fill changing to cut.

level where fill changes to cut. The trapezoidal rule (or possibly the prismoidal rule) can be used to calculate cut and fill elements of portion B separately.

Example 8.3

Figure 8.12 shows a perspective view and dimensioned cross-sections

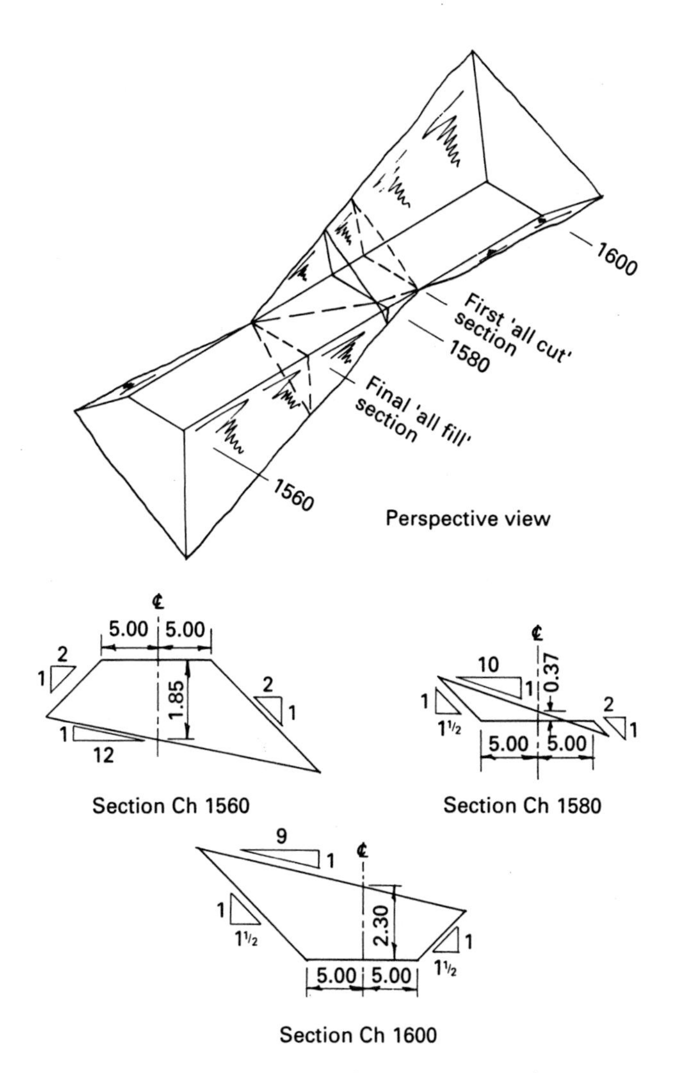

Fig. 8.12 Fill changing to cut: example.

for a length of road changing from embankment to cutting and cross-sections at Chainages 1560, 1580 and 1600.

Find the cut and fill volumes between Chainages 1560 and 1600.

Solution

It is necessary to locate the chainages of the final all fill and the first all cut sections. Assume that original ground has a constant rise between Chainages 1560 and 1580, and between Chainages 1580 and 1600.

Depths between formation and original ground at formation edges are:

Chainage (m)	Left	Right
1560	1.85 − 5/12 = 1.433 m fill	1.85 + 5/12 = 2.267 m fill
1580	0.37 + 5/10 = 0.870 m cut	−0.37 + 5/10 = 0.130 m fill
1600	2.30 + 5/9 = 2.856 m cut	2.30 − 5/9 = 1.744 m cut

Final all fill section will be at chainage

$$1560 + 20 \times \frac{1.433}{1.433 + 0.870} = 1572.445\,\text{m}$$

First all cut section will be at chainage

$$1580 + 20 \times \frac{0.130}{0.130 + 1.744} = 1581.387\,\text{m}$$

The dimensions of sections at these chainages must be determined (see Fig. 8.13).
Chainage 1572.445 m

$$h \text{ (depth at centre line)} = 1.85 - \frac{12.445}{20.000}(1.85 + 0.37) = 0.469\,\text{m}$$

$$g \text{ (original ground crossfall)} = \frac{5.000}{0.469} = 1 \text{ in } 10.661$$

Chainage 1581.387 m

$$h = 0.37 + \frac{1.387}{20.000}(2.30 - 0.37) = 0.504\,\text{m}$$

$$g = \frac{5.000}{0.504} = 1 \text{ in } 9.921$$

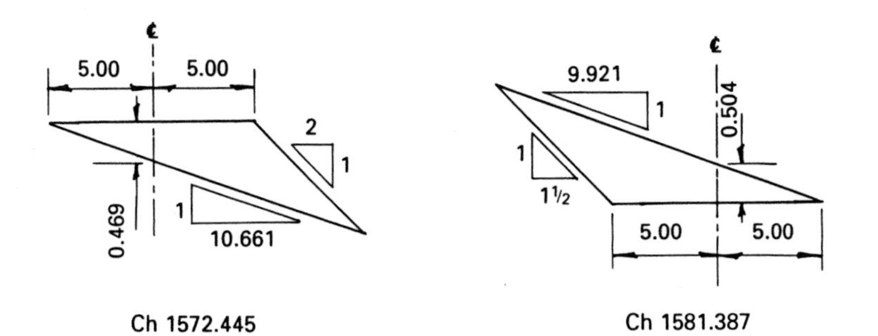

Fig. 8.13 Cross-sections.

Section areas can be calculated using the formulae previously developed:

Chainage	b	h	g	m	n	Fill (m²)	Cut (m²)
1560.000	10.000	1.850	12	2		26.426	–
1572.445	10.000	0.469	10.661	2		5.773	0.000
1580.000	10.000	0.370	10	2	1½	0.106	4.452
1581.387	10.000	0.504	9.921		1½	0.000	5.938
1600.000	10.000	2.300	9		1½	–	32.295

Volume

Volumes between adjacent sections can be evaluated individually by the end areas method. Where the fill/cut tapers out, the formula for a pyramid is used.

Ch 1560 to 1572.445

		Fill (m³)	Cut (m³)
$V = \dfrac{12.445}{2}(26.426 + 5.773)$	=	200.358	

Ch 1572.445 to 1580.00

		Fill (m³)	Cut (m³)
$V_{\text{FILL}} = \dfrac{7.555}{2}(5.773 + 0.106)$	=	22.208	
$V_{\text{CUT}} = \dfrac{7.555 \times 4.452}{3}$	=		11.212

Ch 1580.000 to 1581.387

			Fill	Cut
$V_{\text{FILL}} = \dfrac{1.387 \times 0.106}{3}$	=		0.049	
$V_{\text{CUT}} = \dfrac{1.387}{2}(4.452 + 5.938)$	=			7.205

Ch 1581.387 to 1600.000

			Fill	Cut
$V = \dfrac{18.613}{2}(5.938 + 32.295)$	=			355.815
		Total	222.615	374.232

These values may be rounded off to 222.6 (or 223) m^3 fill and 374.2 (or 374) m^2 cut.

Volume from a grid of levels

After ground levels have been taken at the intersection points of a rectangular grid, depths of cut or fill are calculated. The volume within each grid rectangle can be considered as the product of the plan area and the mean of the corner depths (Fig. 8.14(a) and (b)).

$$\text{Volume within rectangle} = 10.00 \times 12.00(3.35 + 3.80 + 2.74 + 2.91)/4$$
$$= 384\,\text{m}^3$$

By combining all rectangles it can be seen that each depth should be multiplied by a factor equal to the number of rectangles to which it applies, and the formula

$$\text{Volume} = \frac{d_1 \times d_2}{4}\Sigma\,(\text{depth} \times \text{factor})$$

used where d_1 and d_2 are the grid intervals (Fig. 8.14(c)).

The application of the trapezoidal rule for areas and volumes would produce the same factors.

Alternatively, each rectangle may be considered as two triangles (Fig. 8.15(a)).

$$\text{Volume} = \frac{10.00 \times 12.00}{2} \times \frac{3.35 + 2.74 + 2.91}{3}$$
$$+ \frac{10.00 \times 12.00}{2} \times \frac{3.35 + 3.80 + 2.91}{3}$$
$$= 381.2\,\text{m}^3$$

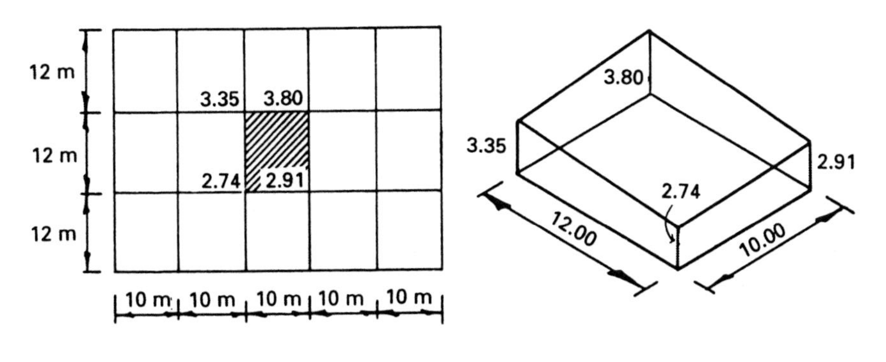

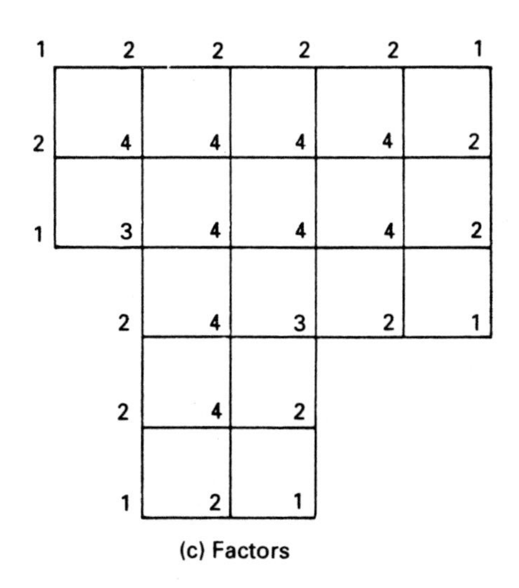

Fig. 8.14 Volume from level grid.

Then the formula

$$\text{Volume} = \frac{d_1 \times d_2}{6} \Sigma \text{ (depth} \times \text{factor)}$$

applies with the factors being the numbers of triangles to which each depth relates (Fig. 8.15(b)). The diagonals to delineate the triangles can be drawn in two directions, giving slightly different values of the final volume.

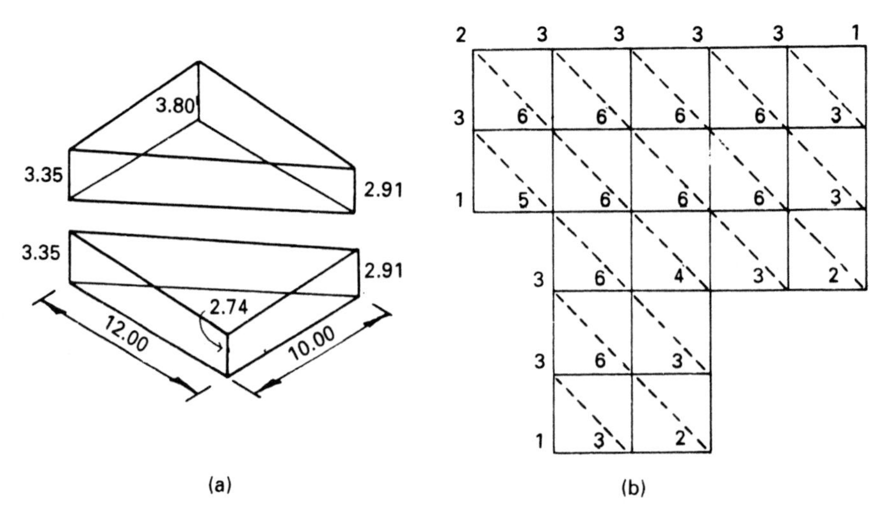

Fig. 8.15 Volume from grid (triangular).

Irregular perimeter and sloping sides

In practice, irregular perimeters and sloping sides are frequently encountered in gridded areas. The plan area between an irregular perimeter and a grid line may be approximated to triangles and trapezia and multiplied by a mean depth.

The volume between a side slope and a vertical edge can be determined by multiplying a triangular section by a length or by taking a number of sections and applying the trapezoidal method.

8.7 CURVED IRREGULAR SOLIDS

Where the centreline of a road or railway curves in plan, the sections will not be parallel to each other and a curvature correction may be required. In many instances, particularly on projects of some length, the corrections due to curvature to right and left will more or less balance each other out, and the resulting error will be insignificant.

Where corrections are necessary, the simplest way of incorporating them is to correct each section. By considering Pappus' theorem and neglecting small quantities, it can be shown that:

Corrected area $= A(1 + e/R)$

where A is the section area, R is the radius of curvature of the road centreline, and e is the eccentricity of the centroid of the section relative to the centreline.

The eccentricity is calculated by taking moments. It is positive if it lies on the other side of the centreline from the centre of curvature and negative if on the same side.

Example 8.4

Figure 8.16 shows a section of road on embankment with a 20.00 m wide formation, 1 in 2 side slopes and a height at the centreline of 12.50 m. Original ground falls at 1 in 10 from right to left and the road is on a right-hand circular curve of 500.00 m. Calculate the section area corrected for curvature.

Solution

From the formulae previously developed:

$$w_1 = 43.75 \qquad w_2 = 29.17 \qquad h_1 = 16.88 \qquad h_2 = 9.58$$
$$A = 588.02$$

To take moments about the centreline, it is necessary to locate the centroids of three triangles, making use of the property that the centroid is 2/3 along a line from an apex to the mid-point of the opposite side. Figure 8.17 shows the relevant dimensions. Taking moments about the centreline:

$$\left(\frac{33.75 \times 13.50}{2}\right) \times \left(\frac{33.75}{3} + 10.00\right) + \left(\frac{2.00 \times 20.00}{2}\right) \times 3.33$$
$$- \left(\frac{19.17 \times 11.50}{2}\right) \times \left(\frac{19.17}{3} + 10.00\right) = 588.02e$$

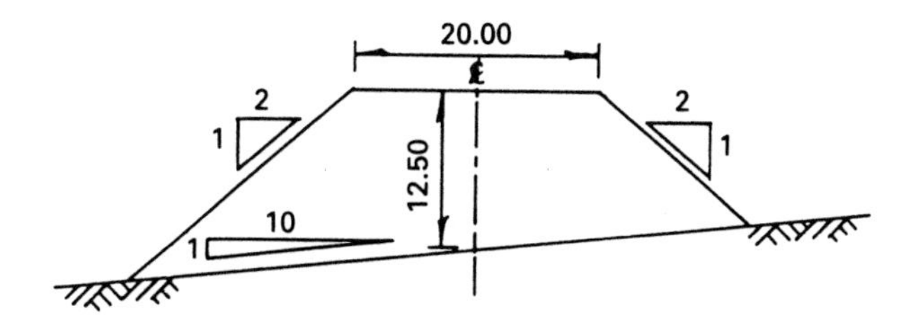

Fig. 8.16 Cross-section.

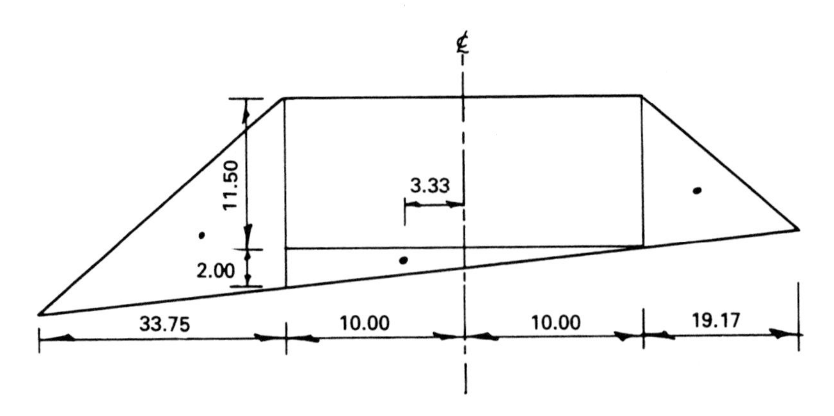

Fig. 8.17 Taking moments.

From which $e = 5.274\,\text{m}$

Then corrected area $= 588.02\left[1 + \dfrac{5.274}{500.00}\right] = 594.22 = 594.2\,\text{m}^2$

8.8 THE MASS-HAUL DIAGRAM

On major road and rail projects, the cost of earthmoving will be considerable, and it will be necessary to determine the most economical way of cutting and filling. Where excavated material is suitable, it may be used for filling, provided that the cost of hauling it does not exceed the combined cost of tipping it off site ('waste') and importing material ('borrow') for the fill sections. The mass-haul diagram allows such comparisons to be made.

The principle is that, in the direction of increasing chainage, the earthwork volume is calculated cumulatively by adding the volume of cut between consecutive sections, and subtracting the volumes of fill. Chainage is plotted horizontally and the cumulative volume vertically, the diagram being drawn under a longitudinal section (Fig. 8.18).

Characteristics are:

(1) A rising gradient represents cut.
(2) A falling gradient represents fill.
(3) The area enclosed by a horizontal line cutting the curve represents cut and fill balancing between the chainages where the line cuts.

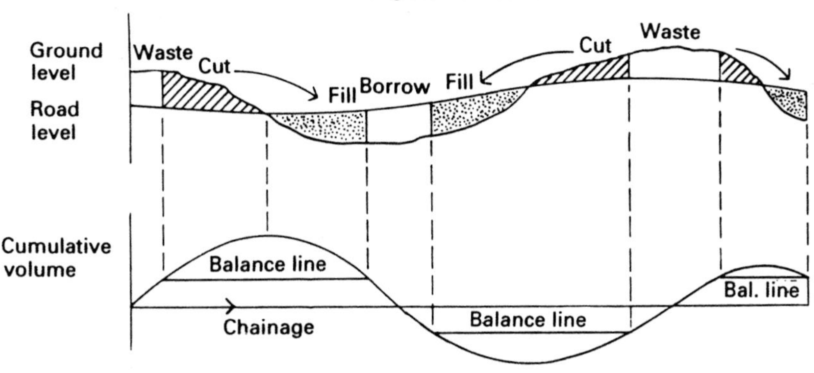

Fig. 8.18 Mass-haul diagram.

(4) The haul distance is the distance between the centroids of cut and fill on the long section.
(5) When the curve is above the axis, haul is in the direction of increasing chainage.
(6) When the curve is below the axis, haul is in the direction of decreasing chainage.
(7) Haul is the product of volume and haul distance, the most useful unit being the 'station metre'; this represents one cubic metre of material being moved a distance of 100 metres.

In pricing earthworks, a freehaul distance is specified, within which the movement of earth is costed by volume. For greater distances (overhaul), the material is still costed at the freehaul rate but an extra over is added, the cost of which is proportional to the product of the amount of material outside the freehaul limit and the extra distance that it is hauled.

The chainage of the centroid of a volume between sections is found on the curve at the mean of the cumulative volumes of the two sections.

Example 8.5

Table 8.1 shows the earthwork volumes (cut positive, fill negative between sections at 100 m chainages for an 1800 m length of motorway. The depths are also given at each section.

Table 8.1

Chainage (m)	Depth (m)	Volume ($m^3 \times 100$)
1000.00	+2.9	
		+120.0
1100.00	+4.0	
		+180.0
1200.00	+4.0	
		+157.5
1300.00	+2.7	
		+45.0
1400.00	0.0	
		−52.5
1500.00	−3.0	
		−150.0
1600.00	−6.1	
		−300.0
1700.00	−6.0	
		−135.0
1800.00	−2.0	
		−7.5
1900.00	−0.9	
		−22.5
2000.00	−1.8	
		−67.5
2100.00	−2.3	
		−94.5
2200.00	−2.0	
		0.0
2300.00	+1.0	
		+162.0
2400.00	+5.7	
		+315.0
2500.00	+5.9	
		+109.0
2600.00	+1.1	
		−15.0
2700.00	−2.1	
		−94.0
2800.00	−2.9	

Draw a long section and a mass-haul diagram and calculate the volume of material to be moved and the total haul.

Solution

Calculate cumulative volumes in Table 8.2.

By studying the long section and the mass-haul diagram (Fig. 8.19), it can be seen that:

- cut from Ch 1000 to 1400 will fill between Ch 1400 and 1700
- cut from Ch 2280 to 2440 will fill between Ch 1700 and 2280
- by drawing a horizontal balance line at 15 000 m^3 from Ch 2800, the cut from Ch 2500 to 2620 will fill between Ch 2620 and 2800
- there is a surplus of 15 000 m^3 between Ch 2440 and 2500 indicated by the curve finishing above the starting point – this surplus will require carting off site ('waste').

(If the curve had finished lower than the starting point, it would have indicated a shortage and 'borrow' material would have to have been imported.)

Table 8.2

Chainage (m)	Volume (m^3)
1000.00	0
1100.00	+12 000
1200.00	+30 000
1300.00	+45 750
1400.00	+50 250
1500.00	+45 000
1600.00	+30 000
1700.00	0
1800.00	−13 500
1900.00	−14 250
2000.00	−16 500
2100.00	−23 250
2200.00	−32 700
2300.00	−32 700
2400.00	−16 500
2500.00	+15 000
2600.00	+25 900
2700.00	+24 400
2800.00	+15 000

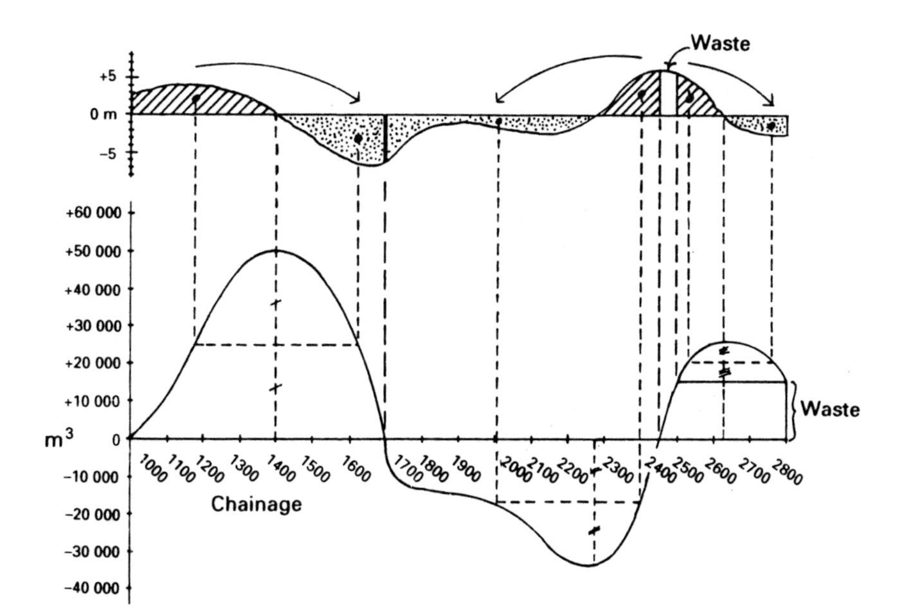

Fig. 8.19 Mass-haul example.

The haul distances between the centroids of balancing cut and fill portions must be found. Between Ch 1000 and 1700, the curve has a maximum of 50 250 m^3 at Ch 1400. (This is the volume to be shifted between these chainages.) A vertical ordinate is drawn at this chainage and bisected by a horizontal line. The centroids are at the intersections of this line and the curve; i.e. at chainages 1170 and 1620.

Haul = 50 250 m^3 (vol) × 450 m (dist) = 226 125 station metres

This value and others are shown in Table 8.3.

Table 8.3

Chainages (m)	Volume (m^3)	Centroids (m)	Haul distance (m)	Haul (station metres)
1000–1700	50 250	1170–1620	450	226 125
1700–2440	34 000	2010–2400	390	132 600
2440–2500	15 000	–	To tip	–
2500–2800	11 500	2540–2750	210	24 150
	Σ110 750			382 875

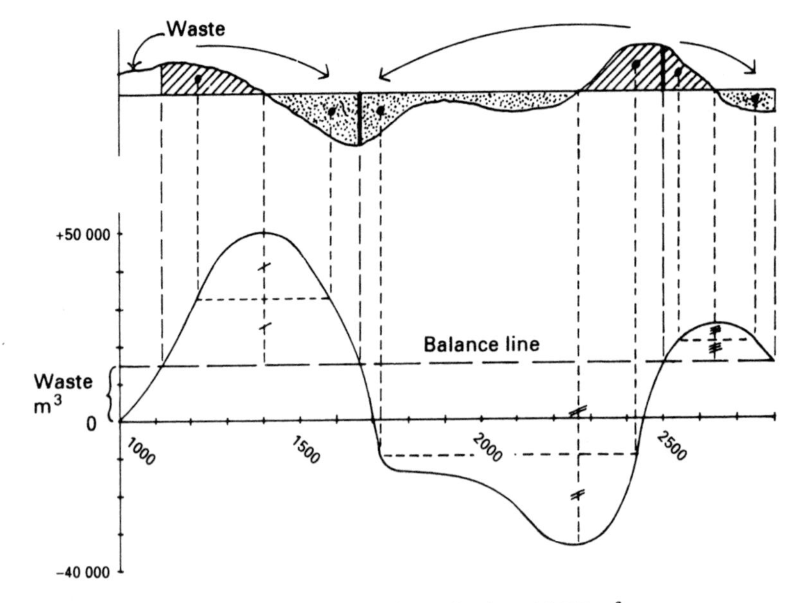

Fig. 8.20 Balance line raised to 15 000 m^3.

Table 8.4

Chainages (m)	Volume (m^3)	Centroids (m)	Haul distance (m)	Haul (station metres)
1000–1120	15 000	–	To tip	–
1120–1660	35 250	1215–1585	370	130 425
1660–2500	49 000	1720–2420	700	343 000
2500–2800	11 500	2540–2750	210	24 150
	Σ110 750			497 575

An alternative approach would be to raise the balance line to give zero volume at the end of the project (Fig. 8.20). The surplus then occurs at the start of the project between chainages 1000 and 1120. The volume to be moved would still be constant, but the haul distance would be altered (Table 8.4). Clearly, this is inferior to the previous scheme.

In pricing, as has been mentioned, rates appropriate to freehauled and overhauled quantities are developed. The next example demonstrates this.

Example 8.6

Using the earthwork figures from the previous example with the following rates:

- freehaul distance = 300 m
- freehaul rate = 72 p/m^3
- overhaul rate = 18 p/station metre
- cost of tipping = 65 p/m^3
- cost of borrowing = 95 p/m^3

investigate the cost of earthmoving

(a) using no borrowed material;
(b) limiting the overhaul to 200 m.

Solution

(a) Figure 8.21 is drawn:

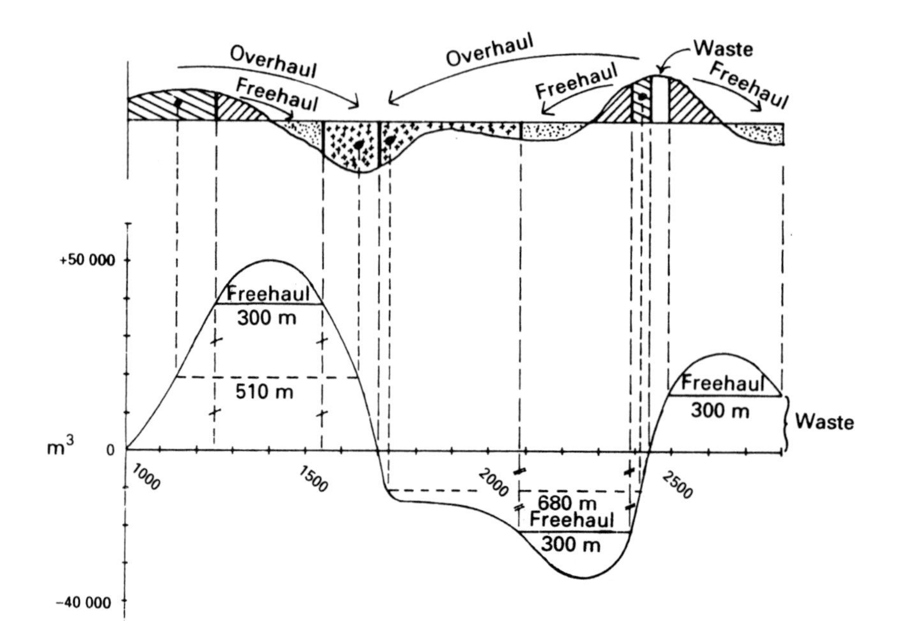

Fig. 8.21 No borrowed material.

Ch 1000 to 1700

			£
Freehaul cost	= 50 250 × 0.72	=	36 180
Overhaul volume	= 38 500		
Overhaul distance	= 510 − 300 = 210		
Overhaul cost	= 38 500 × 210/100 × 0.18	=	14 553
			50 733

Ch 1700 to 2440

Freehaul cost	= 34 000 × 0.72	=	24 480
Overhaul volume	= 21 500		
Overhaul distance	= 680 − 300 = 380		
Overhaul cost	= 21 500 × 380/100 × 0.18	=	14 706
			39 186

Ch 2440 to 2500

Waste (to tip) (15 000 m)	= 15 000 × 0.65	=	9750

Ch 2500 to 2800

All freehaul: cost	= (26 500 − 15 000) × 0.72	=	8280
		Total =	107 949

(b) The diagram is redrawn as Fig. 8.22.

Ch 1000 to 1700

Plot the freehaul line (300 m) and the line representing the freehaul + overhaul (500 m).

The volumes corresponding to these lines are 38 500 m^3 and 20 000 m^3, the difference being 18 500 m^3.

Then the balance line at the limit of overhaul corresponds to a volume of 20 000 − 18 500 = 1500 m^3.

		£
Freehaul cost = (50 250 − 1500) × 0.72	=	35 100
Overhaul cost = (38 500 − 1500) × 200/100 × 0.18	=	13 320
Waste 1500 × 0.65	=	975
Borrow 1500 × 0.95	=	1 425
		50 820

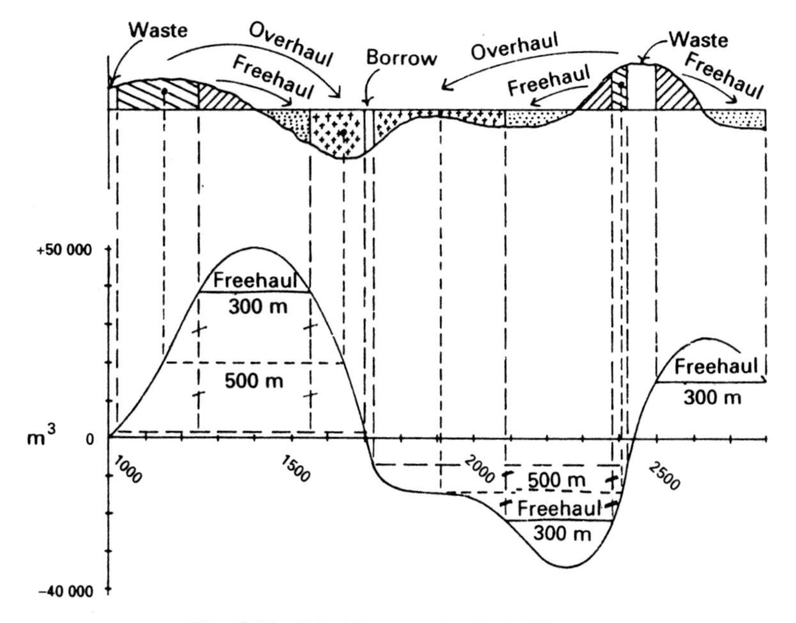

Fig. 8.22 Limiting overhaul of 200 m.

Ch 1700 to 2440

The balance line for the limit of overhaul is at $-7000\,m^3$.

Freehaul cost $= (34\,000 - 7000) \times 0.72$	=	19 440
Overhaul cost $= (21\,500 - 7000) \times 200/100 \times 0.18$	=	5 220
Waste 7000×0.65	=	4 550
Borrow 7000×0.95	=	6 650
		35 860

Ch 2440 to 2800

No change. Costs	=	9 750
		+ 8 280
		18 030
Total	=	104 710

Comment

Overall the second alternative is cheaper, although between chainages 1000 and 1700 it would be disadvantageous to limit the overhaul.

It can be seen that each loop of the mass-haul diagram must be examined separately by a trial and error process to determine the most economic method of earthmoving in all regions of a project.

Other factors

Bulking/shrinking of earth

To determine the volume of excavated material which can be placed in a fill area, it is often necessary to apply a bulking or shrinkage factor, depending on the nature of the material.

Unsuitable material

Not all material excavated is suitable for filling.

Access

Site access and the location of haul roads may affect the planning of earthmoving.

Gradients

It is easier for fully laden machines to travel downgrade than upgrade.

8.9 EXERCISES

Exercise 8.1

Figure 8.23 is a sketch plan of an area of parkland surveyed by taping offsets. From the data in Table 8.5, calculate the area within the perimeter which curves from A through B, C and D to E and runs straight to A. Lines AE and BD are parallel.

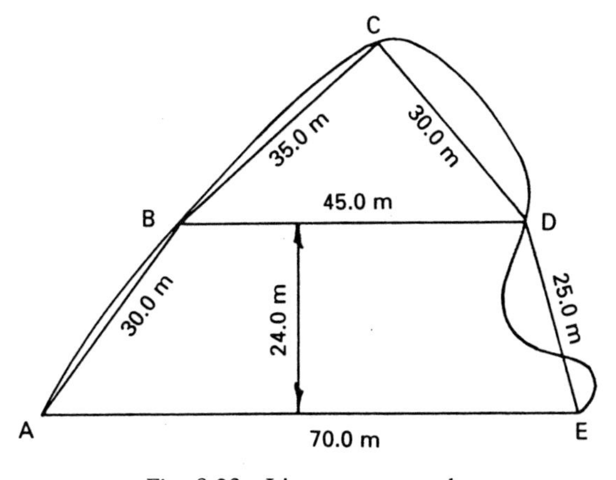

Fig. 8.23 Linear survey: plan.

Table 8.5

Offsets (m) at 5.0 m intervals			
Line AB	Line BC	Line CD	Line DE
A 0.0	B 0.0	C 0.0	D 0.0
1.4	0.8	4.4	−2.6
1.8	1.6	6.2	−6.1
2.0	2.1	6.4	−2.8
2.0	2.3	5.8	3.0
1.2	1.7	4.0	E 0.0
B 0.0	0.9	D 0.0	
	C 0.0		

Exercise 8.2

Figures 8.24(a), (b) and (c) are perspective, plan and end elevations of a motorway embankment. Original ground has a 1 in 10 cross-fall and a 1 in 25 longitudinal fall. Side slopes are 1 (vertically) to 2 (horizontally). The level formation is of 20.00 m width and the embankment tapers out at one end, with a vertical face at the other end. The greatest depth is 20.00 m. Calculate the volume of the embankment.

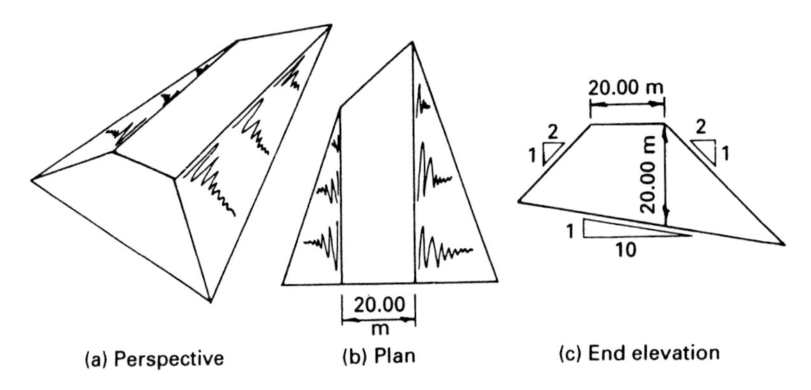

Fig. 8.24 Motorway embankment.

Exercise 8.3

The line of a new slip road involves a length in cutting. Reduced levels (m AOD) were found for five points at each section at 20 m chainage intervals. Table 8.6 shows the values. The road centreline was straight in plan.

The road formation is to be 10.00 m wide and level at 10.00 m AOD. Side slopes are to be formed at 1 to 1. Calculate the volume of excavation between chainages 320 and 400.

Table 8.6

Chainage	Left side offsets		Centre line	Right side offsets	
	20 m	10 m		10 m	20 m
320.00	8.30	9.75	11.10	12.10	13.00
340.00	10.00	11.60	12.50	12.90	13.50
360.00	12.50	13.70	14.60	15.10	15.75
380.00	15.00	16.20	16.80	17.50	18.20
400.00	17.50	18.50	19.20	19.50	20.00

Exercise 8.4

A contoured plan has been prepared of a site for an impounding reservoir. A planimeter has been used to determine the area within the contours of the ground and the upstream face of the dam. Table 8.7 shows the computed areas. Calculate the volume of the reservoir when full with a top water level of 250.0 m.

Draw a graph of water level against volume as the reservoir fills, and from it estimate the water level when the reservoir is half full.

Table 8.7

Level (m)	Area (m^2)
250.0	96 100
245.0	73 600
240.0	57 240
235.0	41 650
230.0	28 060
225.0	19 550
220.0	10 530
215.0	3 380
210.0	880
205.0	10

Exercise 8.5

Figure 8.25 is a plan of a 10 m square level grid over a building site. Excavation has to be taken down to 32.55 m AOD with the sides kept vertical. Calculate the volume of material to be removed.

Further excavation has to be taken down to 30.80 m with sides sloping at 1 to 1. The shaded area shows the bottom of the extra excavation. Calculate its volume.

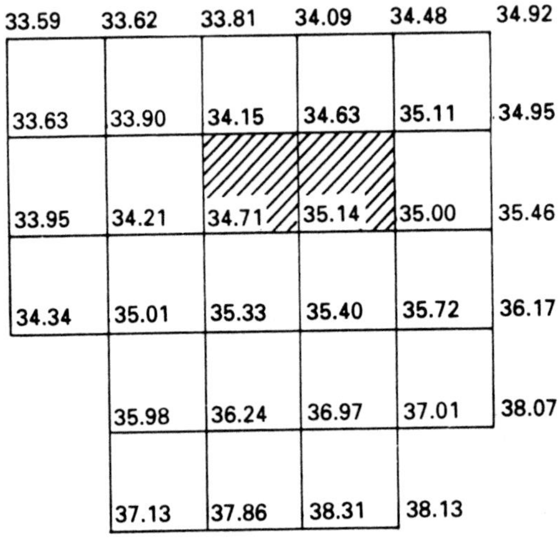

33.59	33.62	33.81	34.09	34.48	34.92
33.63	33.90	34.15	34.63	35.11	34.95
33.95	34.21	34.71	35.14	35.00	35.46
34.34	35.01	35.33	35.40	35.72	36.17
	35.98	36.24	36.97	37.01	38.07
	37.13	37.86	38.31	38.13	

Fig. 8.25 Level grid.

Exercise 8.6

A mass-haul diagram is to be prepared for earthwork planning for a length of road partly on embankment and partly in cutting. Table 8.8 shows the earthwork volumes between sections 100 m apart.

(a) Calculate the cumulative earthwork volumes, and plot the mass-haul curve.
(b) Evaluate the cost of earthmoving for a freehaul distance of 500 m, a freehaul rate of £1.10/m^3 and an overhaul rate of 25p/station-metre.
(c) Investigate the effect of limiting the overhaul to 700 m (500 m freehaul plus 200 m) and borrowing material at £2.00/m^3 and tipping waste at 90 p/m^3. State which of the two strategies examined should be recommended.

Table 8.8

Chainage (m)	Volume between sections (m^3) Cut	Fill
3000		
		16 000
3100		
		12 500
3200		
		9 500
3300		
		7 000
3400		
		3 000
3500		
	4 000	
3600		
	11 000	
3700		
	15 000	
3800		
	18 000	
3900		
	16 500	
4000		
	2 500	
4100		
		10 500
4200		
		8 500
4300		

9 Setting Out – an Introduction

9.1 PREAMBLE

Setting out frequently provides the young engineer's first experience of work in industry. It involves the placing of markers to enable construction works to be located correctly in plan and in elevation. The site engineer must combine surveying skill with ingenuity to provide setting out data precisely and rapidly; this data must be easily understood by the site operatives, for setting out is an exercise in communication. Markers must be established to indicate to the workforce the designer's intention concerning location, size, shape and level of the finished construction as shown on the working drawings.

Notes

(1) In this and the following chapers, the surveyor, engineer or builder responsible for the setting out will be referred to as the site engineer or simply as the engineer.
(2) Control setting out refers to the establishment of 'permanent' markers clear of the works from which pegs, profiles, sight rails and other markers will be set out during the course of the project.
(3) Setting out refers to such pegs, profiles, etc. which will indicate to the workforce the location of the works.

9.2 THE METHOD

Example 9.1: Location of works

A hole in the ground 5.000 m square in plan and with vertical sides has to be excavated. How should it be set out?

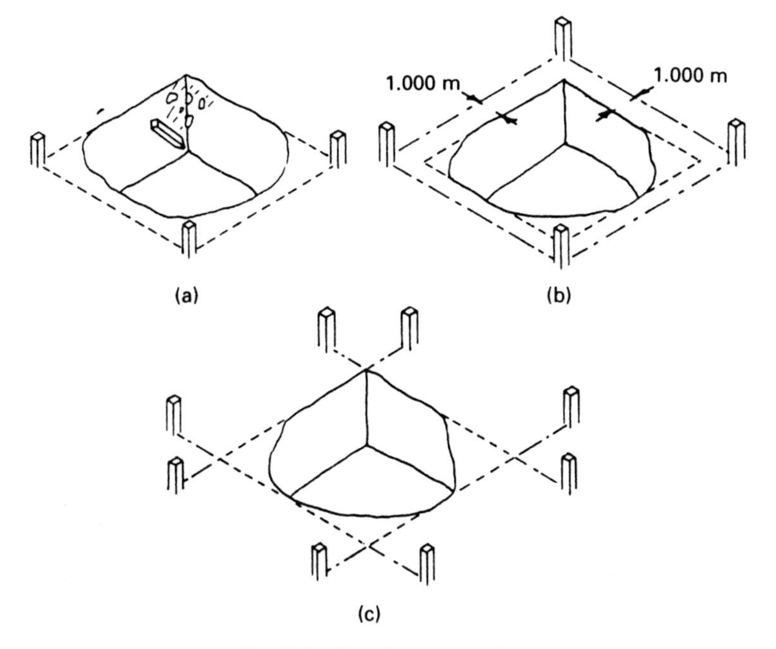

Fig. 9.1 Pegging excavation.

(a) Knock a peg in the ground at each corner (Fig. 9.1(a)).

Advantage	*Disadvantage*
Easily pegged out.	Pegs will be lost as corners are excavated.

(b) Set corner pegs offset 1.000 m back from edge of excavation (Fig. 9.1(b)).

Advantage	*Disadvantages*
Pegs should remain in position.	Longer to set out. How may we be sure that 1.000 m clearance will be observed?

(c) Set eight pegs clear of works on extensions of the side lines (Fig. 9.1(c)).

Advantages	*Disadvantages*
Pegs should remain in position. Limits of excavation can be ranged (interpolated) by eye.	Longer to set out. How may we be sure that corresponding pairs of pegs are used for ranging a side?

Comments

(1) It is likely that a tape and a theodolite (to set right angles and to range accurately along lines) will be used in setting out.
(2) Method (a) will be satisfactory where the pegs can be retained long enough for the sides to become defined adequately during excavation – usually where high accuracy is not required.
(3) For greater accuracy, methods (b) or (c) should be employed, a nail being knocked in the top of each peg on the required line in plan to define it within 5 or even 3 mm.
(4) For method (c), a more robust marker than a peg would be a profile – two pegs connected by a horizontal cross-piece at right angles to the required line; again a nail or saw-cut will give accurate definition.

Levelling of works

Now that we can locate the excavation in plan, we also need to define the level of the bottom. Unless the ground is exceptionally level, it will not be sufficiently accurate to work to a given depth. Boning in should be employed.

Boning in

In method (b) for plan control, a line (or vertical plane) was offset horizontally. Boning in requires the establishment of a plane (horizontal or gently sloping) offset vertically, usually above the construction surface. This plane is define by *sight rails*. The usual form of sight rail is a horizontal timber cross-piece nailed on to one or two upright stakes.

A movable *traveller* or *boning rod* is used to establish whether the construction surface is too high, correct or too low. By looking over one sight rail to another, with the traveller in between, an operative can bone in, a satisfactory level being achieved when the three cross heads form a constant grade (Fig.9.2). The sight rails are set at levels convenient for an operative to sight over and to suit a traveller length which is a multiple of 0.5 m.

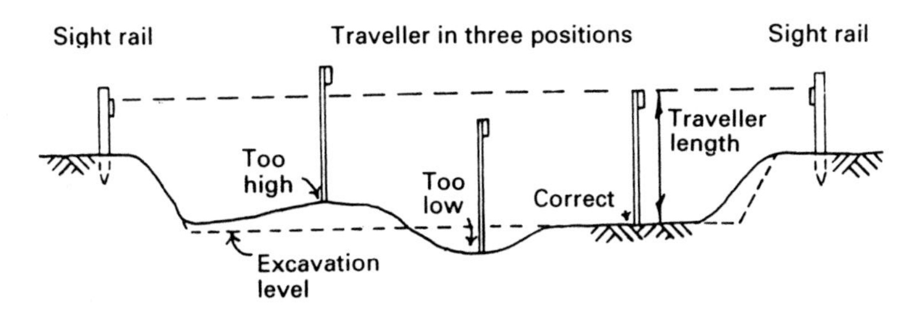

Fig. 9.2 Boning in excavation.

Notes

(1) Sight rails are usually called *profiles* on site.
(2) One marker may be used for plan and level control.

The setting out engineer must:

(1) Position the necessary pegs, profiles and sight rails.
(2) Give details of offset distances and traveller lengths to the site operatives.
(3) Check that the operatives are correctly interpreting the setting out data.
(4) Periodically check the setting out markers in case of disturbances.

9.3 ORGANIZATION OF CONSTRUCTION WORKS

In the United Kingdom, civil engineering works are customarily carried out under ICE Conditions of Contract, building works under RIBA Conditions of Contract.

There are usually three parties to such contracts:

(1) The Promoter/Client/Employer who:
- requires the construction of some works
- funds the project
- becomes the eventual owner or operator
- unless having a civil engineering or building section with sufficient experience will appoint:

(2) The Engineer (civils)/the Architect (building) who, for a fee:
- designs the works
- prepares drawings, specification and contract documents
- advises in appointing a contractor

- supervises the construction
- impartially administers the contract
- measures the works as completed and certifies payment to:

(3) The Contractor who:

- constructs the works in accordance with the drawings and documents
- is paid for the work performed (usually the product of the measured work and the rate for it in the tender)

It is the contractor's responsibility under both ICE and RIBA forms of contract to perform the setting out. Construction errors from faulty setting out must be rectified at the contractor's own expense. Although the engineer/architect may check the setting out, this does not relieve the contractor of any responsibility.

9.4 THE SITE ENGINEER

Duties

(1) To arrange the provision of surveying equipment and to set up a store of consumables.
(2) To train chainmen.
(3) To evaluate the required accuracy and the desired speed in setting out.
(4) To select suitable techniques and equipment for the setting out tasks.
(5) To organize access and transport to facilitate setting out.
(6) To provide all necessary setting out markers accurately, quickly and while under pressure.
(7) To communicate verbally and in writing to ensure the correct interpretation of setting out.
(8) To keep a constant check on the setting out and the alignment of construction.
(9) To perform other engineering tasks required such as planning, supervision and materials testing.

Attributes

(1) Competence in surveying.

(2) Knowledge of construction techniques.
(3) Ability to communicate with, cooperate with, and achieve cooperation from:
- site agent
- resident engineer's or architect's staff
- general foreman
- site foremen
- tradesmen
- drivers
- labourers
- other engineers

(4) A confident, but not dogmatic, attitude.
(5) Quick-wittedness.
(6) Flexibility.

9.5 GOOD RELATIONS

There is a lot of personal contact on construction sites. In addition to liaising with immediate colleagues, the engineer will need to deal with other parties, particularly the general foreman, the trades foremen, the staff of the resident engineer or architect, plant operators and subcontractors.

The general foreman

The engineer may find contact with the general foreman (GF) to be difficult. Pedigree specimens are 2 m high, 2 m in diameter, red-faced and bad tempered. It appears to the engineer, whose preoccupation is with accuracy, that the main concern of the GF is with speed.

In fact the GF is a most important member of the contractor's team, his personality, drive and knowledge of construction techniques being vital to the progress of a project. Cooperation must be achieved, the engineer respecting the knowledge and experience of the GF, while not being afraid to impose his or her will where alignment is involved.

Resident engineer's staff

Although all parties to a construction project hope for safe, sound and efficient construction, the resident engineer's prime concern is for quality while the contractor, to succeed in business, must make a profit. Differences in interpretation will inevitably occur. In setting out, to prevent disputes and possible delay to construction, it may be worthwhile arranging for the RE's engineer to accompany the site engineer and agree levels and positions.

Where faulty setting out is reported by the RE's staff, it should be investigated; if the site engineer is satisfied, work should proceed – the contractor, of course, takes all responsibility for the setting out.

Construction plant

Setting out markers can easily be damaged by construction plant. It is understandable for engineers to think (wrongly of course!) that such acts are deliberate. The engineer must give some thought to the suitable location of pegs and profiles, and, although in theory not dealing directly with the drivers, may achieve cooperation by establishing a rapport with the machine operators.

To minimize the loss of markers, the engineer should

(1) Plan the earthworks setting out with the foreman concerned.
(2) Keep markers well clear of the works.
(3) Liaise with plant drivers, diplomatically 'educating' them.
(4) Reference all setting out to enable the swift replacement of disturbed markers.

The engineer can best develop his or her influence by meeting the majority of key persons on site as soon as possible, by communicating verbally (and confirming in writing) and taking advantage of occasional opportunities to socialize with the site operatives and the RE's (architect's) staff. Such contact should not become so intimate as to cause an adverse effect on judgement or performance at work.

Occasionally it may be necessary for the engineer to disagree quite strongly with other personnel. Speaking one's mind on a construction site is no bad thing; rifts are quickly healed and grudges rarely borne.

9.6 ORGANIZATION AND PLANNING

At the start of a project, the contractor should appoint a sufficient number of setting out engineers with clearly defined responsibilities.

A small contract may require only a single site engineer who will be expected to perform engineering, supervisory and clerical tasks in addition to setting out.

On large contracts there should be a structured team of engineers, ultimate responsibility for controlling setting out resting with a chief engineer, or senior section engineer. The field work is likely to be delegated to the juniors.

The engineer responsible should:

(1) Be familiar with the contract programme and all amendments.
(2) Attend all planning and progress meetings on site.
(3) Be in possession of all the latest revisions of all drawings.
(4) Be in possession of the specification and any schedules.
(5) Plan the setting out at least one week ahead (well, try to anyway).

Note

When other engineering tasks prevent the site engineer from performing the setting out quickly enough, or when a high precision control survey is required, it may be sensible to employ a specialist surveyor. Road centreline establishment and underground control are tasks justifying such action.

9.7 ESTABLISHMENT AND MAINTENANCE OF EQUIPMENT

Surveying equipment

Usually this equipment is provided at the start of a contract by the head office. The engineer should ensure that there are a sufficient number of:

- levels – engineer's tilting or automatic
- levelling staves – 4 m 'E' pattern telescopic or folding

- tapes – 30 m steel, white faced (plus tension grips and thermometer if necessary)
  – 3 m pocket steel tape
- theodolite – 1″, 6″, 10″ or 20″ to suit required precision
- EDM outfits, add-on or integral with theodolite – consider use of data logger and computers
- GPS receivers (for large projects)
- lasers – type according to use
- ranging rods
- optical squares
- arrows
- automatic optical plummets

Where possible each engineer should be responsible for his or her 'own' equipment.

Facilities for drying, cleaning and storage should be provided.

Equipment should be regularly checked, and repaired or adjusted when necessary. If repairs/adjustments cannot be carried out on site, the faulty instruments should be returned to the suppliers for attention and replacement equipment obtained. A number of suppliers offer a regular 'check and adjust' service; take advantage of this facility, particularly on contracts of lengthy duration.

Tools and tackle

Some of these are provided at the start of a contract, others must be bought.

- sledgehammers – 14 lb
- clawhammers
- plumb bobs
- picks
- crowbars (large)
- hand augers
- spirit levels
- paint brushes
- wax crayons
- felt-tip markers

Keep these items separate from similar tackle used for construction.

Consumables

The engineer must organize the provision of consumable materials.

- pegs/stakes – 50 × 50 × 400 mm
  – 75 × 75 × 600
  – 75 × 75 × 1200
- flat timbers – 12 × 100 × 400
  – 12 × 100 × 1200
- nails
- paint
- steel pins/rods – 25 mm diameter reinforcement bars cut into 500 mm lengths

Keep a plentiful supply, liaising with the storekeeper and joiners.

9.8 TRAINING CHAINMEN

The chainman is the engineer's assistant in the field. Duties include holding the staff, holding the zero end of the tape, positioning ranging rods and pegs to the engineer's instructions, carrying equipment, maintaining and cleaning equipment and generally helping the engineer.

The role of chainman is best filled by a second engineer. However, chainmen are usually drawn from the labourers' ranks. Some contractors employ specialist chainmen (trained by engineers), offering payment and conditions commensurate with their duties.

Less enlightened organizations employ no permanent chainmen; the engineer has to use a labourer as and when required. For this task, foremen find it difficult to resist picking operatives with the lowest intelligence and the least physical coordination. In days of high employment, students (usually of philosophy or oriental languages) could obtain vacation jobs as labourers, often becoming chainmen. Such circumstances no longer prevail.

The engineer must impress on site management the importance of the chainman's job. Where experienced chainmen are not available, the engineer must find a suitable candidate and be allowed to keep him and train him. The result will be beneficial to accuracy, speed and morale.

9.9 ACCURACY

Among publications listed in Section 9.17 some give recommended tolerances for setting out operations.

Ideally, the engineer should appreciate the accuracy obtainable using various instruments for various tasks, be aware of the tolerances required in the contract specification and carry out error analysis of the field work. On construction sites there will rarely be time for error analyses and the engineer must adopt an extremely methodical approach. In particular:

(1) Suitable equipment must be used.
(2) The likely accuracy of measurements with the equipment selected must be known.
(3) Skill and care must be used in operating the equipment.
(4) Mistakes must be eliminated.
(5) Systematic errors must be corrected.
(6) Work must be checked by an independent method if possible.

With experience, the engineer should develop a feel for the method and precision to employ.

Accumulation of errors

In addition to errors both in establishing control and in setting out markers, there will be errors in interpretation, i.e. in lining between profiles or in boning in.

Accordingly, markers should generally be set within 3 mm of correct position. Some relaxation of this limit may be allowed for earthworks, and on a large project, such as a motorway, it is often more important that markers are consistent with themselves, or with work completed to date, than being within 3 mm of 'absolute' position. Where prefabricated units are to be installed, and particularly in tunnelling projects, smaller tolerances are frequently required.

Inevitably, during the construction of a large project, inaccuracies will become apparent. In the worst cases the work will have to be demolished and reconstructed. Often, however, the errors will be such that subsequent work can be constructed slightly out of 'true' position, producing a structure still fit for its designed purpose. The engineer should obtain approval for the changes from the RE or architect and may have to recalculate some of the setting out data. A bridge incorrectly aligned in plan (faulty setting out) or too high (residual hogging of beams) may require the realignment of the carriageway in plan or level.

Equipment – type and size of error

These values assume that the equipment has been standardized/adjusted and is correctly set up. They are for guidance; actual corrections should be calculated in the field.

Condition	Error
Steel taping at 30 m	
Slope 1° or rise 0.5 m	+0.004 m
Tension 20 N from standard (13 mm wide tape)	±0.0006 m
Sag at standard tension (13 mm wide tape)	+0.0002 m
Temperature 10°C from standard	±0.003 m
Reading	±0.003 m
EDM at 100 m	
Standard error	±0.005 m
Atmospheric variation by 25 mm of mercury (or 300 m in elevation)	±0.001 m
Temperature variation by 10°C	±0.001 m
Engineer's tilting level, 50 m sight	
Telescope bubble not centred	varies
Staff 3° from vertical, reading at 3 m height	+0.004 m
Staff 6° from vertical, reading at 3 m height	+0.015 m
Lasers – pipeline/rotating, 100 m sight	±(5 to 10) mm
Theodolite	
Centring error of 0.003 m;	
90° angle; both sights 30 m	0 to ±29 seconds
as above but 180° angle	0 to ±41 seconds
Target sighting/centring error	
of 0.003 m at 30 m distance	0 to ±20 seconds
Single circle reading (20″ instrument)	0 to ±10 seconds
(6″ instrument)	0 to ±5 seconds
(1″ instrument)	0 to ±3 seconds

Relation between angular and linear errors

There are 206 265 seconds in one radian. Table 9.1 shows the relationship between angular and transverse (linear) measurements.

Table 9.1

Angular error	Transverse error	In length
1″	0.001 m	200 m
10″	0.005 m	100 m
20″	0.010 m	100 m
1′	0.030 m	100 m

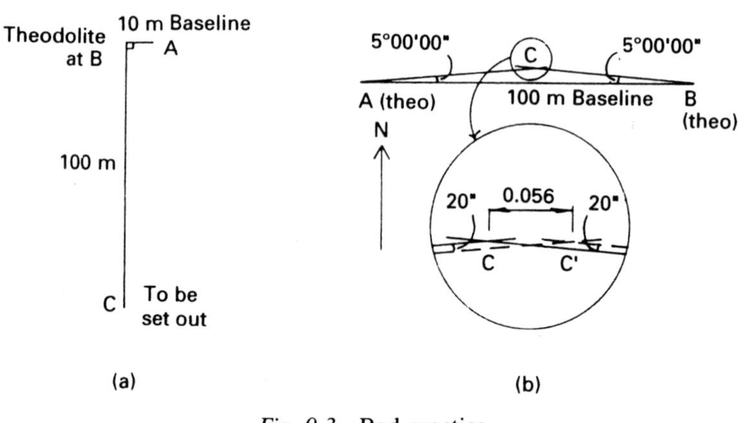

Fig. 9.3 Bad practice.

Bad practice

Methods that will magnify observational errors should be avoided.

Example 9.1
In Fig. 9.3(a) any error of centring at B or sighting at A will be multiplied by 10 in setting out C.
Set out distances shorter than the datum or reference length.

Example 9.2
In Fig. 9.3(b):

Angle CÂB = 5°00′00″ Setting error 20″ too small
Angle AB̂C = 5°00′00″ Setting error 20″ too large
Error at C = 0.056 m east

Keep intersecting rays as close to 90° as possible.

Where these types of unsound methods cannot be avoided (in tunnelling work, for example), greater precision must be used, the engineer selecting more precise equipment, taking more care in setting up and reading, and taking the mean of a greater number of observations.

9.10 SELECTION OF METHOD

Some construction operations have fairly standard methods of setting out (drain installation), but others may have several alternatives. For the accuracy required, a road on a housing estate could be set out entirely by sight rails. A trunk road, however, would require steel pins setting out to give line and level for kerbs.

Factors are:

- accuracy required
- equipment available
- time available
- method comprehensible to operatives (consult the foreman)
- permanency of setting out
- space available
- whether checks on setting out can be incorporated
- whether checks on construction can be carried out
- whether setting out can be easily replaced if accidentally removed

The general programme of setting out should be prepared at least a week in advance by the senior engineer. Site engineers should check daily with the agent, general foreman and planning engineer for any alterations to the construction programme.

9.11 TYPES OF MARKERS

Requirements

- permanent
- prominent
- function clearly marked
- clear of works and access
- easily understood and used

Plan control station

A plan control station can be a steel pin driven into the ground, concreted in and surrounded by a wooden fence. Hacksaw cuts can indicate precisely the position. A wooden peg with a nail or a piece of steel angle can also be used (Fig. 9.4).

When sufficient of the works are complete, stations can be established on parts of the structure. This is done by painting a circular mark on the structure; when it has dried, lines are scribed through it to indicate the position. For a more rapid but less permanent mark, wax crayon may be used instead of paint (Fig. 9.5).

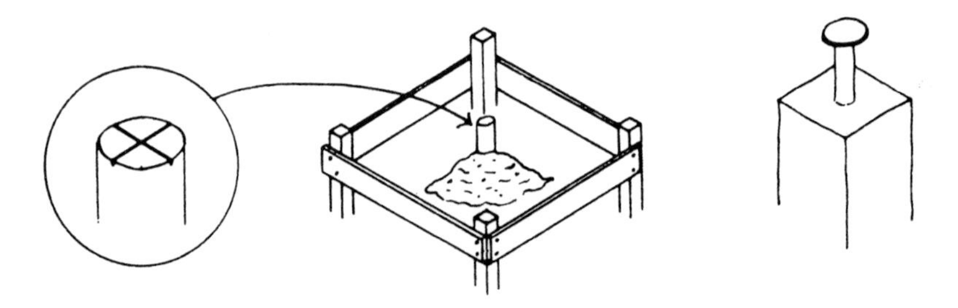

Fig. 9.4 Plan control station.

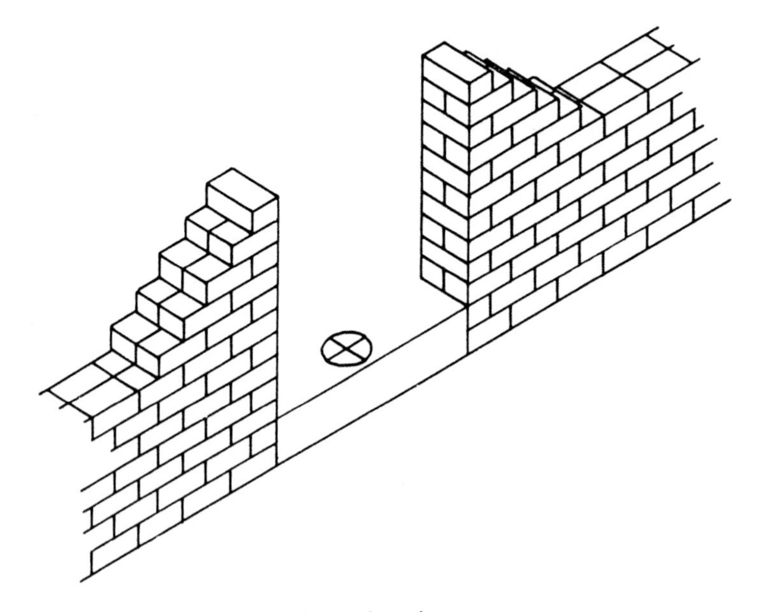

Fig. 9.5 Control station on structure.

Site temporary bench mark

Again a steel pin, suitably protected, is satisfactory; a plan control station can be used. An existing or partially complete structure may also be used by scribing a horizontal line through a paint or crayon mark on a vertical surface. This method is often favoured in building works for setting datum levels 1.000 m above finished floor level (FFL).

Where a horizontal surface is used, it should be truly flat or convex and the position marked distinctly.

Setting out – plan

Pegs are satisfactory for work of low accuracy – top of cutting, toe of embankment, manhole position.

Pegs with nails are needed for more precise marking – however, pegs are likely to have a short life.

Profiles are used for more permanent marking. A horizontal cross-piece is fixed to two pegs. A nail or saw-cut in it indicates the line of construction (Fig. 9.6(a)). Profiles are set back from the construction and a builders' line may be tied between pairs of them. Points above or below the line can be located by plumbing up or down.

Corner profiles are set up outside the corners of buildings to indicate two lines at right angles. More robust than single profiles, they require setting up close to the corners, with a risk of disturbance (Fig. 9.6(b)). If set too far back, the length of cross-pieces may make their use undesirable.

Steel pins may be used to indicate the position of piles or placed at the edge of pavement construction where road forms or piano wire are to be employed.

Batter rails (or rules) resemble sloping profiles. They are set out typically 10 or 20 m apart to indicate the position and gradient of a cutting or embankment side slope (Fig. 9.6(c)).

Setting out – level

The method of boning in is extensively used for earthworks levelling, for the levelling of imported granular material for road and other pavement slabs and for drain installation.

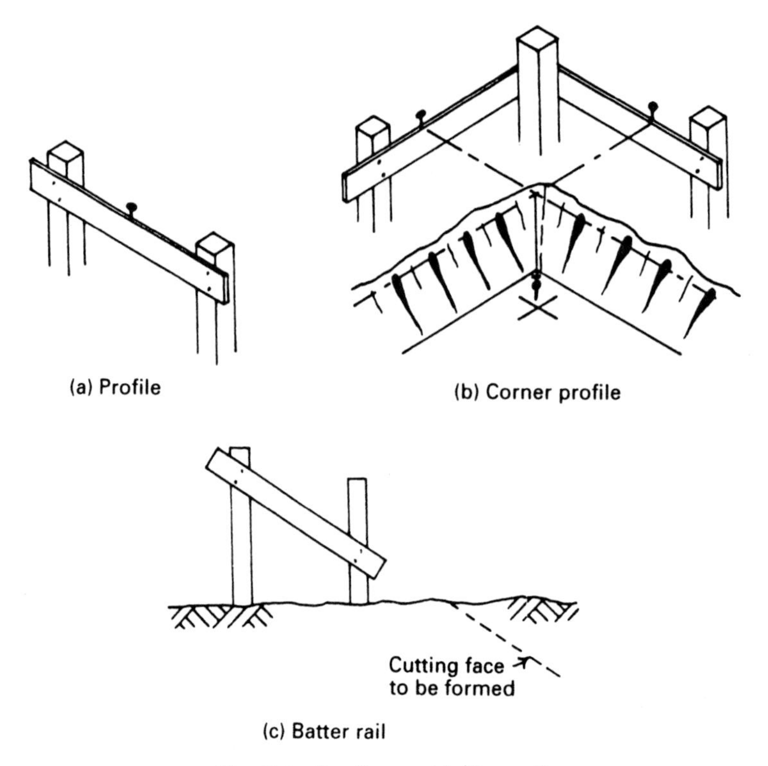

Fig. 9.6 Profiles and batter rails.

Sight rails (profiles) are described in Section 9.2. For finished construction (e.g. in brick, concrete or steel) boning in is not sufficiently accurate and more precise definition is required.

Steel pins for marking the edges of pavements can be set to indicate the pavement level. Blinding concrete at the bottom of excavations can be levelled from steel pins.

Datum marks have been described previously in 'Site temporary bench marks'.

9.12 LASERS

Lasers are setting out instruments and can be used instead of engineers' levels and theodolites for a number of tasks on site. The instrument provides an optical line if fixed, or an optical plane if rotating. A helium-neon gas laser produces an intense narrow beam of red light; it can be used as a reference for horizontal, vertical and sloping control (Fig. 9.7(a)). Although site instruments are at the lower powered end of the laser range, operatives should not look along the beam, nor search for it with a level or theodolite. Warnings should be posted on sites and reference may be made to British Standard BS 4803, 'Safety and the Use of Lasers'.

The beam can be seen where it meets an obstruction. Lining in and boning can be carried out using a template or traveller of suitable dimensions. Alternatively a photoelectric detector (usually on a pole) can be used at the position to be aligned. When alignment is nearly correct, the detector will indicate the amount of further movement required. Lasers are now used extensively in pipelaying and frequently in building construction and tunnelling.

Site lasers are usually self-contained units, although they require a power source. Laser attachments to theodolites and levels are also available such that the beam can be projected through the telescope.

Ground levelling

A rotating laser mounted on a tripod or bracket can sweep out a horizontal plane or one at a specified grade a fixed distance above the trimmed ground. Pole mounted photoelectric detectors can be used to determine the amount of cut or fill required at strategic points. Detectors can also be fitted to excavating and grading machines, with a display indicating to the driver whether cut or fill is required (Fig. 9.7(b)).

Comments

(1) The site engineer will be responsible for aligning the laser and frequent checking.
(2) Most lasers switch off if they are disturbed, preventing setting out in accordance with a faultily aligned beam.
(3) Over long distances, the accuracy may not be as great as that achievable by levelling or by theodolite and tape methods. Accordingly, lasers find their best applications for buildings, drains and earthworks of limited areas. For tunnelling, shaft plumbing

Fig. 9.7 Lasers: (a) Leica Wild LNA10 set for vertical rotation; (b) Laser Alignment LB-4 horizontally rotating laser and beam detector.

and control of multistorey structures, the engineer should check that the accuracy is adequate.

(4) The uses of lasers in setting out pipelines, structures and tunnels are detailed in Chapters 12, 13 and 14 respectively.

9.13 CONTROL SETTING OUT

Control stations should be set out at the commencement of a project, and their location agreed between the contractor's engineer and the RE's or architect's engineer. The stations should be so positioned and protected as to survive the contract. At any stage of the works, the setting out can then be taken from, or checked back to, the original stations, giving consistency of alignment.

The stations should be referenced by taking measurements from features unlikely to be disturbed, and cross-referenced with each other. By establishing more stations than appear necessary, it should be possible to replace one if it is accidentally removed.

The traditional location of control stations has been at specific key points, on grid lines, on centrelines or at fixed offsets from grid or centrelines. Such practice is geared to manual calculations and the use of a theodolite and a 30 m steel tape in the field.

Where EDM and computer assistance for calculations are available, there is not the same need to situate control stations at particular predetermined points. The stations can be arbitrarily fixed and located subsequently (usually with rectangular coordinates) by traversing, triangulation, trilateration, intersection, resection or network methods.

One or more site (temporary) bench marks must be established. On a large project, it will be convenient to have several bench marks. A precise levelling survey must be carried out, so that site levelling can be run between them, the engineer having confidence in their values.

Siting of stations

Stations should be:

- on the construction site (or on land where the stations will neither be liable to removal nor give offence to the landowner)
- clear of anticipated construction
- accessible for setting up instruments

- easily sighted from the works
- convenient for referencing and cross-referencing
- positioned to cover the whole site during all stages of construction
- tied in to Ordnance Survey stations if possible

The determination of levels, angles and lengths should follow the principles in Chapters 2, 3 and 4. Detailed setting out information is given in subsequent chapters.

9.14 WHILE IN THE FIELD

The engineer should:

- check setting out by eye; if suspicious of error, check using instruments
- explain the setting out to the appropriate site operatives
- replace any disturbed markers
- check that the workforce is working correctly to the setting out
- take any necessary measurements of work partially or totally complete for weekly costing, site bonus, subcontractor's measurement or monthly valuation measurement

Remember

Site managers are usually drawn from the ranks of site engineers on the basis of perceived potential. An ambitious engineer must have a demonstrable interest in the technical, organizational and financial aspects of site work, while being employed primarily to set out.

9.15 SKETCH OF SETTING OUT INFORMATION

Returning to the office on completion of a setting out task, the engineer should record the setting out details. A triplicate (carbon copy) book is required. A sketch of the setting out, usually a plan, should be made showing the position and purpose of all markers. Traveller lengths, offset distances, construction details, and type and class of material should be given, as should any other relevant information, including the date and name of the engineer (Fig. 9.8).

The top copy should be given to the foreman in charge of the work,

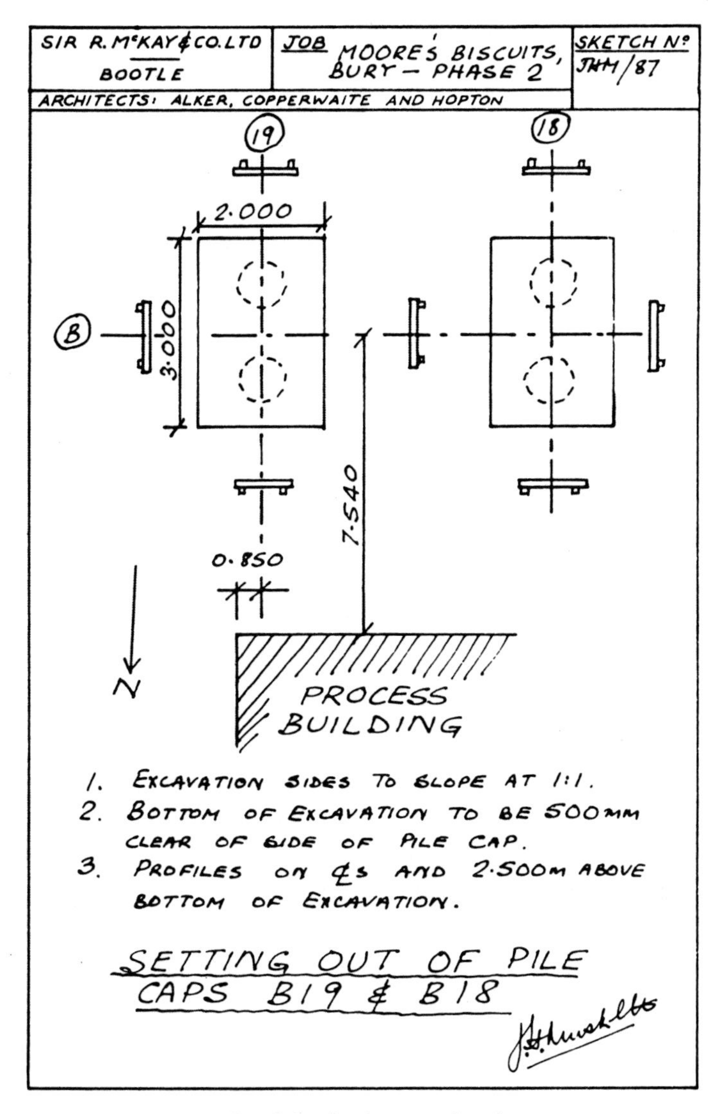

Fig. 9.8 Setting out sketch.

the second to the senior engineer or other responsible party, while the third remains in the book as a record. When, occasionally, mistakes in the alignment are discovered, the engineer will be blamed whether culpable or not. The third copy will convince no-one else, but the engineer, if innocent, may gain reassurance from the record and sleep more soundly at night.

9.16 RECORD DRAWINGS

It is usually the engineer's job to prepare and update record drawings. These may be 'as constructed' drawings of the works, site drawings and bar charts coloured in as work proceeds, or network analysis charts.

Such drawings are to assist in the measurement of particular items of work, and to enable the progress of a section or of the whole of the project to be monitored. It may then be possible to compare the actual value of work so far completed with the budgeted value.

It is no accident that, when the 'top brass' are due to visit a site, a profusion of such drawings appear on the walls, rubbing shoulders with the more common type of adornment found in site cabins.

9.17 FURTHER READING

British Standards Institution (1990) *B.S. 5606:1990 Guide to Accuracy in Buildings*. B.S.I.

British Standards Institution (1990, 1992) *B.S. 7334:1990 (pts 1–3), 1992 (pts 4–8). Building Construction – Measuring Instruments – Methods for determining accuracy in use*. B.S.I.

Building Research Establishment (1977) *Digest 202:1977 Site Use of the Theodolite and Surveyor's Level.* B.R.E.

Building Research Establishment (1980) *Digest 234:1980 Accuracy in Setting Out*. B.R.E.

Institution of Highway and Transportation/Institution of Civil Engineers Joint Working Party (1982) *Survey Standards, Setting Out and Earthworks Measurement*. IHT/ICE.

Price, W.F. and Uren, J. (1988) *Laser Surveying*. London: Van Nostrand Reinhold.

10 Roadworks I – Curve Calculations

10.1 INTRODUCTION

Modern roads are aligned to provide smooth changes of direction and grade. Horizontal curves are generally circular, often with entry and exit spirals to effect the transition between straight and circular sections. Vertical curves are parabolic. Basic design factors incorporate the safe and comfortable progress of motorists and the provision of sight distances adequate for stopping, overtaking and headlight projection. In the United Kingdom, design should be in accordance with Departmental Standard TD 9/93 *Road Layout and Geometry: Highway Link Design*, published by the Department of Transport in 1993.

Railway alignment follows the general principles for road design, and other curved works (pipelines, concrete structures and dams) may be set out in accordance with some of the principles in this chapter.

10.2 HORIZONTAL CIRCULAR CURVES: NOMENCLATURE

In initial road design, the straights or tangents are located to suit the topography and function of the road; curves are then provided to accommodate the changes in direction and grade. Figure 10.1 shows the layout of a circular curve.

Intersection point 'I': the point of intersection of consecutive straights.

Intersection angle 'Δ': the angle at the intersection point between the straights, measured between the first straight projected forward and the second straight; it is also called the *deflection angle*.

Tangent points 'T_1 and T_2': the start and finish of a curve, referred to as entry and exit tangent points.

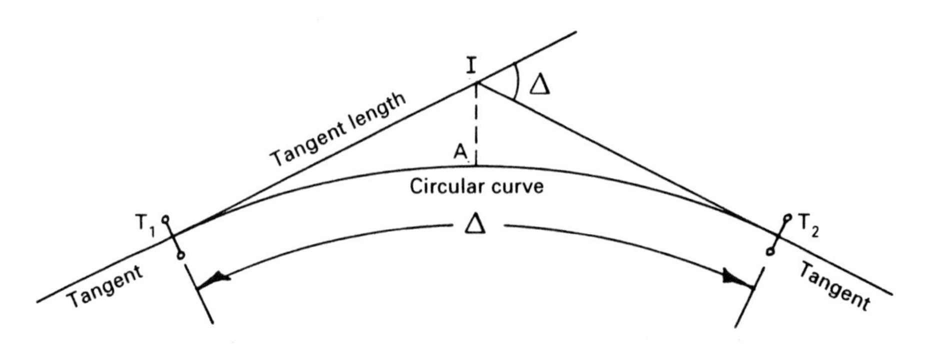

Fig. 10.1 Circular curve geometry.

Tangent length 'IT_1, IT_2': the plan distance between the intersection and tangent points for a curve.

Curve length 'T_1T_2': the length measured around the curve.

For a radius R

$$IT_1 = R\tan\frac{\Delta}{2} \qquad \text{Tangent length}$$

$$T_1T_2(\text{curve}) = \frac{\pi R\Delta}{180} \qquad \text{Curve length}$$

$$IA = R\left(\sec\frac{\Delta}{2} - 1\right) \qquad \text{Apex length}$$

$$T_1T_2 = 2R\sin\frac{\Delta}{2} \qquad \text{Long chord length}$$

Degree of curvature 'D': the deflection of a 100 m arc; sometimes used instead of radius.

$$D = \frac{180 \times 100}{\pi \times R} = \frac{5729.58}{R}\text{(degrees)}$$

A curve is defined if two of the following are known:

- intersection angle
- radius (or degree of curvature)
- tangent length
- curve length
- apex length

Design and site considerations will determine the curve parameters.

Chainage: the distance along the centreline of a road. In the days of imperial measure, the chainage referred to the number of chains (of 100 feet) from the origin of a project to the point in question. In SI units, the corresponding distance is quoted in metres, to which the term chainage is commonly (if inappropriately) applied.

On a particular contract, which may be only part of an overall scheme, it is unlikely that the origin will appear; the starting chainage may be in hundreds or thousands of metres.

Chainages are measured around curves; intersection points may be given *forward chainages* to locate them, these being measured forwards along the first tangent. Note that the chainage of the second tangent point is the sum of the chainage of the first tangent point and the curve length, *not* the sum of the IP forward chainage and the tangent length. For dual carriageways the chainage is usually measured along each offside kerbline.

Centreline or kerbline setting out

The sequence of setting out is:

(1) Locate the straights.
(2) Locate the intersection points.
(3) Measure the intersection angles.
(4) Calculate the curve data.
(5) Locate the tangent points.
(6) Set pegs on the centreline (or kerbline) at regular chainage intervals (20 m or 10 m).

As it is not possible to lay out a curved measurement, a series of short chords are set out. If the length of these chords does not exceed $R/20$, there is a negligible difference in length between the arc and the chord joining consecutive points.

10.3 HORIZONTAL CIRCULAR CURVES: LARGE RADIUS

Theodolite and tape (Rankine's) method

This is the most common method of setting out large radius curves. It makes use of a theodolite set up at a tangent point to swing the deflection angle from the forward tangent to each point to be set out, the distances being measured with a steel tape as successive short

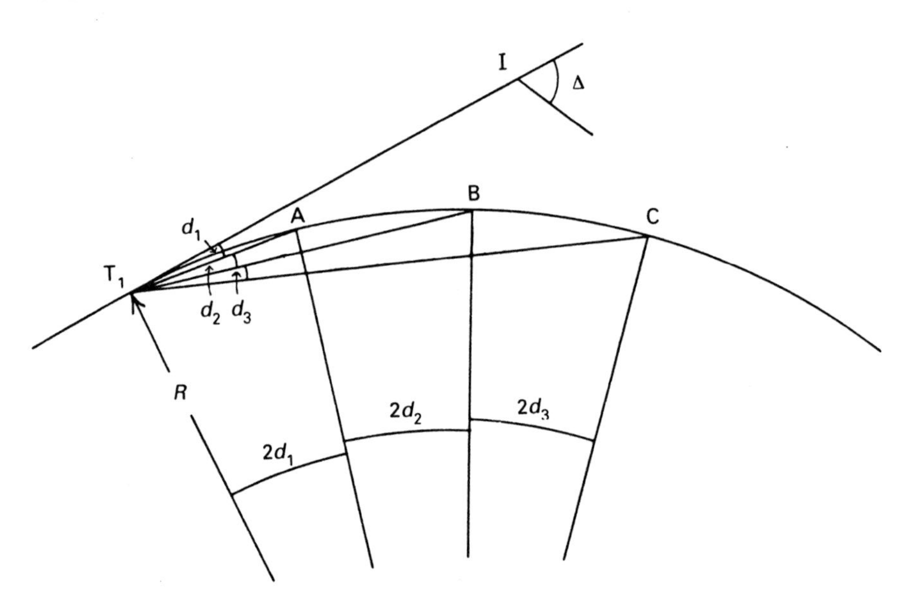

Fig. 10.2 Deflection angles.

chords. It is also known as the *deflection angle* or the *tangential angle* method.

Referring to Fig. 10.2 consider the first three points, A, B and C, to be pegged out on the curve from tangent point T_1. The chords T_1A, AB and BC should not exceed $R/20$.

Let the deflection angles $I\hat{T}_1A$, $A\hat{T}_1B$, $B\hat{T}_1C$ be d_1, d_2, d_3.

Applying the theorems relating the angles at the centre and circumference of a circle, and the angle in the alternate segment, it can be seen that the angles subtended at the centre of the curve, or angles consumed, by arcs T_1A, AB and BC are $2d_1$, $2d_2$, and $2d_3$.

In degrees:

$$\frac{2d_1}{360} = \frac{T_1A(\text{arc})}{2\pi R}$$

$$d_1 = \frac{90 \times T_1A}{\pi R}$$

$$= 28.648\frac{T_1A}{R}$$

Similarly:

$$d_2 = 28.648\frac{AB}{R}$$

and generally:

$$\text{Deflection angle} = 28.648\left(\frac{\text{short chord}}{\text{radius}}\right)(\text{degrees})$$

The reverse sexagesimal key of a calculator can be used to convert decimal degrees to degrees, minutes and seconds.

For the full curve, the sum of these deflection angles should equal half the intersection angle. This gives an arithmetic check. A small angular closing error (due to rounding off values) can be removed by making adjustments to some of the calculated angles. A larger error will indicate a fault in the arithmetic.

The *total deflection angles* for each point A, B, etc. can now be calculated as a cumulative sum of the individual deflection angles; the final value should be equal to half the intersection angle.

Points on the curve will normally be required at specified intervals of through chainage. An 'entry sub-chord' is required which will bring the first curve point to a suitable chainage. The remaining points will be set out to form equal short chords, with an 'exit sub-chord' required between the last curve point and the exit tangent point. A closure check should be made here, or at the middle if the curve has been set out from both ends (see Chapter 11).

Left-hand curves

The curves considered so far have been right-hand curves, deflecting to the right as the chainage increases. A left-hand curve deflects to the left, and the total deflection angles must be calculated by successive subtraction of the individual angles from 360 degrees.

Example 10.1

A 600.000 m radius right-hand circular curve is to join two straights intersecting at 22°30′00″ at forward chainage 3525.600 m. Prepare data for setting out the curve with a theodolite and tape at 20.000 m intervals of through chainage.

Solution

$$\text{Tangent length} = 600.000 \tan\left(\frac{22°30'00''}{2}\right) = 119.347\,\text{m}$$

$$\text{Curve length} \quad = \frac{\pi \times 600.000 \times 22.50}{180} \quad = 235.619\,\text{m}$$

(intersection angle converted to degrees and decimals)

Short chord <600/20 = 30; 20.000 m is therefore satisfactory.

Calculate the chainages of the tangent points by subtracting the tangent length from the forward chainage of the intersection point, and then adding the curve length.

3525.600 − 119.347	= 3406.253 m T_1 chainage		
3406.253 + 235.619	= 3641.872 m T_2 chainage		
Entry sub-chord	= 3420.000 − 3406.253	=	13.747 m
Short chords (3640 − 3420)/20 =	220/20	=	11 No. 220.000 m
Exit sub-chord	= 3641.872 − 3640.000	=	1.872 m
	Curve length =		235.619 m (check)

Figure 10.3 is a plan.

Deflection angles

$$\text{Entry sub-chord, } d = 28.648\left(\frac{13.747}{600.000}\right) = 0.65637 = 00°39'22.9''$$

$$\text{Short chord, } \quad d = 28.648\left(\frac{20.000}{600.000}\right) = 0.95493 = 00°57°17.8''$$

$$\text{Exit sub-chord, } \quad d = 28.648\left(\frac{1.872}{600.000}\right) = 0.08938 = 00°05'21.8''$$

Table 10.1 is prepared.

Each deflection angle is calculated to one decimal place of a second; this is carried forward to the total angle and rounding off to the nearest second is pcrformed by inspection.

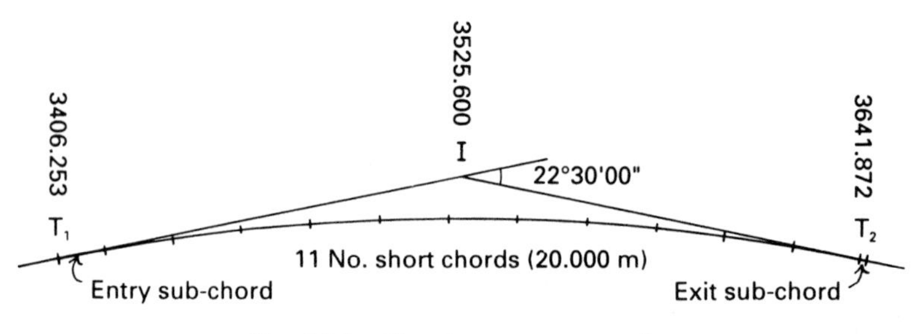

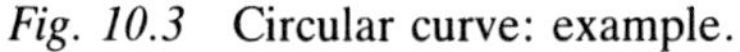
Fig. 10.3 Circular curve: example.

Table 10.1

Chainage	Chord length	Deflection angle °	′	″	Total deflection angle °	′	″
3406.253					00	00	00
3420.000	13.747	00	39	22.9	00	39	23
3440.000	20.000	00	57	17.8	01	36	41
3460.000	20.000	00	57	17.8	02	33	59
3480.000	20.000	00	57	17.8	03	31	16
3500.000	20.000	00	57	17.8	04	28	34
3520.000	20.000	00	57	17.8	05	25	52
3540.000	20.000	00	57	17.8	06	23	10
3560.000	20.000	00	57	17.8	07	20	28
3580.000	20.000	00	57	17.8	08	17	45
3600.000	20.000	00	57	17.8	09	15	03
3620.000	20.000	00	57	17.8	10	12	21
3640.000	20.000	00	57	17.8	11	09	39
3641.872	1.872	00	05	21.8	11	15	00
	Σ235.619	11	15	00.5			

Change of instrument position

Sometimes it may be necessary to move the theodolite to one of the points already set out on the curve. Further curve points may be sighted using the angles previously calculated by setting the theodolite to read 180°00′00″ when sighting the tangent point just vacated (Fig. 10.4(a)).

It can be seen that if the total deflection angle to set out P is a, and the deflection angle for PQ is d, the total angle for Q is $(a + d)$ at P from T_1P projected.

Large radius curves – other methods

Two theodolite method

In this method, theodolites are set up at both tangent points, and pegs are set on the curve by intersecting rays. The total deflection angles are calculated as before for the theodolite at T_1. These values may also be used by the observer at T_2 provided that the instrument has been 'zeroed' while sighting T_1 (Fig. 10.4(b)).

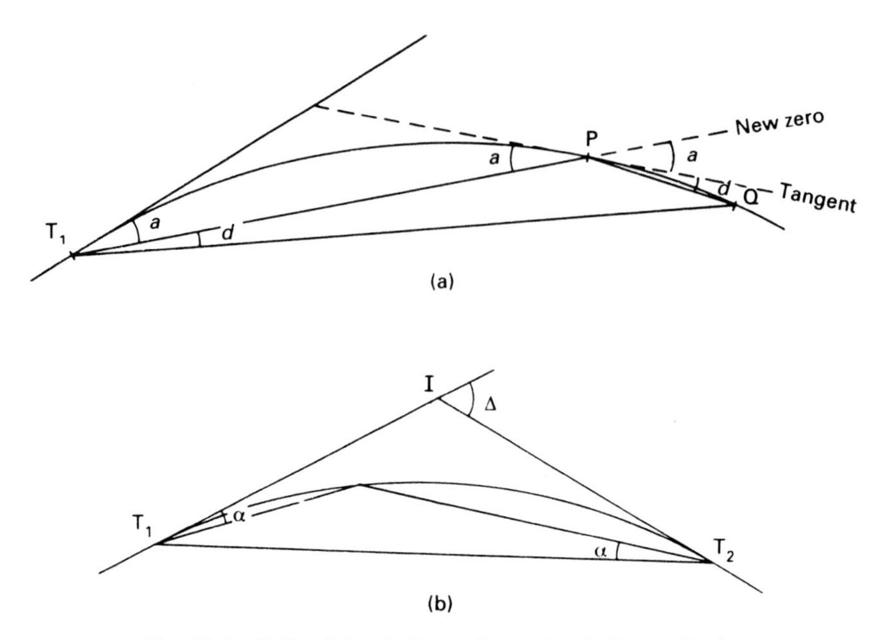

Fig. 10.4 Shift of theodolite and two theodolite methods.

The method saves taping around the curve, but has the following disadvantages:

(1) The risk of readings not being synchronized, especially if T_1 and T_2 are far apart.
(2) The 'weak fix' of points when rays intersect at a flat angle.
(3) The lack of an automatic closing check. A check may be applied by taping between consecutive curve points and ranging by eye.

The method may be useful where the line of a curve is required, but the points do not have to be at specific chainages, for example in a reconnaissance survey. Deflection angles need not then be calculated, but can be chosen for convenience of setting out.

Long chord method

Total deflection angles are calculated as before, but an EDM outfit at T_1 is used to measure directly to each point. Each long chord so set out equals $2R \sin \alpha$ for a total deflection angle of α. Check as for the two theodolite method.

EDM off line

An EDM outfit is set up at a suitable control station of known coordinates. The curve data is computed for the theodolite and tape method, and rectangular coordinates of the curve points are hence calculated. From the coordinate differences between the control station and the curve points, polar coordinates are computed.

There are several advantages to this method:

(1) An instrument station more suitable than the tangent points can be selected.
(2) Parts of a curve (rather than the whole curve) can be set out or re-established.
(3) Obstacles between curve points can be avoided.
(4) Cumulative setting out errors are eliminated.

Against these advantages must be weighed the possible difficulties in setting out with EDM. There is also the risk that, as the method is not a cumulative process, individual errors may pass unnoticed. For this reason, checks should be carried out by taping between consecutive curve points and by ranging around the curve by eye.

Much of the calculation can be carried out by computer.

Obstructions to the theodolite and tape method

Intersection point obstructed

Where the intersection point is obstructed, an alternative method must be employed for measuring the intersection angle and setting out the tangent lengths.

Two points, X and Y, are selected, one on each tangent, so that the distance between them can be measured. The angle between the tangent and the line XY is also measured at each point. Referring to Fig. 10.5(a):

$$\text{Intersection angle} = 360^\circ - (Y\hat{X}P + Q\hat{Y}X)$$

$$XT_1 = IT_1 - \frac{XY \sin Q\hat{Y}X}{\sin(Y\hat{X}P + Q\hat{Y}X - 180^\circ)}$$

$$YT_2 = IT_2 - \frac{XY \sin Y\hat{X}P}{\sin(Y\hat{X}P + Q\hat{Y}X - 180^\circ)}$$

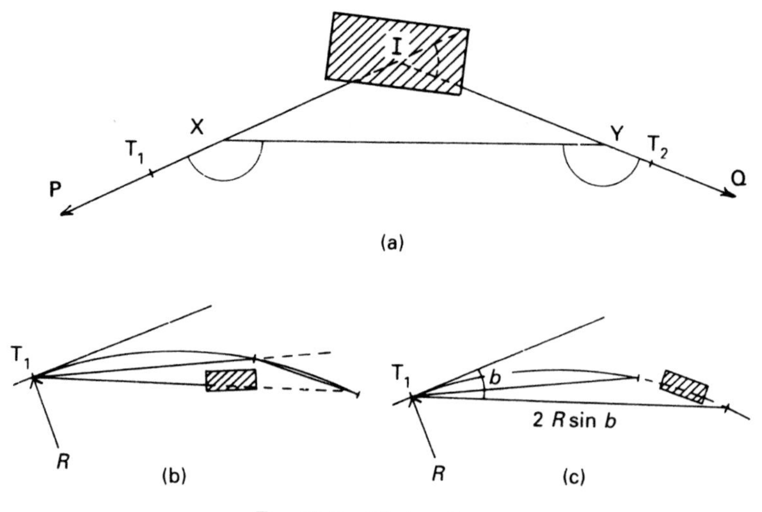

Fig. 10.5 Obstructions.

Then T_1 and T_2 can be set out. A check can be provided by repeating the process for two different points on the tangents.

Obstruction to sighting a deflection angle

An obstruction, although not on the line of the curve, may prevent the sighting of some of the curve points from the tangent point (Fig. 10.5(b)).

Either:

(1) Move the theodolite to the point last established on the curve. Or:
(2) Move the theodolite to the other tangent point. (Find a method of checking closure.)

Obstruction to taping a short chord

In this instance, an obstruction lies on the actual curve line (Fig. 10.5(c)).

Either:

(1) Use the 'two theodolites' method. Or:
(2) Set out the long chord ($= 2R \sin b$.) (Check closure.)

10.4 HORIZONTAL CURVES: SMALL RADIUS

Small radius curves may be set out by the methods described for large radius curves, or by any of the following methods, as appropriate.

Swinging a radius

The most obvious way of setting out a curve, swinging the tape can be employed only when the radius is not longer than the tape. The centre is pegged and a radius swung with one tape, while another can be used to measure short chords if points at particular distances or chainages are required.

Offsets from the long chord

The long chord is marked out (length = $2R\sin\Delta/2$), the middle of it located and the middle offset calculated (Fig. 10.6(a)).

Middle offset $Y_0 = R(1 - \cos\Delta/2)$

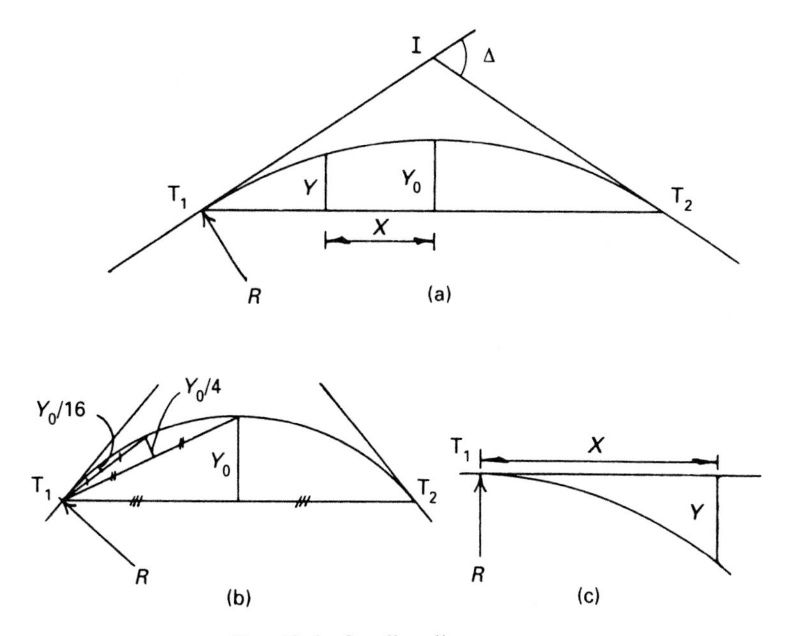

Fig. 10.6 Small radius curves.

Then offsets at points X from the middle are:

$$Y = Y_0 - R + (R^2 - X^2)^{1/2}$$

The method is not suitable for setting out points at particular distances or chainages.

Halving and quartering

Again, the long chord is located and the middle offset calculated and measured as in the previous method. The middle offset from the chord joining the tangent point and the curve mid point is then:

$$Y = Y_0/4 \text{ (Fig. 10.6(b))}$$

The process is continued; as the arc is halved, the mid offset is quartered. The method is not suitable for setting out points at particular distances or chainages.

Railway curves are often checked by this method, the offset being referred to as the *versine*; strictly, the offset is the product of the radius and the versine of the deflection angle for the chord.

Offsets from a tangent

Offsets from the tangent at distances X from the tangent point are:

$$Y = R - (R^2 - X^2)^{1/2} \text{ (Fig. 10.6(c))}$$

To prevent errors occurring from poorly aligned long offsets, the curve should be set out from each end towards the middle.

The method is not suitable for locating points at particular distances or chainages.

10.5 TO PASS A HORIZONTAL CURVE THROUGH A GIVEN POINT

With the tangents fixed, it may be necessary to design a curve to pass through a given point, the position of which is known relative to the tangents. The problem is to find the radius of the curve.

In Fig. 10.7, let the point P be a distance l from the intersection point I, and let IP make an angle of ϕ with IT_1.

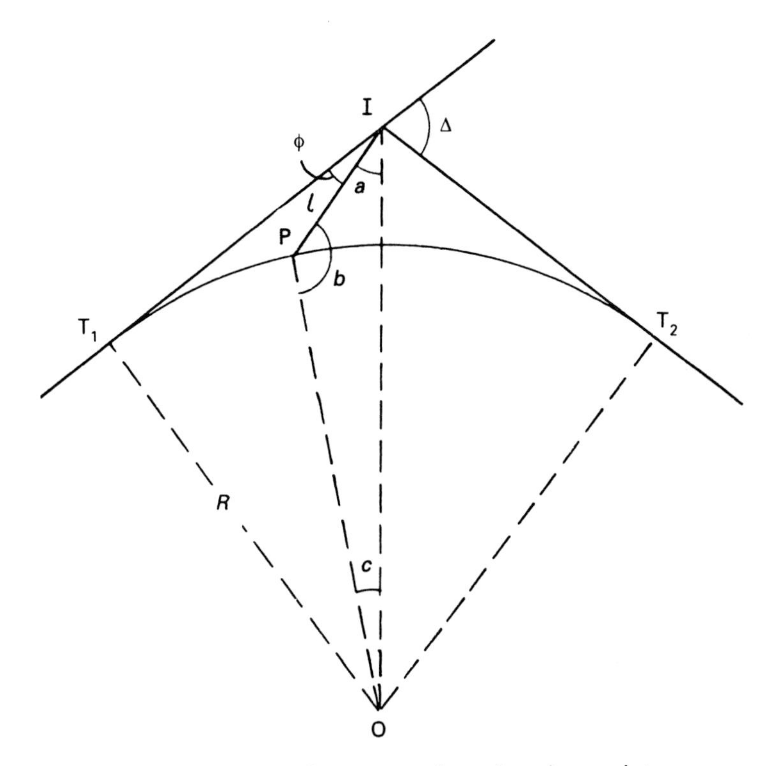

Fig. 10.7 Passing a curve through a given point.

With O as the centre, constructing triangle IPO and labelling its angles as a, b and c, the sine rule can be applied twice to deduce the radius, R.

$$a = 90° - \Delta/2 - \phi$$

$$\frac{\sin b}{\text{IO}} = \frac{\sin a}{\text{OP}}; \qquad b = \sin^{-1}\left(R \sec\frac{\Delta}{2} \times \frac{\sin a}{R}\right)$$

$$= \sin^{-1}\left(\frac{\sin a}{\cos \Delta/2}\right) \quad \text{Take value between } 90° \text{ and } 180°$$

$$\text{Also } c = 180° - a - b$$

$$\frac{\text{OP}}{\sin a} = \frac{\text{IP}}{\sin c} \qquad \therefore R = \frac{l \sin a}{\sin c}$$

Example 10.2

Two straights intersect at an angle of 24°30′00″. A circular curve is to be provided to join these straights and pass through a point which is 5.000 m from one of the straights at a position 55.000 m from the intersection point along the tangent. Calculate the radius.

Solution

Using the rectangular to polar conversion, and referring to Fig. 10.7:

$$\text{IP} = 55.2268 \quad \text{and PIT}_1 = 5°11'39.9''$$

$$\text{Then } a = 90°00'00'' - 12°15'00'' - 5°11'39.9'' = 72°33'20.1''$$

$$\text{and } b = \sin^{-1}\left(\frac{\sin 72°33'20.1''}{\cos 12°15'00''}\right) = 77°29'02.9'' \text{ (or } 102°30'57.1'')$$

Take value >90°

$$\text{Then } c = 180° - 72°33'20.1'' - 102°30'57.1'' = 4°55'42.8''$$

$$R = 55.2268\left(\frac{\sin 72°33'20.1''}{\sin 4°55'42.8''}\right) = 613.254\,\text{m}$$

(Working to 0.1″ to give required accuracy)

10.6 HORIZONTAL TRANSITION CURVES

On high-speed curves, and with increasing frequency on medium speed curves, transitions of steadily decreasing radius are used to lead from the tangent into the circular curve. A similarly shaped but handed curve is provided at the end of the circular arc.

There are two purposes:

(1) To achieve a gradual change of direction (and gradual imposition of lateral forces) from straight (radius = ∞) to the designed circular radius.
(2) To permit the gradual application of superelevation. Superelevation is the lateral tilting of a carriageway to counteract the tendency of vehicles to 'run off' on a curve. In railway engineering it is known as 'cant'.

In Fig. 10.8 if the forces acting when the vehicle is travelling at speed V round a curve of radius R are P (lateral force), W (weight) and N (normal reaction):

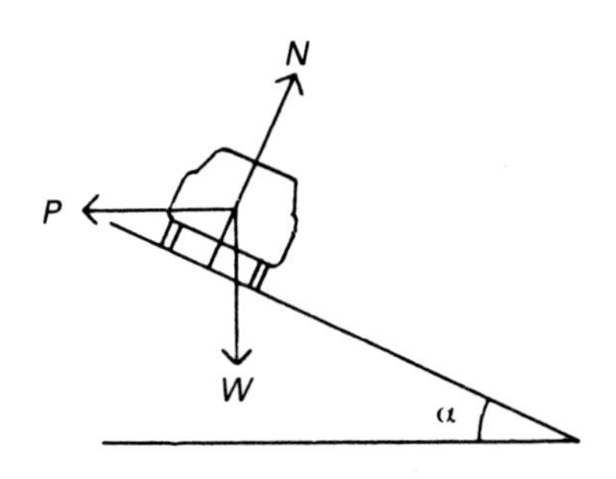

Fig. 10.8 Vehicle on superelevated road.

$$P = \frac{W}{g} \times \frac{V^2}{R} \quad \text{(Newton's second law)}$$

$$\frac{P}{W}\,\text{(centrifugal ratio)} = \frac{V^2}{Rg}$$

If V is in km/h and R is in m:

$$\frac{P}{W} = \frac{V^2}{127R}\,(\text{taking } g = 9.806\,\text{m/s}^2)$$

Circular curve radius

The centrifugal ratio usually has a value in the range 0.21 to 0.25 for roads, and 0.125 for railways.

If a value is taken of 0.25, R can be found for a design speed V:

$$R = \frac{V^2}{0.25 \times 127}$$

In the United Kingdom, design is carried out in accordance with TD 9/93 which, in Table 3, gives minimum radii for different design speeds and a variety of superelevations. The limiting radii correspond to a centrifugal ratio of 0.22; in general, larger radii are recommended.

Length and shape of transition

If the transition length is L, the time taken for a vehicle to travel it at speed V is $L/V = t$.

The radial acceleration increases from zero at the tangent point to V^2/R at the circular arc. If this increase is constant:

$$\frac{V^2/R}{L/V} \text{ is constant } = q$$

$$\therefore \frac{V^3}{RL} = q$$

(Usually $q = 0.3\,\text{m/s}^3$ for roads, though exceptionally 0.45 and 0.6 may be used.)

If V is in km/h and R is in m:

$$\frac{V^3}{(3.6)^3 RL} = q$$

$$\text{or } L = \frac{V^3}{46.7 \times q \times R}$$

At any point along the transition, the product of its distance from the tangent point and the radius is constant, and equals:

$$\frac{V^3}{46.7 \times q} (= rl)$$

From Fig. 10.9,

$$\frac{d\phi}{dl} = \frac{1}{r} = \frac{l}{rl} = \frac{l}{RL} \quad \therefore \phi = \frac{l^2}{2RL} \text{ (by integration)}$$

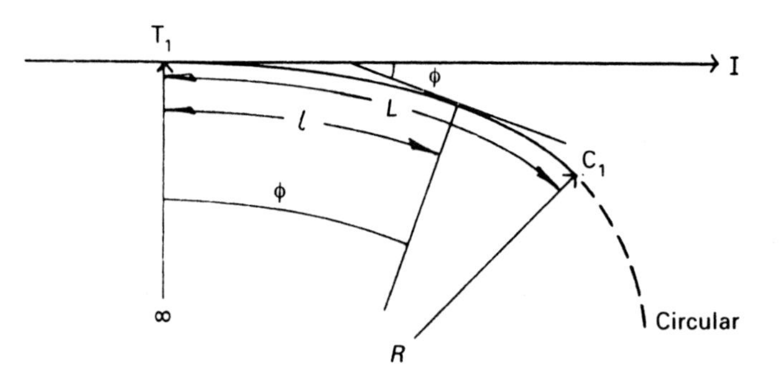

Fig. 10.9 Transition curve.

The curve of such a shape is the clothoid, also known as the true spiral or the Euler spiral. Its equation in intrinsic form is:

$$l = (2RL\phi)^{1/2}$$

Superelevation

Referring to Fig. 10.8, and resolving parallel to the superelevated surface:

$$P\cos\alpha = W\sin\alpha$$

if there is no slipping and no skid resistance.

$$\therefore \tan\alpha = \frac{P}{W} = \frac{V^2}{Rg}$$

To reduce the risk of slipping or overturning at speeds lower than the design speed, only a proportion of the sideways force P is usually resisted by the effect of superelevation. TD 9/93 allows for 45% to be resisted, the remainder being resisted by friction. Hence:

$$\tan\alpha = \frac{0.45P}{W} = \frac{0.45V^2}{Rg}$$

The superelevation is expressed as a percentage:

$$s = 100\tan\alpha = \frac{100 \times 0.45 \times V^2}{R \times g}$$

Expressing V in km/h, and taking $g = 9.806\,\text{m/s}^2$:

$$s = \frac{V^2}{2.824 \times R}$$

(Actually TD 9/93 has a factor of 2.828!)

The maximum permitted value of s is 7%, and normal crossfall (minimum on straight sections) is 2.5%.

The superelevation is gradually applied over the length of the transition: TD 9/93 stipulates that the difference in kerb gradients should not exceed 1% for 'all-purpose' roads and 0.5% for motorways.

A check should be carried out as to whether the transition is of sufficient length to apply the full superelevation without exceeding the permitted kerb gradient difference.

The superelevation is applied by rolling (or warping) the carriageway

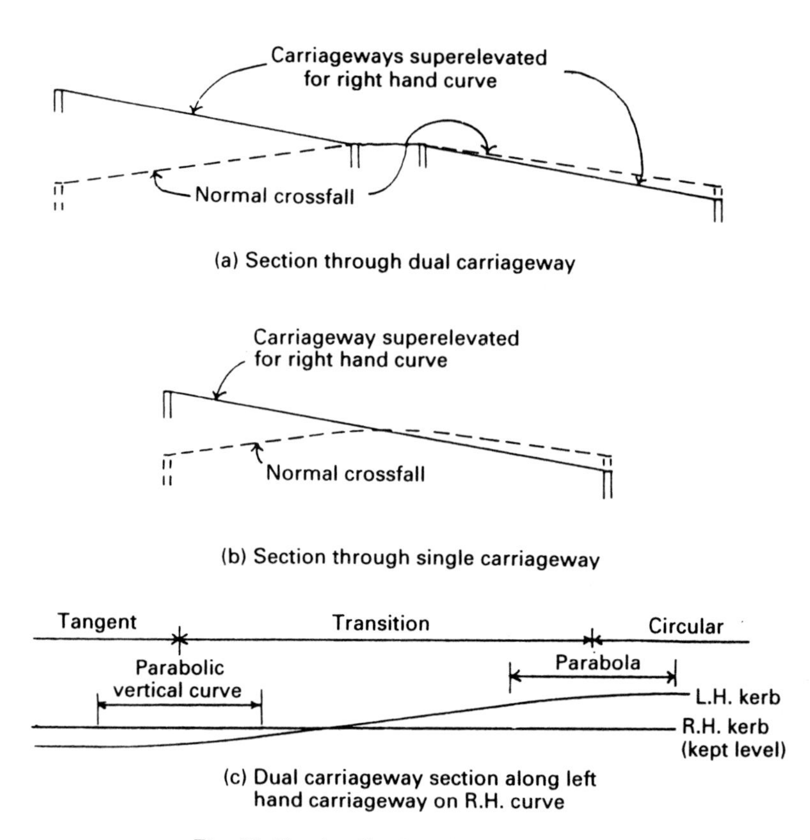

Fig. 10.10 Application of superelevation.

about one kerb, or about its centre (Fig. 10.10(a) and (b)). Where the kerb gradient changes to apply superelevation, short parabolic vertical curves are provided (Fig. 10.10(c)).

Setting out transitions

The intrinsic equation of the clothoid $\phi = l^2/(2RL)$ does not provide dimensions for setting it out. In practice the *Highway Transition Curve Tables* prepared by the County Surveyors' Society are generally used.

Where these tables are not available, and a clothoid is required, use may be made of the following expressions:

$$X = l - \frac{l^5}{40(RL)^2} + \frac{l^9}{3456(RL)^4} - \frac{l^{13}}{599\,040(RL)^6} + \ldots$$

$$Y = \frac{l^3}{6RL} - \frac{l^7}{336(RL)^3} + \frac{l^{11}}{42\,240(RL)^5} - \ldots$$

$$\tan\theta = \frac{\phi}{3} + \frac{\phi^3}{105} + \frac{\phi^5}{5997} - \frac{\phi^7}{198\,700} - \ldots$$

where X is the distance along the tangent from the tangent point, Y is the offset from the tangent, ϕ is in radians and θ, the deflection angle, is in degrees (Fig. 10.11).

Alternatively, for transitions consuming a small angle, the cubic spiral and the cubic parabola are approximations which may be used. Each of these three curves has only one shape; it is the scale and the amount of curve used which vary in individual curve design.

Cubic spiral

A non-standard curve of mixed dimensions, the cubic spiral is:

$$Y = \frac{l^3}{6RL}$$

and θ, the deflection angle, equals $\phi/3$, where ϕ is the angle consumed, The *back angle* therefore equals $2\phi/3$ or 2θ.

The curve has a limited range; beyond $\theta = 2°$, the radius decreases more sharply than that of the clothoid:

1% smaller at $\theta = 2°30'$
5% smaller at $\theta = 6°00'$
10% smaller at $\theta = 9°00'$

At $\theta = 20°$, the radius becomes zero and the curve ceases.

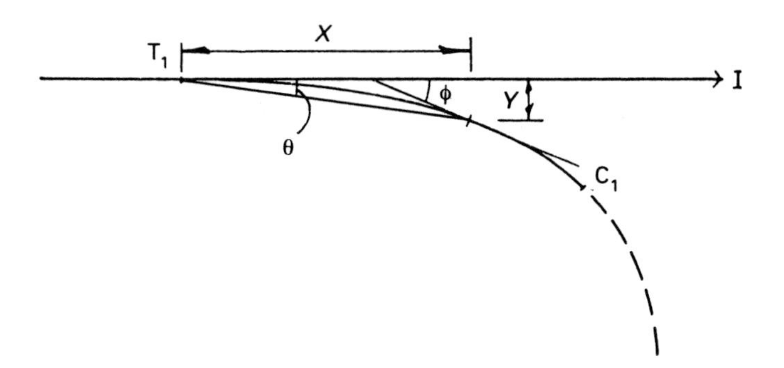

Fig. 10.11 Transition curve: setting out.

The curve may be set out by short chords and deflection angles using a theodolite and a tape, where l, the distance around the curve, is measured as the sum of short chords.

$$\theta = \frac{\phi}{3} = \frac{l^2}{6RL} \text{ radians} = \frac{180 \times l^2}{\pi \times 6RL} \text{ degrees}$$

$$\therefore \theta^\circ = \frac{9.5493 l^2}{RL}$$

Cubic parabola

The equation of the cubic parabola is:

$$Y = \frac{X^3}{6RL}$$

Again, $\theta = \phi/3$, and the *back angle* $= 2\phi/3$.

The cubic parabola also has a limited range; its radius decreases less sharply than the clothoid's:

1% larger at $\theta = 1^\circ 30'$
5% larger at $\theta = 3^\circ 30'$
10% larger at $\theta = 5^\circ 00'$

The shape is most suitable for setting out by offsets from the tangent. The cubic parabola may also be used in the determination of rectangular coordinates.

Clothoid – highway transition curve tables

Table 10.2 shows part of 'Table D' giving θ values for ϕ. These are slightly less than $\phi/3$. Table 10.3 shows 'Table 16', one of 26 tables giving values of R, D, L, ϕ, S, $R + S$, C, Long Chord, X, Y, θ and Back Angle along the transitions for RL values from 572.958 ($q = 0.3\,\text{m/s}^3$, $V = 20\,\text{km/h}$) to 1375098.708 ($q = 0.3\,\text{m/s}^3$, $V = 267.9\,\text{km/h}$).

D is the degree of curvature
S is the shift (see next section)
C is used to find the tangent length

Table 10.2 Highway Transition Curve Tables – metric. (*Reproduced by permission of the County Surveyors' Society*.)

Angle consumed ϕ	$\phi/3$	Deduct	Deflection angle
2°	0°40′	NIL	40′ 0″
3°	1°	0.1″	59′59.9″
4°	1°20′	0.2″	1°19′59.8″
5°	1°40′	0.4″	1°39′59.6″
6°	2°	0.7″	1°59′59.3″
7°	2°20′	1.0″	2°19′59.0″
8°	2°40′	1.6″	2°39′58.4″
9°	3°	2.3″	2°59′57.7″
10°	3°20′	3.1″	3°19′56.9″
11°	3°40′	4.1″	3°39′55.9″
12°	4°	5.4″	3°59′54.6″
13°	4°20′	6.8″	4°19′53.2″
14°	4°40′	8.5″	4°39′51.5″
15°	5°	10.5″	4°59′49.5″
16°	5°20′	12.7″	5°19′47.3″
17°	5°40′	15.2″	5°39′44.8″
18°	6°	18.1″	5°59′41.9″
19°	6°20′	21.3″	6°19′38.7″
20°	6°40′	24.8″	6°39′35.2″
21°	7°	28.8″	6°59′31.2″
22°	7°20′	33.1″	7°19′26.9″
23°	7°40′	37.8″	7°39′22.2″
24°	8°	43.0″	7°59′17.0″
25°	8°20′	48.6″	8°19′11.4″
26°	8°40′	54.7″	8°39′ 5.3″
27°	9°	1′ 1.2″	8°58′58.8″
28°	9°20′	1′ 8.3″	9°18′51.7″
29°	9°40′	1′16.0″	9°38′44.0″
30°	10°	1′24.1″	9°58′35.9″
31°	10°20′	1′32.8″	10°18′27.2″

Table D Interpolated Deflection Angles
For any point on a spiral where Angle Consumed = ϕ, the true Deflection Angle is $\phi/3$ minus the correction tabled above.
The Back Angle is $2\phi/3$ plus the same correction.
The series for Tan (Deflection Angle) as given under 'Formulae' gives slight errors when ϕ is large. These have been adjusted in this table.

Table 10.3 Highway Transition Curve Tables – metric. *(Reproduced by permission of the County Surveyors' Society.)*

Radius R metres	Degree of curve D °	′	″	Spiral length L metres	Angle consumed ϕ °	′	″	$R + S$ metres	
22 918.3118	0	15	0.0	5.00	0	0	22.5	0.0000	22 918.3118
11 459.1559	0	30	0.0	10.00	0	1	30.0	0.0004	11 459.1563
7 639.4373	0	45	0.0	15.00	0	3	22.5	0.0012	7 639.4385
5 729.5780	1	0	0.0	20.00	0	6	0.0	0.0029	5 729.5809
4 583.6624	1	15	0.0	25.00	0	9	22.5	0.0057	4 583.6680
3 819.7186	1	30	0.0	30.00	0	13	30.0	0.0098	3 819.7285
3 274.0445	1	45	0.0	35.00	0	18	22.5	0.0156	3 274.0601
2 864.7890	2	0	0.0	40.00	0	24	0.0	0.0233	2 864.8122
2 546.4791	2	15	0.0	45.00	0	30	22.5	0.0331	2 546.5122
2 291.8312	2	30	0.0	50.00	0	37	30.0	0.0455	2 291.8766
2 083.4829	2	45	0.0	55.00	0	45	22.5	0.0605	2 083.5434
1 909.8593	3	0	0.0	60.00	0	54	0.0	0.0785	1 909.9379
1 762.9471	3	15	0.0	65.00	1	3	22.5	0.0999	1 763.0469
1 637.0223	3	30	0.0	70.00	1	13	30.0	0.1247	1 637.1470
1 527.8875	3	45	0.0	75.00	1	24	22.5	0.1534	1 528.0408
1 432.3945	4	0	0.0	80.00	1	36	0.0	0.1862	1 432.5807
1 348.1360	4	15	0.0	85.00	1	48	22.5	0.2233	1 348.3593
1 273.2395	4	30	0.0	90.00	2	1	30.0	0.2651	1 273.5046
1 206.2269	4	45	0.0	95.00	2	15	22.5	0.3117	1 206.5387
1 145.9156	5	0	0.0	100.00	2	30	0.0	0.3636	1 146.2793
1 091.3482	5	15	0.0	105.00	2	45	22.5	0.4209	1 091.7691
1 041.7414	5	30	0.0	110.00	3	1	30.0	0.4839	1 042.2254
996.4483	5	45	0.0	115.00	3	18	22.5	0.5529	997.0013
954.9297	6	0	0.0	120.00	3	36	0.0	0.6282	955.5579
916.7325	6	15	0.0	125.00	3	54	22.5	0.7101	917.4425
881.4735	6	30	0.0	130.00	4	13	30.0	0.7987	882.2722
848.8264	6	45	0.0	135.00	4	33	22.5	0.8944	849.7208
818.5111	7	0	0.0	140.00	4	54	0.0	0.9975	819.5086
790.2866	7	15	0.0	145.00	5	15	22.5	1.1082	791.3948
763.9437	7	30	0.0	150.00	5	37	30.0	1.2268	765.1705
739.3004	7	45	0.0	155.00	6	0	22.5	1.3535	740.6539
716.1972	8	0	0.0	160.00	6	24	0.0	1.4887	717.6859
694.4943	8	15	0.0	165.00	6	48	22.5	1.6326	696.1269
674.0680	8	30	0.0	170.00	7	13	30.0	1.7854	675.8534
654.8089	8	45	0.0	175.00	7	39	22.5	1.9475	656.7564
636.6198	9	0	0.0	180.00	8	6	0.0	2.1191	638.7388

Values continued up to 15°15′00″ degree of curve

Gain of accn. m/s^3	0.30	0.45	0.60
Speed value km/h	117.0	133.9	147.4

Table 10.3

C metres	Long chord metres	Coordinates X metres	Coordinates Y metres	Deflection angle from origin °	′	″	Back angle to origin °	′	″
2.5000	5.0000	5.0000	0.0002	0	0	7.5	0	0	15.0
5.0000	10.0000	10.0000	0.0015	0	0	30.0	0	1	0.0
7.5000	15.0000	15.0000	0.0049	0	1	7.5	0	2	15.0
10.0000	20.0000	20.0000	0.0116	0	2	0.0	0	4	0.0
12.5000	25.0000	25.0000	0.0227	0	3	7.5	0	6	15.0
15.0000	30.0000	30.0000	0.0393	0	4	30.0	0	9	0.0
17.5000	35.0000	34.9999	0.0624	0	6	7.5	0	12	15.0
20.0000	39.9999	39.9998	0.0931	0	8	0.0	0	16	0.0
22.4999	44.9998	44.9996	0.1325	0	10	7.5	0	20	15.0
24.9999	49.9997	49.9994	0.1818	0	12	30.0	0	25	0.0
27.4998	54.9996	54.9990	0.2420	0	15	7.5	0	30	15.0
29.9998	59.9993	59.9985	0.3142	0	18	0.0	0	36	0.0
32.4996	64.9990	64.9978	0.3994	0	21	7.5	0	42	15.0
34.9995	69.9986	69.9968	0.4989	0	24	30.0	0	49	0.0
37.4992	74.9980	74.9955	0.6136	0	28	7.5	0	56	15.0
39.9990	79.9972	79.9938	0.7446	0	32	90.0	1	4	0.0
42.4986	84.9962	84.9916	0.8931	0	36	7.5	1	12	15.0
44.9981	89.9950	89.9888	1.0602	0	40	30.0	1	21	0.0
47.4975	94.9935	94.9853	1.2469	0	45	7.5	1	30	15.0
49.9968	99.9915	99.9810	1.4542	0	50	0.0	1	40	0.0
52.4960	104.9892	104.9757	1.6834	0	55	7.4	1	50	15.1
54.9949	109.9864	109.9693	1.9355	1	0	29.9	2	1	0.1
57.4936	114.9830	114.9617	2.2115	1	6	7.4	2	12	15.1
59.9921	119.9789	119.9526	2.5126	1	11	59.9	2	24	0.1
62.4903	124.9742	124.9419	2.8398	1	18	7.3	2	36	15.2
64.9882	129.9686	129.9293	3.1942	1	24	29.8	2	49	0.2
67.4858	134.9621	134.9147	3.5769	1	31	7.2	3	2	15.3
69.9829	139.9545	139.8976	3.9889	1	37	59.6	3	16	0.4
72.4797	144.9458	144.8780	4.4314	1	45	7.1	3	30	15.4
74.9759	149.9358	149.8555	4.9054	1	52	29.4	3	45	0.6
77.4716	154.9243	154.8298	5.4119	2	0	6.8	4	0	15.7
79.9667	159.9113	159.8005	5.9521	2	7	59.2	4	16	0.8
82.4612	164.8965	164.7673	6.5269	2	16	6.5	4	32	16.0
84.9550	169.8799	169.7299	7.1376	2	24	28.8	4	49	1.2
87.4479	174.8612	174.6878	7.7850	2	33	6.1	5	6	16.4
89.9401	179.8402	179.6406	8.4702	2	41	58.4	5	24	1.6

Increase in degree of curve per meter = D/L = 0°3′0.0″
RL constant = 114 591.559
Degree of curvature based on 100 m standard arc

Composite curves

Consider a composite curve consisting of a circular curve connected to the tangents by transitions as in Fig. 10.12(a) and (b).

It can be seen that the circular curve is shifted in from the tangents by an amount S. At the point where a radius (from the circular curve

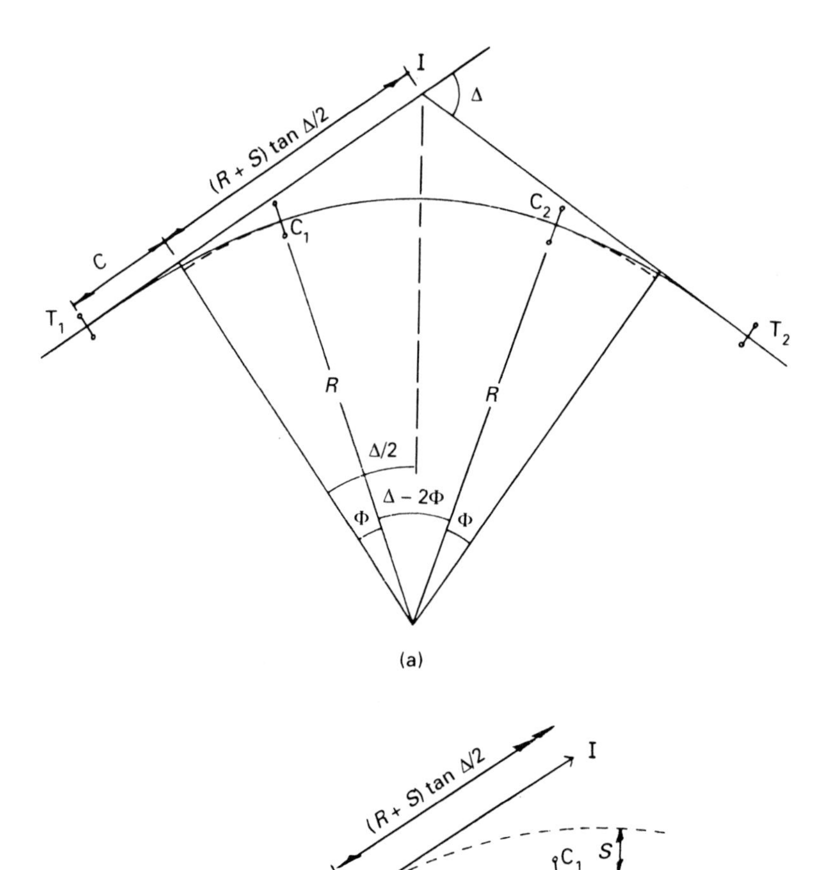

Fig. 10.12 Composite curve.

centre) perpendicular to the tangent intersects the transition, the transition and the shift mutually bisect each other, with negligible error.

The tangent length = $(R + S)\tan\Delta/2 + C$. Either:

$$S = \frac{L^2}{24R} \quad \text{and } C = L/2$$

may be used with negligible error, or the values can be read from the *Highway Transition Tables*.

The angle consumed by the whole transition

$$\Phi = \frac{L}{2R} \text{ radians}$$

(and the final deflection angle is Θ).

For a composite curve with equal transitions, the angle consumed by the circular curve

$$= \Delta - 2\Phi$$

and the curve length

$$= \frac{R(\Delta - 2\Phi)}{180} \text{ (angles in degrees).}$$

The circular curve can be set out using a theodolite set up at C_1 or C_2 and a steel tape. For a right-hand curve, the instrument circle is set by sighting T_1 while reading $180° - (\Phi - \Theta)$ as shown in Fig. 10.13, the corresponding value for a left hand curve being $180° + (\Phi - \Theta)$.

With transitions of cubic spiral/cubic parabola shapes:

$$\Phi - \Theta = 2\Phi/3$$

For a clothoid, the back angles $(\phi - \theta)$ are given in the tables.

A reading of zero then represents the forward tangent at C_1; the customary deflection angle method can then be used to set out the circular arc by theodolite and tape.

The second transition should be set out from T_2; the whole composite curve should close either at C_2 or at the middle of the circular curve if this has been set out from C_1 and C_2.

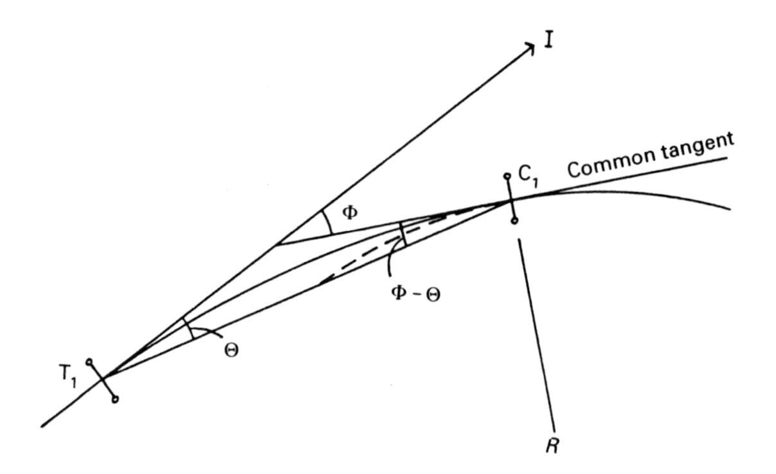

Fig. 10.13 Use of back angle.

10.7 COMPOSITE CURVE EXAMPLE

Example 10.3

A composite curve is to be designed to allow a 7.300 m wide single carriageway with a double crossfall to negotiate a right-hand curve with an intersection angle of 30°00′00″. The design speed is 120 km/h, the intersection point forward chainage is 2148.500 m and the permissible rate of gain of radial acceleration is to be 0.3 m/s^3. Prepare data to set out points at 20.000 m intervals of through chainage using a theodolite and tape.

Two solutions will be given, the first showing the manual calculations for a cubic spiral and the second using the transition tables for a clothoid.

Solution – using a cubic spiral

Radius

From TD 9/93, minimum radius for a maximum superelevation of 7% is 720 m for a design speed of 120 km/h.

Transition length

The nearest speed value in the transition tables is 117 km/h; to allow a comparison to be made between methods, this value has been taken for both solutions.

$$L = \frac{117^3}{3.6^3 \times 0.3 \times 720} = 158.926\,\text{m}.$$

Round up to 160.000 m. Keep $R = 720.000$ m.
Then RL (constant) = 115 200.

Before the transition length is accepted, it should be checked for the application of superelevation.

Superelevation

$$s = \frac{V^2}{2.828R} = 6.72\%$$

As this nearly equals 7%, the criteria for selection of the radius are satisfactory.

For 7.300 m carriageway width, with 6.72% superelevation, the outer kerb must be laid 7.3 × 0.0672 m (=0.491 m) above the inner kerb on the circular arc.

Over 160 m, the kerb gradient difference

$$= 0.491 \times 100/160 = 0.307\%$$

which is less than the limit of 1%, so the 160.000 m transition length is adequate.

Chainage

$$\text{Shift, } S = \frac{160^2}{24 \times 720} = 1.481\,\text{m}$$

$$C = 160.000/2 = 80.000\,\text{m}$$

Tangent length

$$= (720.000 + 1.481)\tan 15° + 80.000 = 273.320\,\text{m}$$

Angle consumed by transition = $L/2R$ (radians)

$$= \frac{160}{2 \times 720} \times \frac{180}{\pi} = 6°21'58''$$

Angle consumed by circular arc = 30°00′00″ − 2(6°21′58″)

$= 17°16'04''$

Length of circular arc

$$= \frac{\pi \times 720 \times 17°16'04''}{180}$$

$= 216.993\,\text{m}$

$T_1 = 2148.500 - 273.320 = 1875.180\,\text{m}$

$C_1 = 1875.180 + 160.000 = 2035.180\,\text{m}$

$C_2 = 2035.180 + 216.993 = 2252.173\,\text{m}$

$T_2 = 2252.173 + 160.000 = 2412.173\,\text{m}$

First transition

Table 10.4

Chainage	Short chord	Distance along transition l	Deflection angle °	′	″
1875.180	0.000	0.000	00	00	00
1880.000	4.820	4.820	00	00	07
1900.000	20.000	24.820	00	03	04
1920.000	20.000	44.820	00	09	59
1940.000	20.000	64.820	00	20	54
1960.000	20.000	84.820	00	35	47
1980.000	20.000	104.820	00	54	39
2000.000	20.000	124.820	01	17	59
2020.000	20.000	144.820	01	44	19
2035.180	15.180	160.000	02	07	19
	Σ160.000				

Deflection angle $= \dfrac{9.5493 \times l^2}{RL}$. Table 10.4 is prepared.

As a check, the final angle should equal $\Phi/3 = (6°21'58'')/3$

$= 2°07'19.3''$

which provides the check.

Circular arc

Back angle at C_1 = 2(6°21′58″)/3 = 4°14′39″
Set theodolite at C_1 sighting T_1 with a reading of

180° − 4°14′39″ = 175°45′21″

on the horizontal circle. Table 10.5 is prepared.

Table 10.5

Chainage	Chord length	Deflection angle °	′	″	Total deflection angle °	′	″
2035.180							
2040.000	4.820	00	11	30.4	00	11	30
2060.000	20.000	00	47	44.8	00	59	15
2080.000	20.000	00	47	44.8	01	47	00
2100.000	20.000	00	47	44.8	02	34	45
2120.000	20.000	00	47	44.8	03	22	30
2140.000	20.000	00	47	44.8	04	10	14
2160.000	20.000	00	47	44.8	04	57	59
2180.000	20.000	00	47	44.8	05	45	44
2200.000	20.000	00	47	44.8	06	33	29
2220.000	20.000	00	47	44.8	07	21	14
2240.000	20.000	00	47	44.8	08	08	59
2252.173	12.173	00	29	03.7	08	38	02
	Σ216.993	08	38	02.1			

The sum of the deflection angles should equal half the angle consumed by the circular curve

= (17°16′04″)/2 = 08°38′02″

which checks.

Second transition

Theodolite set up at T_2. Table 10.6 shows the data.

Table 10.6

Chainage	Short chord	Distance along transition l	Deflection angle °	′	″	Instrument angle °	′	″
2412.173	0.000	0.000	00	00	00	00	00	00
2400.000	12.173	12.173	00	00	44	359	59	16
2380.000	20.000	32.173	00	05	09	359	54	51
2360.000	20.000	52.173	00	13	32	359	46	28
2340.000	20.000	72.173	00	25	54	359	34	06
2320.000	20.000	92.173	00	40	15	359	19	45
2300.000	20.000	112.173	01	02	35	358	57	25
2280.000	20.000	132.173	01	26	53	358	33	07
2260.000	20.000	152.173	01	55	10	358	04	50
2252.173	7.827	160.000	02	07	19	357	52	41
	Σ160.000							

Solution – using a clothoid

Radius

As in previous solution:

$$R = 720.000\,\text{m}$$

Using Table 16 in the Highway Transition Curve Tables, the nearest speed value to 120 is 117 km/h.

$$RL = 114\,591.559 \text{ (from the table)}$$

Transition length

As before, length L = 158.926 m; again, round up to 160.000 m. With a fixed value of RL as in the tables, the radius should be modified, and will become 114591.559/160.000 = 716.1972 m. (Alternatively L = 155.000 m and R = 739.3004 m could be taken.)

From the table:

$$S = 1.4887\,\text{m}$$

$$C = 79.9667\,\text{m}$$

$$\Theta = 2°07'59''$$

$$\Phi = 6°24'00''$$
$$\text{Back angle} = 4°16'01''$$

Superelevation

As before.

Chainage

Tangent length

$$= (717.6859)\tan 15° + 79.9667 = 272.2700$$

Angle consumed by circular arc

$$= 30°00'00'' - 2(6°24'00'')$$
$$= 17°12'00''$$

Length of circular arc

$$= \frac{\pi \times 716.1972 \times 17°12'00''}{180}$$
$$= 215.000\,\text{m}$$

$$T_1 = 2148.500 - 272.270 = 1876.230\,\text{m}$$
$$C_1 = 1876.230 + 160.000 = 2036.230\,\text{m}$$
$$C_2 = 2036.230 + 215.000 = 2251.230\,\text{m}$$
$$T_2 = 2251.230 + 160.000 = 2411.230\,\text{m}$$

First transition

Where the required chords are not tabulated, there are two methods that can be used.

(1) Calculate $\theta = \phi/3 = \dfrac{9.5493 \times l^2}{RL}$

(2) Calculate $\theta = \phi/3 = \dfrac{\Phi}{3} \times \dfrac{l^2}{L^2}$

and in both cases deduct the appropriate correction given in Table D; usually small, often negligible.

For example, consider the deflection angle at 83.770 m:

$$\theta = \frac{9.5493 \times l^2}{114591.559} = 00°35'05''$$

$$\text{or } \theta = \frac{6°24'00''}{3} \times \frac{83.770^2}{160.000^2} = 00°35'05''$$

From Table D, correction is zero. Table 10.7 is prepared.

Table 10.7

Chainage	Short chord	Distance along transition l	Deflection angle °	′	″	Correction and corrected angle
1876.230	0.000	0.000	00	00	00	
1880.000	3.770	3.770	00	00	04	
1900.000	20.000	23.770	00	02	50	
1920.000	20.000	43.770	00	09	35	
1940.000	20.000	63.770	00	20	20	
1960.000	20.000	83.770	00	35	05	
1980.000	20.000	103.770	00	53	50	
2000.000	20.000	123.770	01	16	36	
2020.000	20.000	143.770	01	43	21	−0.4″
2036.230	16.230	160.000	02	08	00	−0.8″/02°07′59″
	Σ160.000					

Circular arc

The back angle of 4°16′01″ is used to enable the theodolite to be set up at C_1. The method follows that previously described.

Second transition

Deflection angles from T_2 are calculated.

10.8 TRANSITION ORIGIN INACCESSIBLE

It may happen that a transition has to be set out from a point other than its origin. One method is to calculate the coordinates of points to be set out, using X and Y values either taken from the tables or calculated for a cubic parabola. Polar coordinates from a suitable instrument station can be calculated and theodolite and tape or EDM used in the field.

10.9 COMPOUND CURVES

In addition to the composite curves already described, there may be occasions when a compound curve is required. Such a curve is likely to combine circular arcs of differing radii (joined by a transition) or transitions of differing characteristics. Slip road construction frequently requires the provision of compound curves.

If the principles associated with circular and transition curves are understood, the engineer should be able to design and set out such curves with success.

10.10 COMPUTER DESIGN

In many instances curve design will be carried out by computer and a 'printout' will be supplied to site. If the site engineer is fortunate, all the data necessary for setting out horizontal curves will be available. More frequently, intermediate points on the curves may not be tabulated and it will be the engineer's job to prepare additional calculations so that all the points required on site can be set out.

10.11 VERTICAL CURVES

Vertical curves are used to effect smooth changes between different grades on roads or railways. The basic criterion to be considered in design is that of passenger comfort: a constant rate of change of grade will achieve this. In addition, the sight distance along a crest (summit) curve must be considered, as must be headlight distance for adequate illumination along a road sag curve.

A parabola is generally used for a vertical curve, satisfying the requirement for the rate of change of grade to be constant.

Conventions

Figure 10.14 refers.

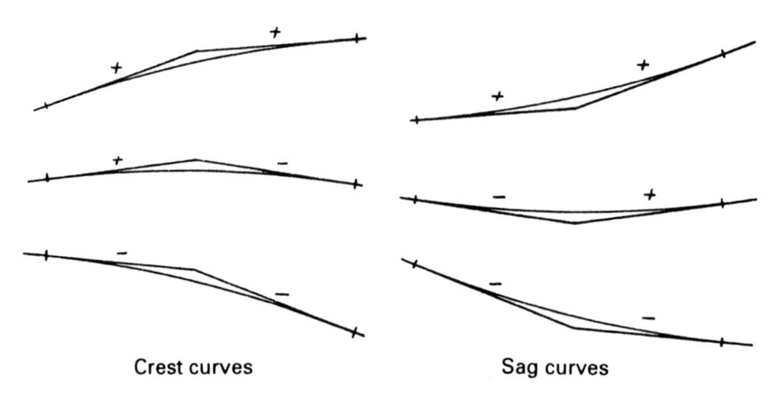

Fig. 10.14 Vertical curves.

(1) Grade rising as chainage increases is *positive*.
(2) Grade falling as chainage decreases is *negative*.
(3) Long sections are drawn with chainage increasing from left to right.
(4) When algebraic grade value is to be decreased, a *crest* (or summit) curve is provided.
(5) When algebraic grade value is to be increased, a *sag* curve is provided.
(6) Grades are expressed as percentages.

Approximations in curve design

Grade values are small and vertical curves are flat. In design, the following approximations are acceptable:

(1) All distances along the curve are taken as horizontal.
(2) Curve offsets are vertical.
(3) The curve and the vertical line joining the intersection point to the long chord mutually bisect each other.

Thus, in Fig. 10.15:

$$T_1I = IV = IT_2 = T_1P = PT_2 = T_1Q = QT_2 = \frac{T_1U}{2}$$

$$\text{and } IP = PQ = \frac{VT_2}{4}$$

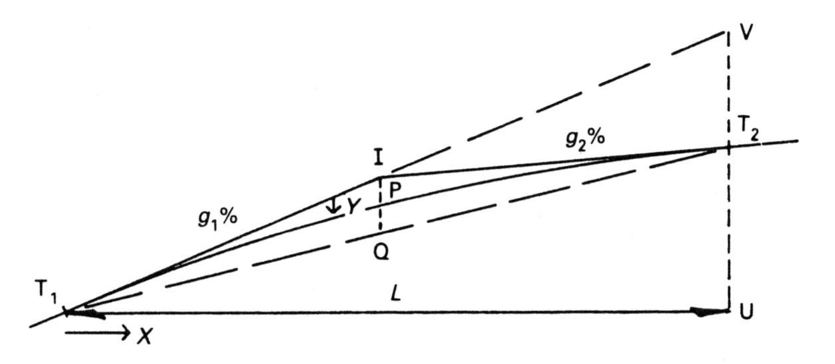

Fig. 10.15 Vertical curve geometry.

Note that these approximations do not relieve the engineer of the responsibility for applying slope corrections on site.

Curve equation

The entry grade is g_1 (%), the exit grade is g_2 (%), the curve length is L, the horizontal distance is X and the curve offset (vertical) is Y. It can be shown that the curve equation is:

$$Y = \frac{(g_1 - g_2)X^2}{200L}$$

The origin is taken as the start of the curve T_1, and the offsets are measured vertically from the entry grade (projected when $X > L/2$).

The same equation can be applied taking the curve mid-point (P) as the origin with offsets measured from the tangent at this point. The formula for crest sight distance can be derived using this form of the equation.

$(g_1 - g_2)/(200L)$ will be positive for a crest curve and negative for a sag curve. The offset must be subtracted from the entry grade level.

Design procedure

Curve equation

Road design in the UK should be carried out in accordance with Department of Transport Standard TD 9/93 which specifies desirable maximum gradients. When the grades have been determined, Table

10.8 ('Table 3' in TD 9/93) should be used to obtain K, values for crest and sag curves being given at a variety of vehicle speeds. The K value enables the length of the curve to be determined as:

$$L = K(g_1 - g_2)$$

On single carriageways, full overtaking sight distance should be provided on crest curves. Where the overtaking criterion is not applicable (e.g. on a dual carriageway), K values for Vertical Curvature should be taken; they provide adequate stopping distances. Sag curve K values in Vertical Curvative give suitable headlight distances. All values are such that the vertical acceleration never exceeds $0.3\,m/s^2$.

Grade levels

Reduced levels should be calculated at the required chainages along the entry grade for the full length of curve, and at the end of the curve (on the exit grade). It is advisable to prepare a table of the calculations.

Offsets

The curve offsets can now be computed using the equation:

$$Y = \frac{(g_1 - g_2)X^2}{200L}$$

A check on the arithmetic can be carried out at this stage: provided that the chainage intervals are constant, the second differences of the offset values should be constant (constant change of grade).

Curve levels

The offsets should be subtracted from the entry grade level, taking account of the sign. The level at the end of the curve, calculated in this way, should equal the value previously computed.

Alternatively, the curve levels may be calculated directly if the reduced level at the start of the curve is known:

$$\text{Curve level} = RL \text{ at start} + g_1X/100 - (g_1 - g_2)X^2/(200L)$$

Table 10.8 Road curve data (Table 3 in D.o.T. Highway Link Design Standard TD 9/93).
(Reproduced with the permission of the Controller of Her Majesty's Stationery Office.)

Design Speed km/h	120	100	85	70	60	50	V^2/R
STOPPING SIGHT DISTANCE m							
Desirable Minimum	295	215	160	120	90	70	
Absolute Minimum	215	160	120	90	70	50	
HORIZONTAL CURVATURE m							
Minimum R * without elimination of Adverse Camber and Transitions	2880	2040	1440	1020	720	510	5
Minimum R * with Superelevation of 2.5%	2040	1440	1020	720	510	360	7.07
Minimum R * with Superelevation of 3.5%	1440	1020	720	510	360	255	10
Desirable Minimum R with Superelevation of 5%	1020	720	510	360	255	180	14.14
One Step below Desirable Minimum R with Superelevation of 7%	720	510	360	255	180	127	20
Two Steps below Desirable Minimum R with Superelevation of 7% at sites of special difficulty (Category B Design Speeds Only)	510	360	255	180	127	90	28.28
VERTICAL CURVATURE							
Desirable Minimum * Crest K Value	182	100	55	30	17	10	
One Step below Desirable Minimum Crest K Value	100	55	30	17	10	6.5	
Absolute Minimum Sag K Value	37	26	20	20	13	9	
OVERTAKING SIGHT DISTANCE							
Full Overtaking Sight Distance FOSD m	*	580	490	410	345	290	
FOSD Overtaking Crest K Value	*	400	285	200	142	100	

* Not recommended for use in the design of single carriageways (see Paragraphs 7.25 to 7.31 inclusive)

Example 10.4

A crest curve of 160.000 m length is to lead from an upgrade of 2% into a downgrade of 3%. The two grades if projected would meet at a reduced level of 26.360 m AOD at chainage 2360.000 m. Calculate curve levels at 20 m intervals of chainage.

Solution

$g_1 = +2\%, g_2 = -3\%, L = 160.000\,\text{m}$

The entry grade will rise by 2 × 20/100 = 0.400 m every 20 m and the exit grade will fall by 3 × 20/100 = 0.600 m every 20 m.

The reduced level at the start of the curve will be:

$$26.360 - 2 \times 80/100 = 24.760\,\text{m AOD}$$

while at the end of the curve it will be

$$26.360 - 3 \times 80/100 = 23.960\,\text{m AOD}$$

The curve offset equation will be:

$$Y = \frac{[+2 - (-3)]X^2}{200 \times 160} = 0.000156X^2$$

Start of curve chainage = 2360.000 − 80.000 = 2280.000 m.

End of curve chainage = 2360.000 + 80.000 = 2440.000 m.

Table 10.9 is prepared. The second differences check the reliability of the offsets.

Curve highest or lowest point

The highest point on a crest curve where the grade changes from positive to negative and the lowest point on a sag curve where the grade changes from negative to positive can be found. The location of these points is frequently necessary in road drainage design.

The grade at the highest (or lowest) point is clearly zero; and as the rate of change of grade is constant, the distance of this point is:

$$\frac{g_1 L}{(g_1 - g_2)}$$

from the start of the curve. Thus in the example previously worked:

Table 10.9

		Grade levels			Difference checks		
Chainage	Distance	Entry	Exit	Offsets	First	Second	Curve levels
2280.000		24.760		0.000			24.760
					0.063		
2300.000	20.000	25.160		0.063		0.124	25.097
					0.187		
2320.000	40.000	25.560		0.250		0.126	25.310
					0.313		
2340.000	60.000	25.960		0.563		0.124	25.397
					0.437		
2360.000	80.000	26.360		1.000		0.126	25.360
					0.563		
2380.000	100.000	26.760		1.563		0.124	25.197
					0.687		
2400.000	120.000	27.160		2.250		0.126	24.910
					0.813		
2420.000	140.000	27.560		3.063		0.124	24.497
					0.937		
2440.000	160.000	27.960	23.960	4.000			23.960

$$\text{Distance of highest point} = \frac{(2 \times 160)}{[+2 - (-3)]} = 320/5 = 64.000\,\text{m}$$

$$\text{Chainage} = 2280.000 + 64.000 = 2344.000\,\text{m}$$

$$\begin{aligned}\text{Curve level} &= 24.760 + 2 \times 64.000/100 \\ &\quad - 5 \times 64.000^2/(200 \times 160.000) \\ &= 25.400\,\text{m AOD}\end{aligned}$$

Comfort criterion – vertical curve radius

On most vertical curves, there is little difference between the parabola designed and a circular arc running between the same start and end points. It may sometimes be useful to calculate the radius of this curve, either to allow it to be plotted (using a railway curve) or to determine the vertical acceleration experienced by a vehicle on the curve.

The grade angle $= (g_1 - g_2)/100$ radians. Then, with negligible error, $L = R(g_1 - g_2)/100$.

$$\text{i.e. } R = \frac{100L}{g_1 - g_2}$$

Vertical acceleration at speed V

$$= V^2/R$$

$$= \frac{V^2(g_1 - g_2)}{100L}$$

With the limitation of $0.3\,\text{m/s}^2$ in the UK:

$$0.3 = \frac{V^2(g_1 - g_2)}{(3.6)^2\,100L}$$

Hence:

$$L = \frac{V^2(g_1 - g_2)}{389}$$

with V in km/h and L in m.

Crest sight distances

For overtaking

On single carriageway roads where overtaking is to be permitted, adequate sight distance must be provided. This distance S is measured between points at a height of h above the surface. The distance may be shorter than, longer than or equal to the curve length; consider the case where it is shorter.

Referring to Fig. 10.16:

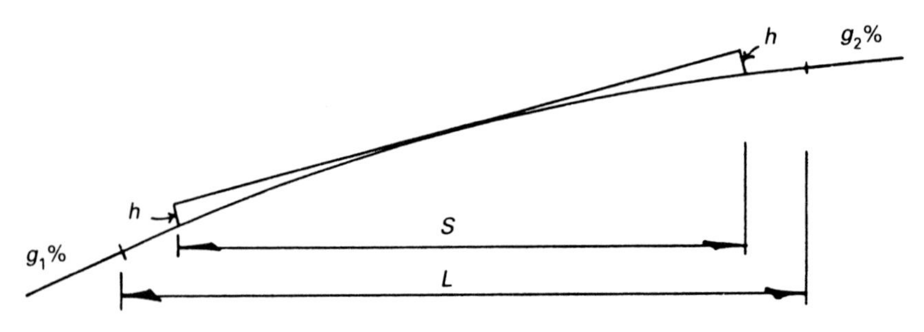

Fig. 10.16 Crest sight distance.

$$h = \frac{(g_1 - g_2)(S/2)^2}{200L}, \text{ thus } L = \frac{(g_1 - g_2)S^2}{800h}$$

Where $S > L$, a different formula applies; for high speed roads S is large and g_1 and g_2 are small, resulting in only a small difference for the value of L calculated by either formula. For low speeds passenger comfort is the ruling criterion. The given formula may therefore be used regardless of the length of S relative to L.

In the UK, h is taken as 1.050 m, representing the height of a sports car driver's eyes above the road and:

$$L = \frac{(g_1 - g_2)S^2}{840}$$

(corresponding to overtaking sight distance values in TD 9/93).

For stopping

The driver's range of vision (to see an object of negligible height on the surface) is $S/2$. Where a sight line is to run between points of different height above the surface, it may be assumed that at its limit it is tangential to the mid-point of the curve.

Then $S = S_1/2 + S_2/2$ where

$$S_1^2 = \frac{800h_1L}{g_1 - g_2} \text{ and } S_2^2 = \frac{800h_2L}{g_1 - g_2}$$

Then $$S^2 = \frac{200L}{g_1 - g_2}(h_1^{1/2} + h_2^{1/2})^2$$

and $$L = \frac{(g_1 - g_2)S^2}{200[h_1 + h_2 + 2(h_1h_2)^{1/2}]}$$

In the UK stopping sight distances (for dual carriageways) are calculated between driver's eye height (1.050 m) and an object 0.260 m high. Hence:

$$L = \frac{(g_1 - g_2)S^2}{471}$$

(corresponding to 'Vertical Curvature' K values)

The values of S are determined from the road design speeds in accordance with Table 3 in TD 9/93. K values are calculated to satisfy comfort and sight distance criteria.

TD 9/93 specifies envelopes of visibility with an upper limit between points 2.000 m high at driver(s)/object (Fig. 10.17(a) and (b)).

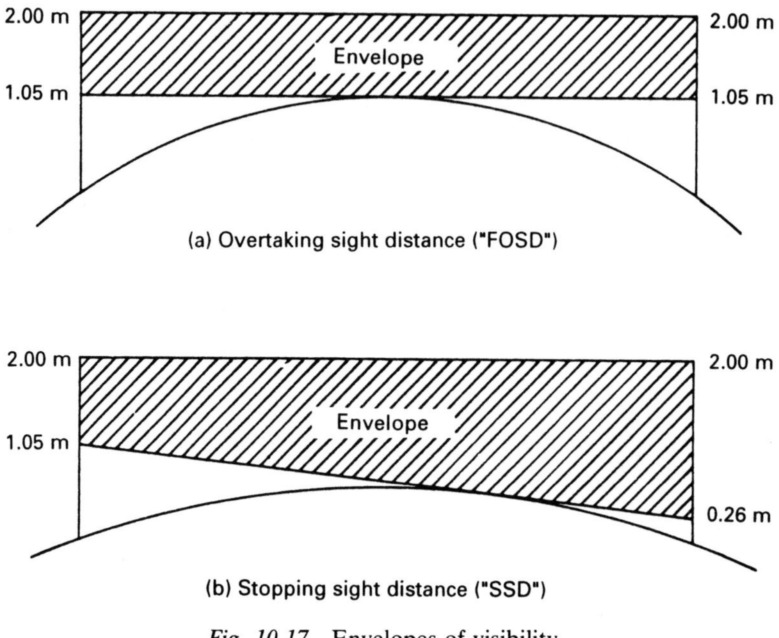

Fig. 10.17 Envelopes of visibility.

Sag headlight distances

On sag curves attention must be paid to the distance that vehicle headlights are projected forwards so that adequate illumination is provided at night. The worst case occurs with the vehicle at the start of the curve. Take the upper limit of the headlight beam of a vehicle on a level surface to have an elevation angle of $\alpha°$ and the height of the headlight above the road to be j and, as with crest curves, consider the sight distance to be less than the curve length, as in Fig. 10.18.

Where the beam 'runs out':

$$j + S\tan\alpha° = \frac{(g_2 - g_1)S^2}{200L}$$

$$\text{Therefore } L = \frac{(g_2 - g_1)S^2}{200(j + S\tan\alpha°)}$$

In the UK, j is taken as 0.600 m and α as 1°.

$$\text{Then } L = \frac{(g_2 - g_1)S^2}{120 + 3.49S}$$

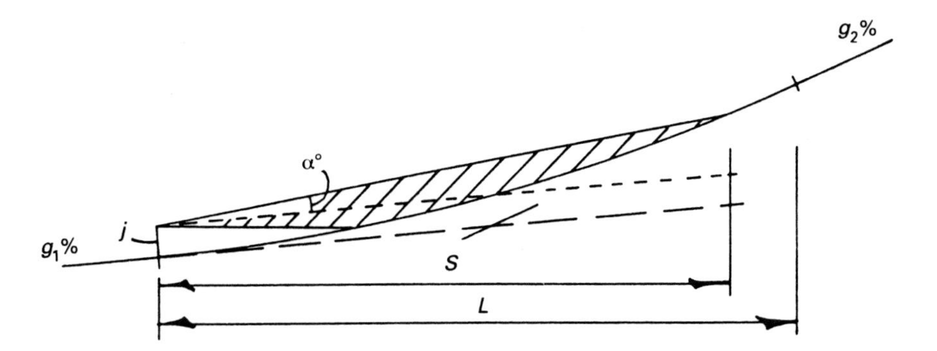

Fig. 10.18 Headlight distance.

Table 10.8 gives K values for sag curves in 'Vertical Curvature'. Because of the variation in headlight strength and alignment, the headlight formula is used for design speeds not exceeding 70 km/h; at higher speeds the comfort criterion:

$$L = \frac{(g_2 - g_1)V^2}{389}$$

although giving lower K values, is used.

10.12 TO PASS A VERTICAL CURVE THROUGH A GIVEN POINT

It is often necessary to pass a curve through a given point either at a road intersection or to provide clearance at a bridge. The usual problem is to determine the curve length if the entry and exit grades are fixed. Two cases occur.

Chainage and reduced level of curve start known

Using the known chainage and reduced level of the given point, calculate the distance and offset of the curve at the given point. Substitute these values in the curve equation:

$$Y = \frac{(g_1 - g_2)X^2}{200L}$$

and solve for L.

Chainage and reduced level of intersection point known

In this case, the distance from the curve start to the given point is not known. Suppose that the intersection point is a distance b beyond the given point. Calculate the curve offset at the given point.

Again, substitute in the curve equation, writing $(L/2 - b)$ for the X value. Solve the resulting quadratic equation for L, taking the larger value.

Example 10.5

As part of a road project, a crest curve is to run between a 2% upgrade and a 3.5% downgrade. The intersection point of the grades (projected) is at chainage 1786.000 m with a reduced level of 53.450 m AOD. At chainage 1804.000 m the road crosses a canal where a clearance of 3.500 m above the water level of 44.600 m AOD has to be provided. The combined construction thickness of bridge beams and road pavement is 1.760 m. Calculate the longest possible length of the curve.

Solution

Figure 10.19 refers.

Road level at the canal

$= 44.600 + 3.500 + 1.760$

$= 49.860\,\text{m AOD}$

Entry grade level at the canal

$= 53.450 + 2(1804 - 1786)/100$

$= 53.810$

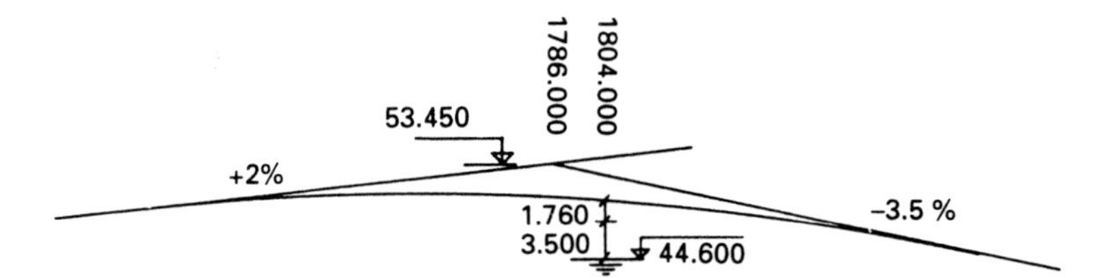

Fig. 10.19 Passing curve through a given point.

Offset at canal:

$$= 53.810 - 49.860 = 3.950\,\text{m}$$

$$\therefore 3.950 = \frac{[2 - (-3.5)](L/2 + 18.000)^2}{200L}$$

$$\therefore L^2 - 502.55L + 1296 = 0$$

Solving, and taking larger value, $L = 499.958$ m. (This would probably be rounded up to 500.000 m.)

10.13 COORDINATED DESIGN

Major roads should be designed so that horizontal and vertical curves are imposed simultaneously; to ensure that both types of curve start and end together, it is often necessary to lengthen one of the curves (more commonly the vertical one) to achieve this coordination.

Much road alignment design is now performed by computer, programmed to carry out a coordinated design.

10.14 FURTHER READING

TD 9/93 (1993) *Highway Link Design Standard*. Department of Transport.

County Surveyors' Society (1969) *Highway Transition Curve Tables (Metric)*. The Carriers Publishing Company Ltd.

10.15 EXERCISES

Exercise 10.1

A right-hand circular curve is to be designed and set out to connect two straights which intersect with a deflection angle of 43°30′00″ so that the apex distance (mid-point of curve to intersection point) is 39.09 m.

Show that the curve radius is 510.000 m and prepare a table of data for setting out the curve using a tape and a theodolite set up at the initial tangent point. The intersection point is at forward chainage 2546.35 m and pegs are required at 20 m intervals of through chainage.

After the peg at chainage 2660.00 m has been set out, the tape is accidentally broken and the remaining points have to be set out using two theodolites. The second theodolite is set up at the final tangent point with the horizontal circle set to read zero when sighting the IP.

Calculate the angles for the second theodolite to set out the remaining points on the curve.

Exercise 10.2

A horizontal circular curve is to start at station A, pass through station B and finish at station C. The chainage at A is 3571.082 m and station coordinates are as follows:

A (1000.000 m E; 1000.000 m N)
B (1134.832 m E; 1033.784 m N)
C (1433.173 m E; 1192.861 m N)

(a) Calculate the curve radius.
(b) A theodolite is set up at A and the horizontal circle is zeroed sighting C. Calculate the angle and distance to set out the intersection point.
(c) Prepare data for setting out the curve at 50 m intervals of through chainage with a tape and the theodolite at A now zeroed on the IP.

Exercise 10.3

The left-hand 11.000 m wide carriageway of a motorway on normal crossfall (2.5% to nearside kerb) is to deflect to the right through 41°00′00″ by means of a 1500 m radius circular curve with cubic spiral transitions at each end applied to the right-hand kerb. The rate of increase of radial acceleration is to be $0.3\,\text{m/s}^3$, the design speed is 120 km/h and the forward chainage of the IP is to be 3396.003 m. Calculate:

(a) A suitable length of transition.
(b) The length of circular curve.
(c) The superelevation to balance 45% of the centrifugal forces at 120 km/h.
(d) The difference in kerb gradients along the transition to apply the superelevation.

(e) Data to set out the first eleven points on the curve at 20 m intervals of through chainage, using a steel tape and a theodolite set up at the straight/transition junction for the transition curve, and at the transition/circular junction for the circular curve.

Exercise 10.4

An existing road has a vertical parabolic crest curve of 160.000 m length joining an upgrade of 4.5% to an upgrade of 0.75%. The curve is to be replaced by a longer parabola such that the mid-point of the curve is lowered by 0.375 m.

(a) Prove that the length of the new curve will be 240.000 m.
(b) Calculate curve levels at 20 m intervals along the new curve, if the level at the start of the original curve was 63.150 m AOD.
(c) Determine, for the worst case, the maximum distance at which a motorist, with eyes 1.05 m above the road surface, can just see an object 0.26 m high on the surface.

Exercise 10.5

A 3.5% rising grade is to be connected by a crest parabolic curve to 0.5% rising grade. The grades, if projected, would intersect at chainage 2130.000 m and level 158.750 m AOD. At chainage 2154.000 m an overbridge is being constructed, beam soffit level being 165.007 m AOD. Calculate:

(a) Minimum length of curve to give 6.500 m clearance at the overbridge.
(b) Curve levels at 20 m intervals of through chainage.
(c) The range of vision along the curve for a motorist with eyes 1.05 m above the surface.

Comment on the suitability of the curve for a dual carriageway with a design speed of 100 km/h.

11 Roadworks II – Setting Out

11.1 INITIAL SURVEY

Major roads

During the feasibility and planning stages of a road, the horizontal alignment is determined by a process of trial plotting on plans usually prepared from aerial photographs. At the same time the vertical alignment must be considered and longitudinal sections are prepared.

Calculations for both alignments are performed using known coordinates and ground levels until the approximate alignment of a preferred route is established. In the United Kingdom, design should be in accordance with Department of Transport Standard TD 9/93. At this stage the line is set out on the ground and modifications made until a final satisfactory route is achieved.

Key alignment stations (*intersection points* and *tangent points*) are likely to be tied in to the National Grid coordinate system, by taking measurements to, or from, points previously coordinated. Details of triangulation points on the National Grid are available from the Director General of the Ordnance Survey, Southampton.

The centreline, or kerbline, is set out along the full length of road, and a final longitudinal section and cross-sections (at 20 m or other convenient intervals) measured and plotted, with the construction alignment superimposed.

Minor roads

While there should be no relaxation of the standards for road design, a minor road is likely to be referenced to adjacent major roads, building lines or industrial structures according to its purpose.

11.2 INITIAL SETTING OUT

At the commencement of roadworks, a convention for the type of marker and colour-coding should be established. It should be the engineer's endeavour that a clear distinction can be made between:

- centreline pegs
- offset pegs
- reference pegs
- curve intersection points
- tangent points
- bridge intersection points
- manhole positions

and between:

- road profiles for different layers or grades
- drainage profiles.

A plentiful supply of differently coloured paints is necessary. The convention should be issued to all engineers and to all foremen.

On occupying a site, the contractor's engineer should 'walk the line' of all roadworks, and carry out a proving survey. This should entail relocating the key alignment stations; use should be made of survey information documented at the design and initial survey stage. Checks should be made of all intersection angles, tangent lengths and straight lengths, and any discrepancies brought to the notice of the Resident Engineer/Architect, so that amendments to the alignment can be made promptly.

At this stage, more immediate referencing of intersection and tangent points should be carried out. Such work will involve placing pegs so that displaced IPs and TPs can be relocated swiftly.

Such reference stations must be established so that they:

(1) Are clear of all works.
(2) Are easily accessible.
(3) Can provide a simple and accurate method of relocating key points.

A satisfactory method is to establish four stations so that the diagonals of the quadrilateral formed intersect at the required point, as shown in Fig. 11.1.

A disadvantage is that two theodolites are needed to relocate the

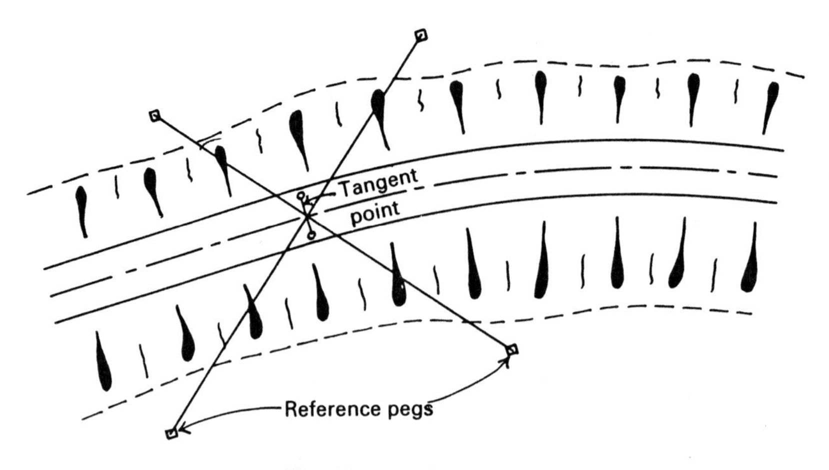

Fig. 11.1 Referencing.

key point; the method is still often preferable to location by distances or angles, where values require precisely recording and setting out.

Curve setting out

Although it may be convenient to set out the tangent points using EDM, the traditional theodolite and tape may be preferred to locate points on a curve, whether circular or spiral. (See Chapter 4 for notes on EDM use.) The usual method of setting out a curve is to knock wooden pegs into the ground, a nail being inserted in the top of each peg to give precise positioning.

Two chainmen are required for short chord taping. In the event (a frequent one!) of a second chainman not being available, the tape can be looped on the nail of the preceding peg, *provided* that:

(1) The peg is not disturbed.
(2) Cumulative errors are not introduced.

Alternatively, the engineer will have the opportunity to get plenty of exercise proceeding from theodolite to pegs and back to hold the tape and set the angle.

To save time and to minimize error accumulation, the curve, unless very short, should be set out from each tangent point towards the middle. In the event of a large misclosure, only half the curve will require repeating (assuming that the errors are all in one half!), and for a small misclosure (a likely occurrence on a long curve) adjustments can be made of decreasing magnitude to the stations proceeding in both directions from the middle. Care should be taken over slope correction, calculating to four places of decimals (of a metre) to prevent cumulative errors in arithmetic and measurement from affecting the chainages.

At this stage of the work, the boundaries of the construction should be defined, and any temporary fencing set out. Boundary and fence location can be performed by taping from the setting out line.

11.3 EARTHWORKS

On a major road project, earthworks can form a significant part of the total cost. There is no permanent construction to guide the earthwork location, and the landscape can change dramatically in a few days. The site engineer is required to set out quickly and accurately (within certain tolerances) and must be prepared to replace damaged or lost markers at a moment's notice. Imagination and quick-wittedness are needed, and the engineer must maintain vigilance; the cost of remedying wrongly aligned earthworks may be considerable and any blame will be laid at the engineer's door, regardless of who is at fault.

Topsoil strip

The limits of topsoil strip should be pegged out at the start of a project. The soil may be required for replacing on side slopes in cuttings or on embankments, and a stockpile on site can be formed.

Usually dimensions for stripping and piling soil can be scaled from drawings, and setting out done by tape (or EDM) and optical square (or theodolite) as appropriate. A measurement of the volume of the topsoil may well be required.

Cuttings and embankments

High-speed roads run in cuttings or on embankments for considerable distances. The top of cutting, or toe of embankment, must be located, the angle of slope indicated, and the bottom of cutting, or top of embankment, positioned.

The slope angle is indicated by *batter rails* (or *rules*), these being sloping boards on two stakes set both sides of the road every 10 or 20 metres or at a convenient distance. For a cutting, the rails are set flush with the slope for boning it directly, whereas for embankments they are installed above the slope, with a traveller (boning rod) being used. This prevents disappearance of the batter rails when overfilling occurs. Figure 11.2 shows sections in cut, fill and part cut/part fill.

For cuttings, excavation can start as soon as the batter rails have been positioned. Frequently, however, before embankment filling commences, pegs (slope stakes) are placed to indicate the limits of deposition. Only when a sufficient height of fill has been placed, compacted and is ready for trimming to shape, is it worthwhile setting up the batter rails; if set out earlier they stand a considerable risk of disturbance.

The process of setting out batter rails is usually a trial and error affair, their exact position depending on the existing ground profile.

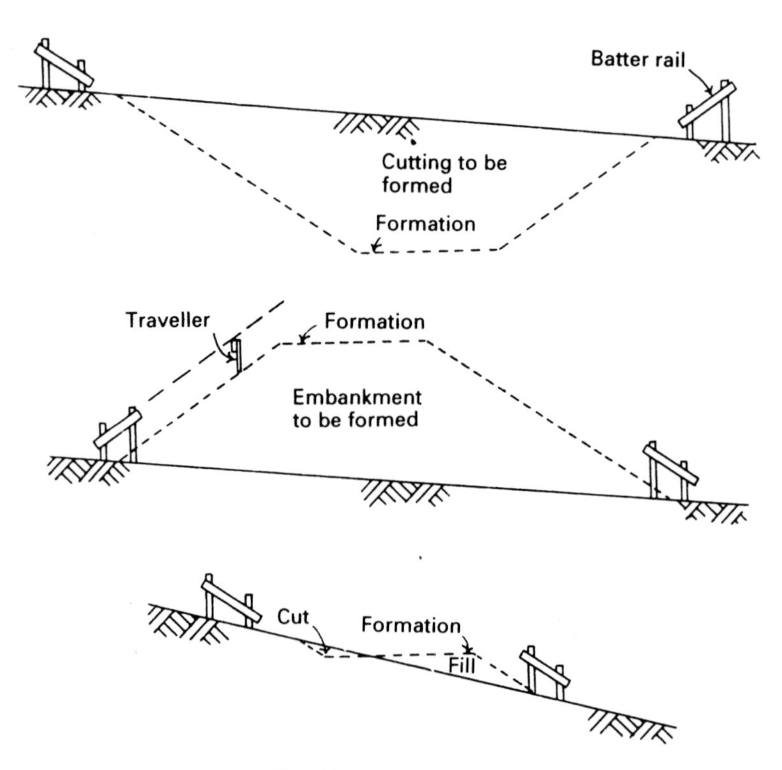

Fig. 11.2 Batter rails.

Example 11.1: setting out batter rails for a cutting

There is no set procedure for establishing batter rails; all methods are variations on the following process.

At each cross-section (as in Fig. 11.3), for one side of a cutting, the engineer records from the cross-section drawing:

- the cross-section (chainage) value
- the offset of the cutting toe
- the reduced level at the cutting toe
- the side slope value
- the offset of the cutting top
- the depth of any topsoil subsequently to be placed

The theoretical ground levels at the cutting top should be calculated (Fig. 11.4(a)). In the field, the engineer will then measure from the centreline peg (assuming that it has not been removed!) the theoretical distance to the cutting top on each side, and take ground levels at these positions (Fig. 11.4(b)). Batter rails must be set back from the cutting top; for the rail to clear the ground by about 300 mm, the front stake should be set back from the edge by 600 mm for a 1 in 2 slope (Fig. 11.4(c)).

It can be seen that the ground is higher on the left, and lower on the right than as indicated on the cross-section (Fig. 11.4(d)). It will be necessary to move away from the centreline on the left, and towards the centreline on the right to position the batter rails.

The new theoretical positions of the front posts should now be calculated as in Fig. 11.5(a).

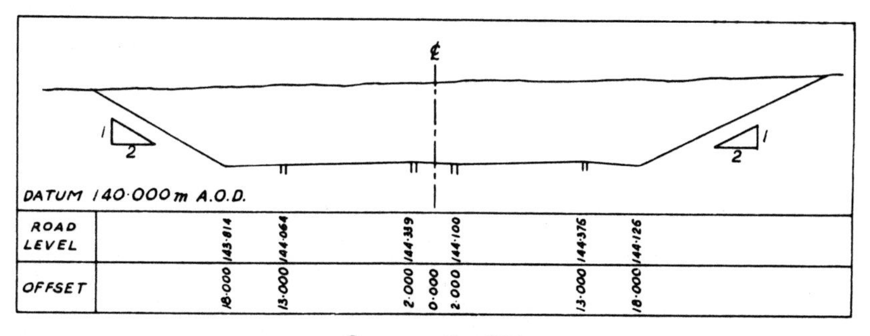

Cross-section 320

Fig. 11.3 Road cross-section.

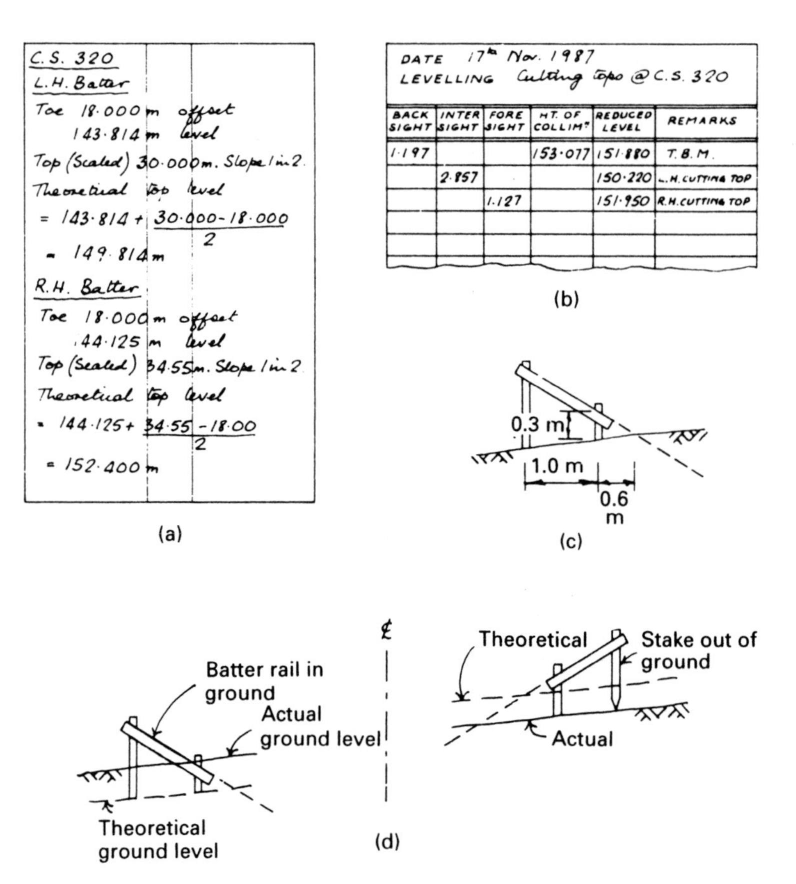

Fig. 11.4 Survey for batter rails.

These distances should be set out, and fresh levels taken on the ground (Fig. 11.5(b)); if these levels are approximately 300 mm below the slope levels, the front stakes should be knocked into the ground and levels taken on their tops. (If a reliable chainman is available, the required staff readings can be calculated, and the slope level marked directly on the stake.)

By taping down the stake, the engineer can mark the rail position (Fig. 11.5(c)). A second stake is then knocked into the ground; the rail level can be marked on it by transferring the level from the first post. This can be done using a spirit-level, a tape and simple proportion (Fig. 11.5(d)). Alternatively, a wooden template of the slope can made up and used with a spirit-level.

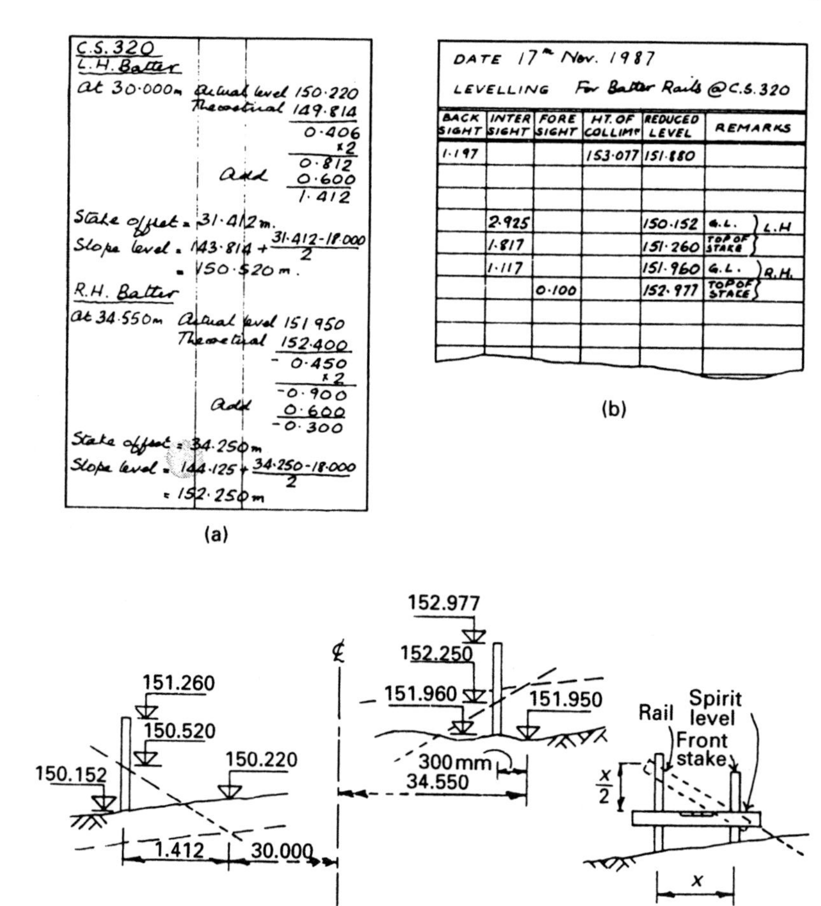

BACK SIGHT	INTER SIGHT	FORE SIGHT	HT. OF COLLIMn	REDUCED LEVEL	REMARKS
1.197			153.077	151.880	
	2.925			150.152	G.L. } L.H
	1.817			151.260	TOP OF STAKE }
	1.117			151.960	G.L. } R.H.
		0.100		152.977	TOP OF STAKE }

Fig. 11.5 Setting out batter rails.

Embankments

A similar trial and error process is employed to set pegs at the foot of an embankment. If the batter rails are not to be set until after an amount of material has been placed, it is worthwhile placing offset pegs at each cross-section clear of the work. This will save relocating centre pegs while filling is proceeding.

The rails should be set to indicate a plane 1.000 m (or other convenient height) above the surface to be formed and a traveller of the same height used.

Deep cuttings/high embankments

For a depth or height greater than 5 or 6 metres, further batter rails should be set out at this level to maintain accuracy. In cuttings, this may require that they are set (1.000 m) above the surface, and a traveller used.

Where topsoil is to be placed and the same batter rails are to be used, the rails should initially be set an amount higher to correspond with the topsoil thickness (say 0.150 m) and a traveller used in cuttings as well as for embankments.

Bottom of cutting/top of embankment

When formation level is approached, the engineer must act quickly to set out sight rails (road profiles). These should be set to allow boning in of the formation, and placed at a fixed offset from the formation edge, so that the toe of the cutting or top of the embankment can be defined and formed to the required accuracy. The chainage and traveller length should be marked on each profile (Fig. 11.6).

11.4 FORMATION AND SUB-BASE

The formation is the top of the sub-soil or the underside of the pavement of a road, i.e. the bottom of a cutting, or the top of placed material which has been excavated, usually from another part of the contract.

On large roadworks, the formation is often left high (between 150 mm and 300 mm). The purpose is to protect the surface from damage by weather and by construction traffic until the contractor is ready to commence pavement construction. The profiles already set out for finishing cuttings and embankments should be used for this second cut or fill.

Road profiles

Road profiles are sight rails set out to control formation level (first and second cut/fill) and stages of the pavement construction.

(*Note:* The term 'pavement' refers to the placing of imported

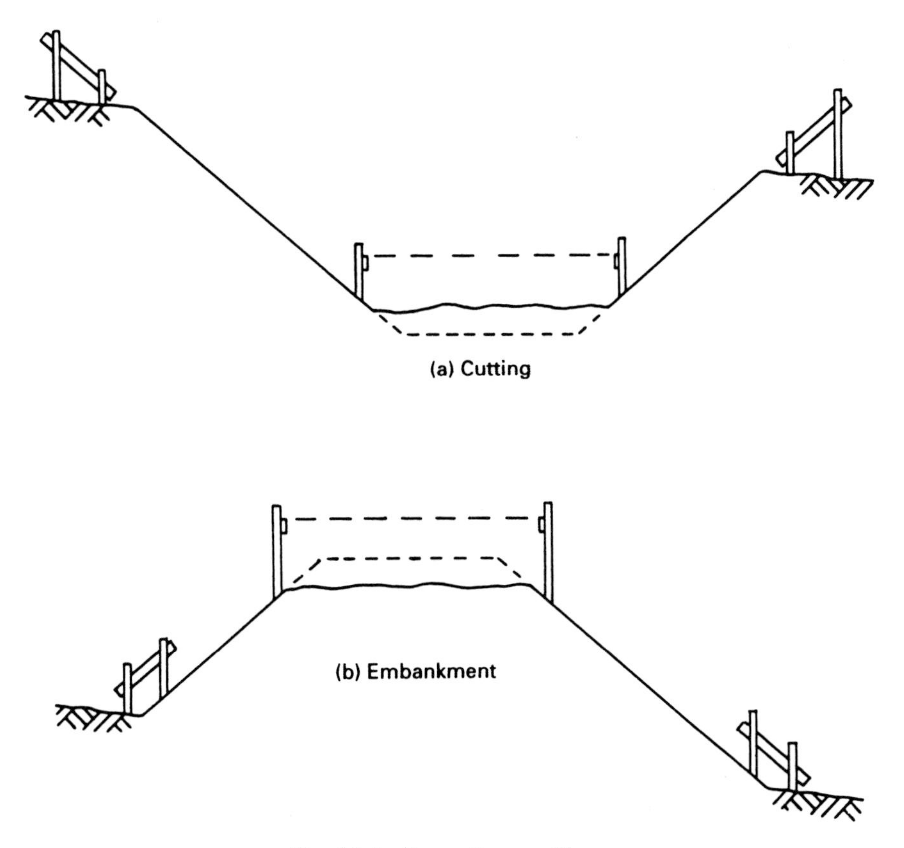

Fig. 11.6 Formation profiles.

material to form a satisfactory surface for vehicular or pedestrian use, for example in construction of airport runways, motorways and shopping walkways. It should not be confused with the layman's term for a footpath in the United Kingdom.)

By being set at fixed distances back from the edges of the road, the profiles define the edge of formation, and the limits of placing pavement material. It is customary to set the cross-heads (on a single stake) parallel to the carriageway line. On major roads (more than two lanes, often dual carriageway) construction is likely to comprise:

- sub-base (selected granular material)
- base (granular material with a bitumen or cement binder)
- surfacing (2 or 3 layers of bitumen bound material)

for a flexible pavement. A rigid pavement will consist of concrete surfacing overlying the sub-base.

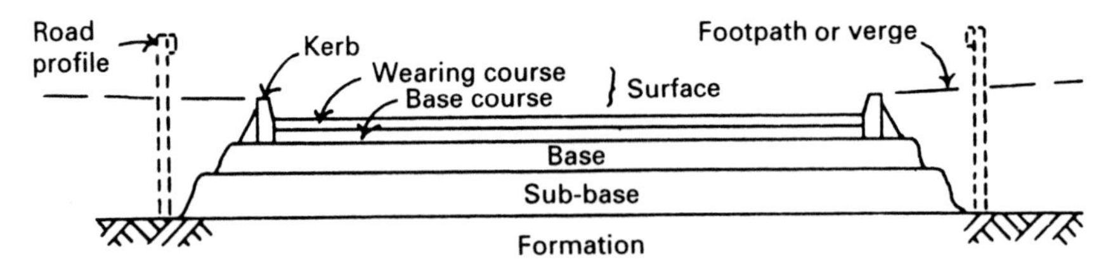

Fig. 11.7 Road construction.

The road profiles can be used for boning in the sub-base and sometimes the base; it is unlikely that boning in will be sufficiently accurate for surface levelling. Figure 11.7 shows a typical road cross-section.

Road profiles are most commonly set 1.000 m above finished surface grade, with the traveller length being adjusted at the various stages. The engineer should supply the foreman with traveller details and should frequently check that these are being complied with. On the profile itself should be written the chainage and the height of the cross-head above the finished road level.

Occasionally it may be decided to use a constant traveller length; this will necessitate several cross-heads on one stake, and it will be the engineer's duty to ensure that the correct cross-heads are being used at every stage. A system of colour coding is essential.

Minor roads will generally have fewer pavement layers (sub-base +2 surfacing courses) than major roads, with narrower carriageways. The road profiles can be used for all stages of construction; to help the kerblayer, the cross-heads may be set perpendicular to the carriageway line. Note that, although boning in may be satisfactory for levelling the kerbs, the stakes will not be precisely enough set out for their plan location.

For a dual carriageway, four profiles at each cross-section will be required, those between offside kerblines being set in the central reservation. The stakes may be set at their correct chainage by using an optical square on straight sections and by using a taped 'construction' on curved sections (Fig. 11.8(a)). In determining the levels of the profiles, the road crossfall must be taken into account (Fig. 11.8(b)).

Double crossfall

On a straight single carriageway the surface will usually fall from the

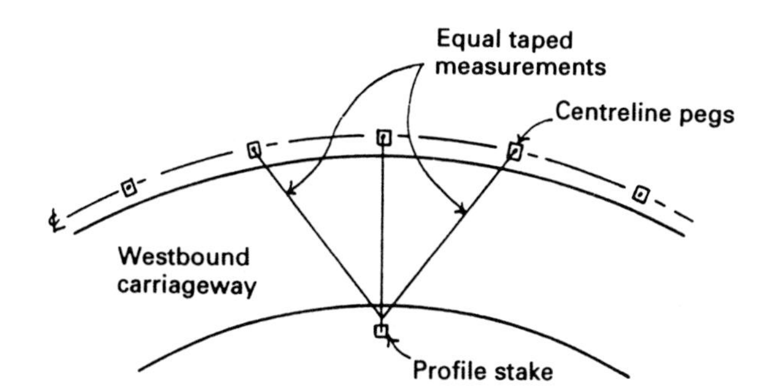

(a) Location plan

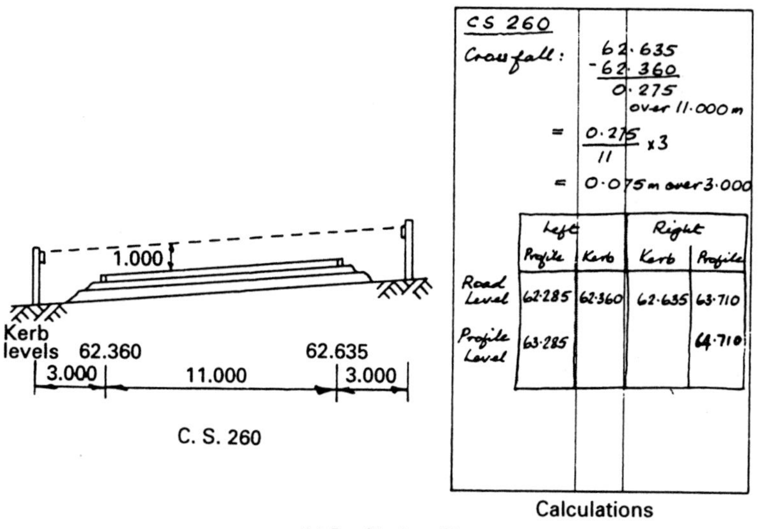

(b) Profile levelling

Fig. 11.8 Road profiling.

centre to the sides, and on dual carriageways there may be double crossfalls where lanes merge or diverge at junctions. There are several methods of profiling such sections.

Profile at change of crossfall

Excavation or filling must be carried out while leaving the central profiles intact (Fig. 11.9(a)); the engineer will have to use all his or her powers to dissuade the machine drivers from prematurely remov-

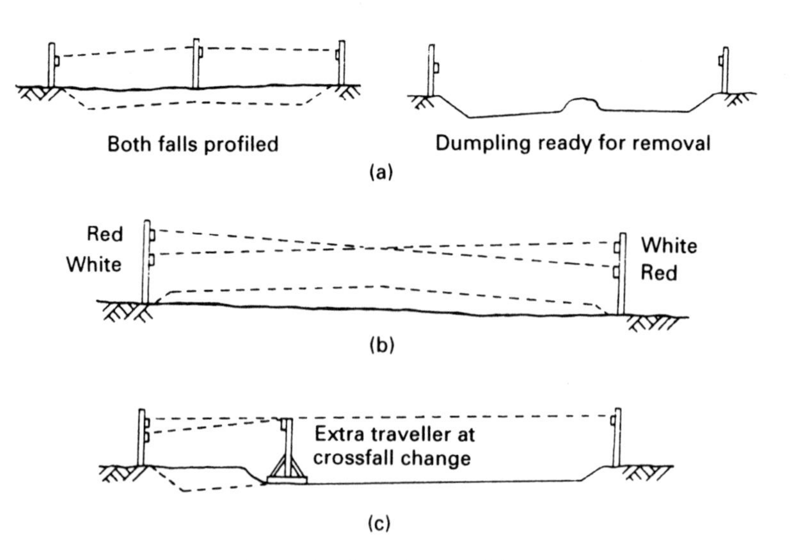

Fig. 11.9 Double crossfalls.

ing these. Only when the two crossfalls have been adequately formed may the centre profiles be taken out, and either the dumpling removed or the trough filled and compacted.

Double cross-heads (1)

Boning is carried out between lower and upper cross-heads which should be colour coded (Fig. 11.9(b)). The engineer will have to guard against misinterpretation by the colour-blind or the witless.

Double cross-heads (2)

This method is suitable where a slip road joins a carriageway. The crossfall of the main carriageway is formed, using the upper (left) cross-head. By setting an additional traveller at the point of crossfall change (dimension given to foreman), the sliproad crossfall is then formed (Fig. 11.9(c)).

11.5 ROAD BASE

The level and position of the base is usually defined by piano wire run between steel pins, or by roadforms secured to the pins.

The pins are set at intervals (20 m; 10 m for sharp curves) either side of the carriageway. Measurements are taped from the centreline, kerbline or offset line, and an optical square (or geometric construction) is used to keep the pins at the correct chainage.

If piano wire is to be tied to the pins, levels must be taken on the pins, and the wire level marked by taping down and pencilling a line over a paint or wax crayon mark, or by making a hacksaw cut, or by fixing coloured adhesive tape.

Where roadforms (screeds) are to be fixed to the pins, the pins are knocked down until their tops are at the required level, the engineer having calculated the necessary staff readings. Alternatively, the roadforms can be levelled directly, the engineer arranging for packings to be placed under them until the required staff readings are observed.

Gully connections

Before kerbs are laid, pipe junctions must be located, and gully pots must be placed and connected to the drains. The engineer must mark their positions and levels, so that raising pieces and gratings can be located to suit the kerb.

11.6 KERBS

The most common form of kerbing is precast concrete, although in situ concrete has been used, and channels can be formed by extruding the asphalt surfacing.

Steel pins should be placed to indicate the line and level of the kerbs. The pins are placed a convenient distance (1.000 m or 0.500 m) outside the kerbline and, after levelling, are knocked down so that the tops indicate the kerb levels. Alternatively, the pins can be left high and a mark (wax crayon or adhesive tape) made on them. They are usually set at 10 m chainage intervals, care being taken to mark out the chainage accurately.

An alternative method can be employed where the road base provides a surface on which marks can be scribed to position the kerbline. Levels are taken at these marks, and the height of kerb above each mark is calculated and written by the mark, or added to a list which is given to the kerblayer.

11.7 SURFACING

The two activities concerning the engineer over surfacing are dipping and setting double crossfalls.

Dipping

A check has to be made that the specified depth of surfacing can be achieved by measuring at several positions at each cross-section. This process of dipping is performed by stretching a line between opposite kerbs at every cross-section (10 m or 20 m) and taping the depth to the road base. The depths should be recorded for passing to the Resident Engineer.

Double crossfall

Where a double crossfall has to be formed between kerbs, the engineer must set a row of steel pins to indicate the position and level at the change of crossfall.

11.8 OTHER WORKS

After the surfacing, there will be a number of other items to be set out, for example:

- road markings/catseyes
- permanent fencing
- footpaths
- topsoil placing
- lighting columns
- signposts and gantries

These can usually be set out by taping from the kerbs.

12 Drains and Pipelines

12.1 INTRODUCTION

The majority of construction projects will require the installation of drains. Surface water must be removed, and in some cases there will be foul sewage to convey to a treatment works. Drain laying, being below ground, is an early item in a construction programme; there are also obvious advantages in being able to drain a site during construction and in providing running drains from the site cabins.

Pipelines may form whole projects, typically designed to carry water, gas, oil and pulverized fuel ash over considerable distances.

Commonly, drain and pipe laying will be carried out in 'open cut', a trench being excavated, pipes laid and jointed and the trench backfilled. The engineer's tasks will be to set out the positions of the drains or pipes, to indicate the trench depths and to ensure that the pipes are laid at the correct levels and to the correct gradients.

There will be other items to set out: gullies, manholes, outfalls, air valves and wash-out chambers. Occasionally pipes may have to be laid in a tunnel (Chapter 14) or by thrust boring or pipe jacking. There will be exposed pipework at chemical and refinery works, at sewage works and at pipe bridges and siphons.

12.2 POSITION IN PLAN

Drains

Drains are designed to run in straight lines and at constant gradients between manholes. If the manholes are set out in plan, the drain positions are fixed. The centre of each manhole should be pegged; dimensions will be given on a drawing, usually relating the manhole to part of the works above ground. Pegging to within 25 mm of correct plan position is usually acceptable and the engineer will find that a tape and theodolite, or two tapes, will be adequate for the job. On many projects, the drains will connect into existing sewers, the position of which must be found and checked. Although manholes should be constructed 'on line', there may be some scope for moving the position upstream or downstream a small amount to save cutting a pipe.

This only applies where the drain runs straight through the manhole; where changes of direction or cross-runs occur, there will be little tolerance for any repositioning. Before any manhole position is moved, the site engineer must get permission from the RE/Architect. The revised position must be recorded and the manhole invert level modified so that the drain gradients are not affected.

Intermediate points on a drain can be pegged by lining in with ranging rods or a theodolite. Manhole positions must be referenced, usually by offsetting, as pegs on the line will be lost when excavation starts.

Junction positions (for gully connections) must be marked; the actual position is sometimes varied slightly from the drawing to save cutting a pipe (purpose made junction pipes are used).

Pipelines

Pipelines usually run in straight lines between 'bends', special pipes giving the required deflection. Gentle curvature in plan can sometimes be applied if the pipes are flexible or if the joints can be 'pulled'. Accordingly, the bends should be pegged, and the pipe position is then defined – intermediate positions must be ranged by theodolite as there may be considerable distances between bends. If a bend is moved to accommodate a whole number of pipes, the engineer must check that alignment after the bend is not adversely affected.

Additional pipework running from wash-out (drain off) chambers may have to be set out.

All pegs on the line must be referenced.

12.3 LEVELLING

Drains

Drain levels and gradients are often quite critical despite the cavalier comments of some site operatives ('twill be a dark night that water won't find its own level')! Flow characteristics are sensitive to variations of gradient, and if a gradient changes along a pipe run there is a risk of scouring or silting.

The ideal procedure of working upstream from the outfall cannot always be followed; it is important that intermediate lengths, subsequently to be connected, are set to the correct levels and gradients.

Boning in is the traditional method for excavation, levelling of bedding material and pipelaying. The accumulation of errors in this method is such that it is only just accurate enough, particularly if the profiles are well to the side of the trench and the excavation is deep. Lasers are now being used in preference to boning in for much drainage work.

Pipelines

Pipelines carry pumped material; gradients are therefore not so critical and pipes are often designed to run at a constant depth below ground. Some are designed to specific levels (particularly in developed areas), and gradients must be adhered to so that air valves are at high points and wash-outs at low points. Boning in is still often preferred to the use of lasers.

Boning in pipes

Drain profiles are set to the side of the trench. For robustness and accuracy two uprights are used for each profile with the cross-head set at right angles to the line of the drain. A traveller with its cross-head extended one side is used.

Figure 12.1(a) shows a cross-section through a drain being installed. The following features should be noted.

(1) The nearer upright should be a constant distance from the drain line and greater than the half-width of the excavator.

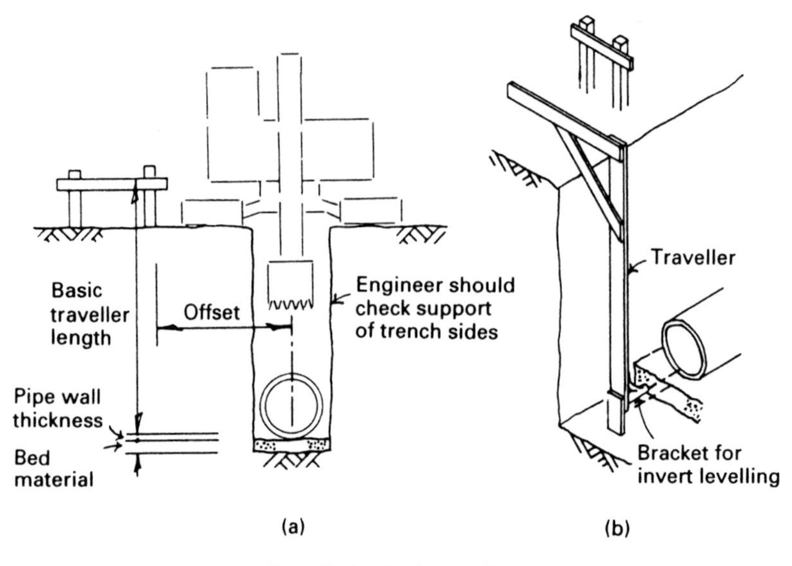

Fig. 12.1 Drain profiling.

(2) On one side of the trench the pipes will be strung out; on the other side the excavated material will be dumped. The engineer may have to seek the foreman's advice over which side to position the profiles.
(3) The basic traveller length is the depth from the cross-head to the pipe invert, usually a multiple of 0.5 m.
(4) The cross-heads should be between 0.8 m and 1.5 m above ground.
(5) Profile spacing should not be greater than 50 m.
(6) From a single set of profiles, excavation, bed and pipes must be levelled. One way of providing a suitable single traveller is to make the length to suit the top of the bedding material, fixing a foot to mark the pipe invert (a shelf bracket can be used) and adding an extension piece for excavation level (Fig. 12.1(b)). Pegs or pins can be boned in to indicate the top of the bed.
(7) Plan alignment is usually achieved by maintaining the excavator on line, keeping its tracks or wheels running along a string line offset from the profile stakes.
(8) The offset distance and traveller length should be marked on the profile (e.g. '3.500 m TO INVERT'). On some sites a distinctive colour for cross-heads is used.

To establish profiles for a drain, the stakes should be knocked in the ground at the required positions. The engineer should take levels on the tops of the stakes and select a traveller length to give a suitable height of cross-head above ground. By adding the traveller length to the pipe invert levels at the profile positions, the engineer can calculate cross-head levels and the distances they should be below (or above) the stake tops. Afterwards the engineer should check:

(1) By eye that the cross-heads appear level.
(2) By eye that the profiles bone in with themselves (there should be three for all but very short drain runs).
(3) By eye that the fall is in the correct direction and of sensible magnitude.
(4) With an instrument that the cross-heads are at the correct levels.

Example 12.1

A drain of 90 m length is to fall at 1 in 150 from manhole 63 (invert level 27.450 m AOD) to manhole 64. Becuase of access problems, the first 40 m downstream from manhole 63 has to be set out and installed ahead of the remainder of the drain. Offset profile stakes have been knocked in the ground opposite each end of the 40 m length.

Table 12.1 shows the engineer's level book with staff readings taken on the ground and on the tops of the stakes. Figure 12.2 is a sketch plan.

Calculate suitable cross-head levels, and find the distances to be measured up or down from the stake tops.

Table 12.1

BS	IS	FS	Reduced level	Remarks
1.585			29.650	TBM.
	1.690			Pos^n 1 Ground level
	0.714			2 Near stake (top)
	0.631			3 Far stake (top)
	1.443			4 Ground level
	0.392			5 Near stake (top)
		0.434		6 Far stake (top)

Solution

Fall over 40 m = 40/150 = 0.267 m

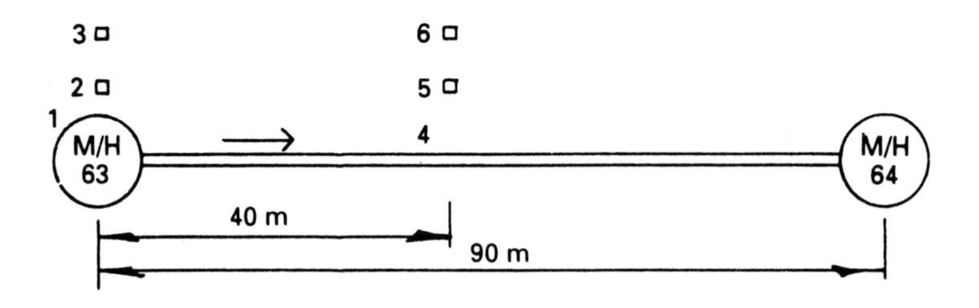

Fig. 12.2 Plan of drain.

Table 12.2

BS	IS	FS	Height of collimation	Reduced level	Remarks
1.585			31.235	29.650	TBM
	1.690			29.545	Pos^n 1 Ground level
	0.714			30.521	2 Near stake
	0.631			30.604	3 Far stake
	1.443			29.792	4 Ground level
	0.392			30.843	5 Near stake
		0.434		30.801	6 Far stake
−0.434		Check		−29.650	
1.151				1.151	

Invert level at 40 m downstream = 27.450 − 0.267 = 27.183 m

The reduced levels are shown in Table 12.2.

At M/H 63, depth to invert = 29.545 − 27.450 = 2.095

40 m downstream, depth to invert = 29.792 − 27.183 = 2.609

A traveller of 3.500 m would set the cross-heads:

3.500 − 2.095 = 1.405 m above ground at M/H 63

3.500 − 2.609 = 0.891 m above ground at 40 m

Therefore with a 3.500 m traveller, cross-head levels will be:

30.950 m AOD at M/H 63

30.683 m AOD at 40 m

Figure 12.3 shows a long section. The setting of the cross-heads should be as follows:

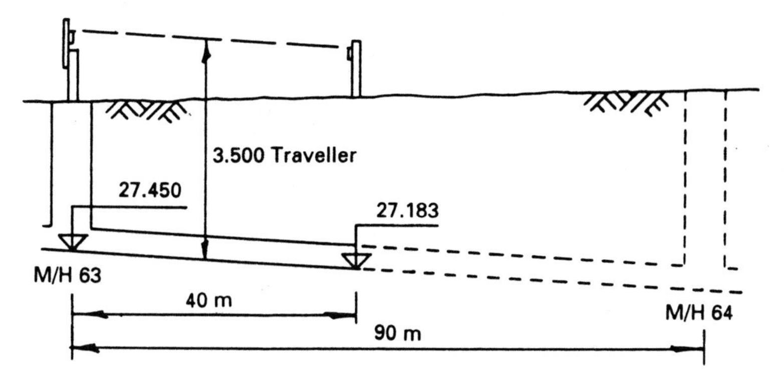

Fig. 12.3 Long section of drain.

Stake	Level	Cross-head	Up or down
2	30.521	30.950	0.429 m UP
3	30.604		0.346 m UP
5	30.843	30.683	0.160 m DOWN
6	30.801		0.118 m DOWN

Sketch

A sketch should be prepared and given to the foreman, the engineer keeping at least one copy. It should show:

- identification and position of manholes
- location of profiles
- profile offset
- traveller length
- pipe material, diameter, class, type of joint
- bed (haunch/surround) details
- any other details (e.g. backdrops)

Laser control

Laser control of pipelaying shares with boning in the establishment of a line parallel to and above the invert of a pipe run.

Laser above the pipe

The laser can be set up at ground level, or in the trench above the pipe. If above ground it can be used to control the whole drainlaying operation; a detector on a pole (corresponding to a traveller) is used to level the trench bottom, bedding and pipes. It may, however, be difficult to provide a sufficiently stable support for the laser above ground.

The engineer must set the laser at a suitable level, determine the 'traveller length' and set the correct grade.

Laser in the pipe

A short length of trench must be excavated, bedding placed and a pipe set at the correct level and grade by the engineer. A laser can then be fixed in the pipe to project a beam through at the required grade (Fig. 12.4). Excavation, bedding and pipelaying can then be controlled visually using a short traveller, or a detector on a short pole.

The engineer should provide the foreman with a sketch, and should make frequent checks that the laser has not been disturbed and that the correct traveller/detector height is being used.

12.4 VARIATIONS IN BONING IN

Double cross-heads

Where the length of the ground necessitates a change in traveller length, or where the pipe diameter changes, it may be necessary to set two cross-heads at different levels on the one profile, as in Fig. 12.5.

The engineer must try to prevent confusion occurring in boning in; colour coding may be employed.

Boning in insufficiently accurate

On occasions boning in may not give adequate accuracy. In deep trenches, an unacceptably large error may occur if the traveller is not held vertically. Where very flat gradients are specified, levels can be quite critical. The engineer will have to determine when boning in, as

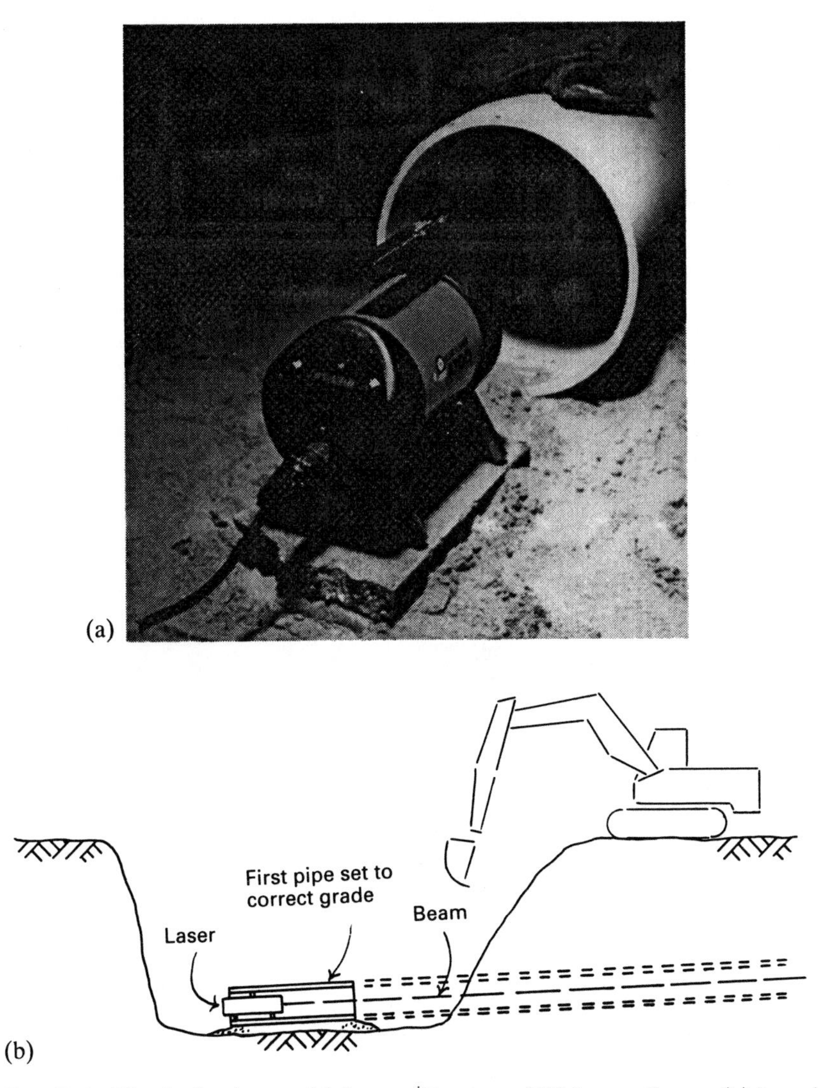

Fig. 12.4 Pipe laying lasers. (a) Laser Alignment 4700 beam aligner; (b) laser in pipe.

described so far, is not suitable. There are several alternative approaches:

(1) A laser, at pipe level, can be used.
(2) Deep excavation is sometimes carried out in two stages; profiles can be set on the 'bench' formed at the first excavation level, as in Fig. 12.6(a).

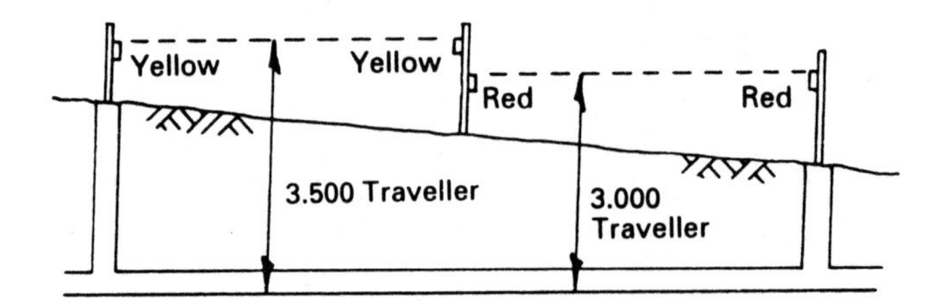

Fig. 12.5 Double cross-heads.

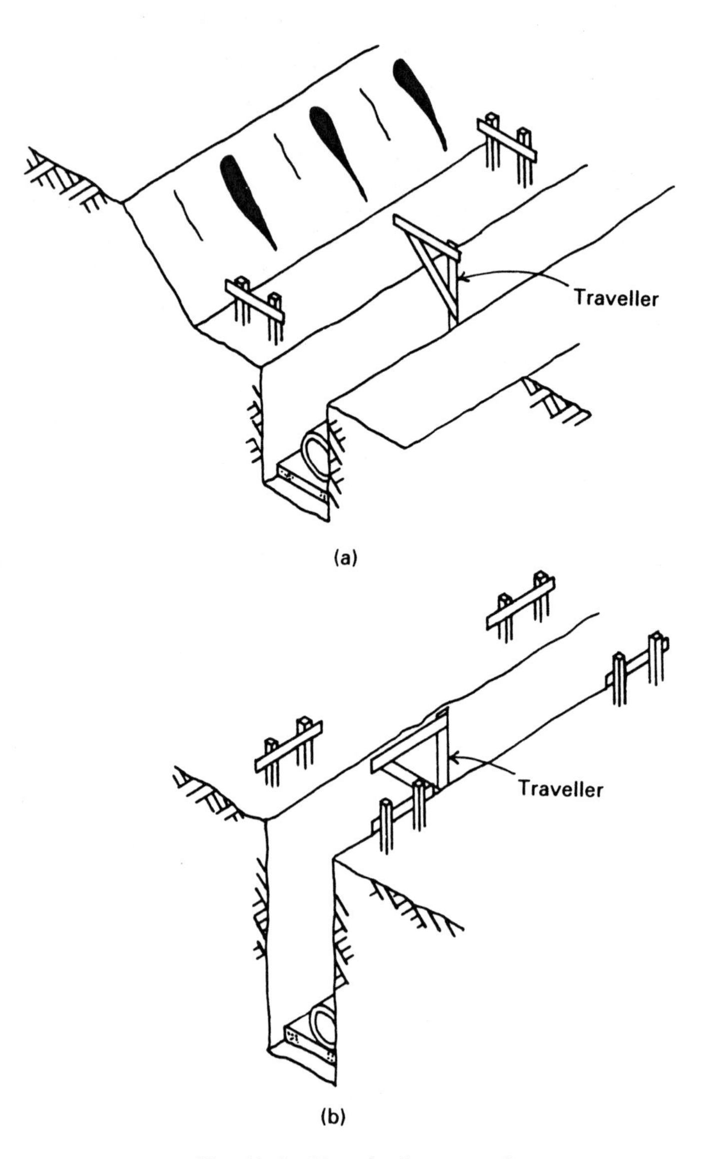

Fig. 12.6 Pipes in deep trenches.

(3) Conventional profiles can be used for excavation, and further profiles then set at the bottom of the trench for levelling, bedding and pipelaying. A length sufficient to allow a number of pipes to be laid would have to be excavated.
(4) Profiles can be set transversely in pairs on opposite sides of the trench (Fig. 12.6(b)). Boning in can then be carried out more accurately, but errors from a non-vertical traveller can still occur.

12.5 MANHOLES, GULLIES, AIR VALVES, WASH-OUTS

Manholes

Manholes may be constructed at the time that drains are laid, or they may be built and furnished later. The engineer should give the foreman details concerning:

- position
- size
- shape
- material
- inlets/outlet
- benching
- backdrops
- steps
- landings
- tapers
- access shaft
- slab and cover

Completion is often near the end of a project so that the cover can be set to the correct level.

Gully connections

The engineer should note the 'as laid' positions of junctions. Gullies are usually installed and connected some time after the drain has been laid and backfilled, and time (and money) can be wasted searching in the wrong place for a gully connection. The positions and levels of gullies have to suit surface features and should be in accordance with the drawings.

Air valves

At high points on pipelines, air valves are installed to enable the pipeline to be bled of air. A pipe with a bleed valve is laid and a small access chamber is built around it. Setting out is straightforward.

Wash-out (drain-off) chambers

Provision has to be made for draining pipelines. A valve at a low point in a dry chamber allows the pipe contents to be drained into a sump. There will be pipework, possibly a pump and a high chamber, to permit disposal. Construction of the chambers is usually in concrete and may be quite deep, possibly involving sheet piling. The setting out should follow the principles in Chapter 13.

13 Foundations, Temporary Works and Structures

13.1 INTRODUCTION

Typically, a construction project will include earthworks, drainage, foundations and structural works. Earthworks are covered in Chapter 11 (Roadworks II) and drainage in Chapter 12. This chapter deals with the setting out of foundations and a variety of structural work. The setting out of temporary works is also covered.

13.2 PILES – LOAD BEARING

General

Where the ground is not sufficiently strong to carry the loads imposed by a structure, a transfer of loads to lower strata capable of withstanding them can be achieved by installing piles. In cohesionless ground, end bearing piles transfer all the load to a lower stratum; they are bored through the ground. In cohesive ground, piles can be driven so that the load is transferred partly to the lowest stratum in end bearing and partly to intermediate strata through skin friction. There are several systems of both bored and driven piles and the engineer will have to adapt setting out techniques to suit the equipment used and the site conditions.

Pile location

Piles are usually located in groups, supporting a pile cap which, in turn, supports a column. Single large diameter piles are also used. A drawing will show the position in plan of all piles.

Before piles are set out, the engineer must be sure that the ground

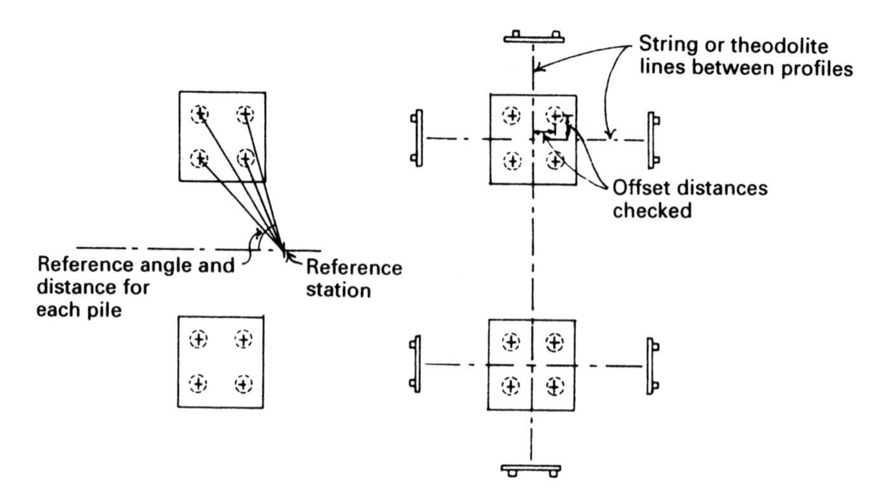

Fig. 13.1 Two methods of referencing piles.

is at a suitable level: too high and there will be unnecessary boring or driving; too low and it will not be possible to form the upper part of an in situ pile, nor, for a driven pile, will the full length for skin friction be achieved.

Each pile position can be located using a theodolite and a (steel) tape and marked with a pin driven into the ground. Because some pins are likely to be disturbed, a method of quick referencing must be employed. The reference system must also allow the piles to be checked during installation – as soon as an auger, casing or preformed pile is positioned, the pin is lost. Piling rigs are large and, in addition to their propensity for dislodging neighbouring pins, can easily obscure a sight line. Ground heave and vibration during installation can also disturb the pins.

Figure 13.1 shows two methods of checking pile positions.

Levels for piles must also be given. The depth of bored piles is usually specified on the drawings; a level peg or profile close to the pile position (beware of ground heave) can act as a datum. Driven piles are usually specified in terms of the 'set': the maximum penetration for a given number of blows of the hammer. The engineer must check this for each pile.

The depth of bore or drive must be checked by the engineer who should beware of tapes with missing middle sections offered by the piling subcontractor! The top level of concrete must be indicated, as must the cut-off level when the concrete has matured and the ground around the pile has been excavated (Fig. 13.2).

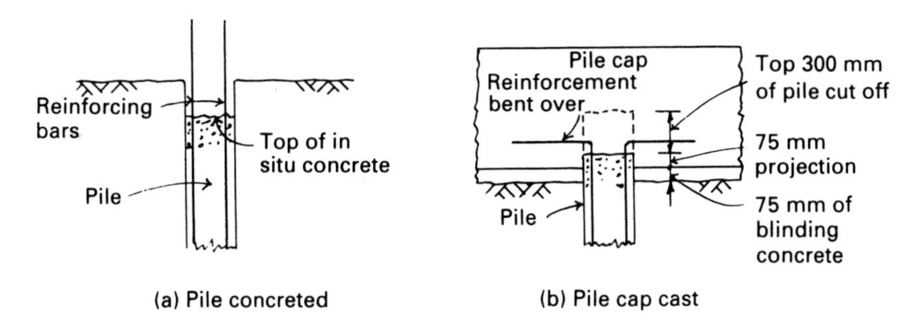

Fig. 13.2 Pile installation.

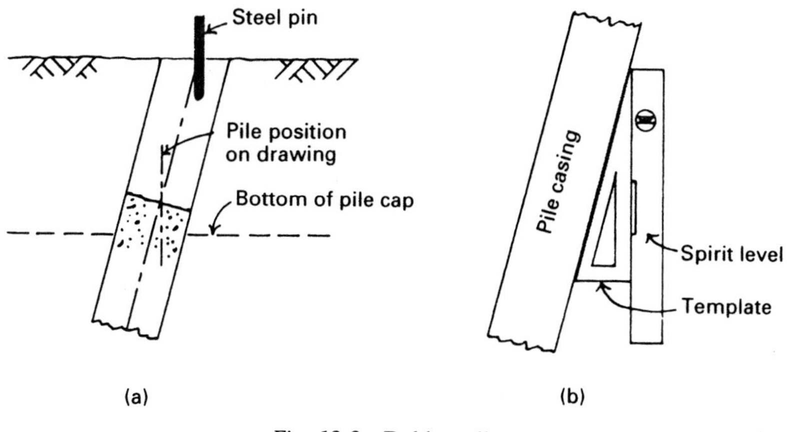

Fig. 13.3 Raking piles.

For raking piles the engineer will have to set out the position where the pile centre meets existing ground; the drawings usually show the position of the top of the cut-off pile (Fig. 13.3(a)). The angle of rake during driving may be checked by using a wooden template and a spirit level as plumb (Fig. 13.3(b)).

13.3 PILE CAPS, COLUMN BASES, FOOTINGS

Pile caps, bases and footings have several requirements so far as setting out is concerned. They must be in the correct plan position and of the correct size; excavation must be taken down to the correct level; the top, when concreted, must be set to the correct level; bolts, pockets, upstands and kickers must be correctly positioned.

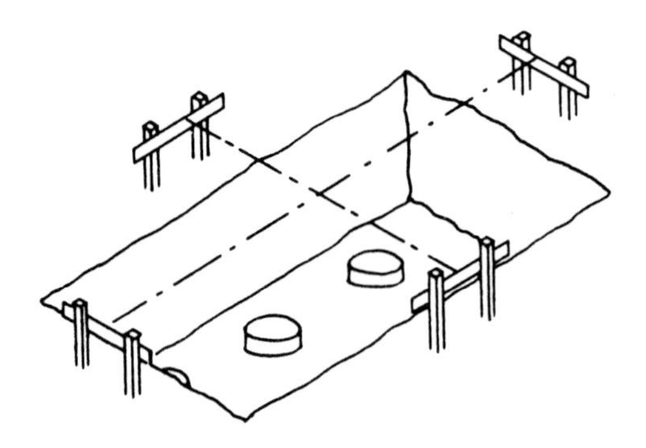

Fig. 13.4 Pile cap.

For position and size, pegs or profiles should be set in the ground. For excavation, corner pegs set back from the edge of the dig will be satisfactory. Profiles defining centrelines can then be set and can double as sight rails for levelling the excavation. Blinding concrete will be required at the bottom of the excavation; steel pins should be set to indicate the level. When excavation is complete, the profiles can be used by joiners to set formwork for the base (Fig. 13.4).

After the formwork has been fixed, the engineer should mark on it at a number of positions the level of the top of the concrete – nails or strips of timber ('grout checks') are satisfactory. Tops of bases are usually set slightly low where prefabricated columns or machinery are to be installed. Some tolerance is thus provided should the concrete be poured higher than the marks.

The site operatives should have enough information to position bolts or to set formwork for pockets, upstands or kickers; the site engineer should check, preferably with instruments.

Strip footings are prepared by excavating a trench in the correct position to the required depth and pouring concrete up to a given level as indicated by pegs, pins or profiles.

13.4 DEEP FOUNDATIONS

Where a suitable stratum is some distance below ground level but not so deep as to warrant piles, and where deep basements are to be constructed, deep excavations will be required. Excavation sides will probably need supporting, very likely by sheet piles, diaphragm walling or contiguous bored piling.

The engineer will be responsible for fairly precise plan location, the maintenance of verticality and correct depth of installation. Methods must be adopted to permit checking during installation and, possibly, monitoring for movement subsequently. It is likely that the engineer will have to transfer alignment to the bottom of the excavation.

13.5 TEMPORARY WORKS

Virtually all construction projects involve temporary works. Scaffolding, column propping, formwork and falsework for in situ concrete, lifting schemes, temporary access and diversion of services are all examples, and the engineer will be responsible for their setting out. Not only is the alignment critical so that the finished structure can perform its function correctly, but wrongly positioned temporary works can impose additional loading to temporary and permanent construction. A number of the most dramatic collapses have occurred during construction, with faulty temporary works being one of the causes.

The following three examples indicate the engineer's role in positioning temporary works.

In situ concrete slab

The same technique is used for a suspended floor slab of a building, for a bridge deck and for the roof of a service reservoir. Proprietary supports will probably be used; these have a limited amount of adjustment for height, usually at the top (Fig. 13.5).

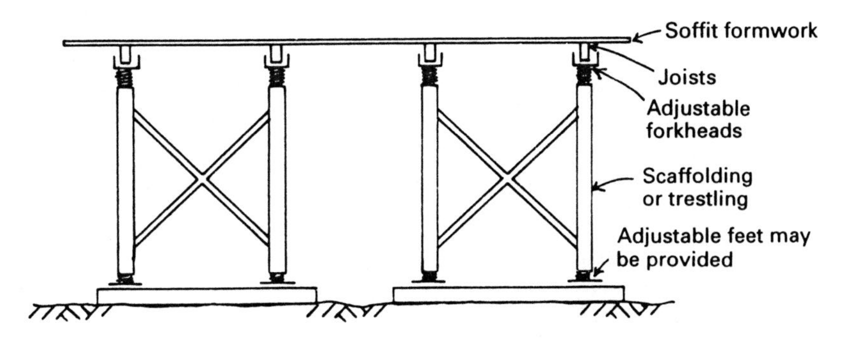

Fig. 13.5 Temporary support for concrete slab.

The engineer must check:

(1) The level of the base on which the trestling/scaffolding will stand.
(2) The plan position of the trestling, particularly so that the forkheads will line up with the joists.
(3) The levels at the forkheads/joists.
(4) The formwork level.

Bridge deck to be rolled into position

Where a bridge is to be constructed under an existing railway, it is often assembled on temporary supports alongside the line and rolled or slid into position during a limited 'possession' of the line. This minimizes disruption to rail traffic.

Permanent foundations will be constructed inside temporary cofferdams or tunnels under the line while the bridge is constructed on the temporary supports. The supports and track for rolling or sliding must be carefully aligned to permit smooth transfer of the bridge from temporary to permanent foundations.

Sheet piling

Interlocking steel sheet piling is driven into the ground to support the earth during excavation or to hold back water. Usually a temporary measure, sheet piling is sometimes permanently installed for dock and harbour works.

The piles are set in a vertical frame in the correct position; the engineer must set out the frame. Where the piles are to form a closed cofferdam, the setting out will be quite critical to obtain closure. The top level of the piles (after driving) must be given to ensure that correct penetration is achieved. When excavation is carried out, care must be taken not to remove too much material, reducing the 'toe' of the piles.

Sheet piling may be supported by internal horizontal framing (in cofferdams) or by horizontal beams and tie rods or ground anchors for walls. It may be necessary for sheet piling to be monitored for movement throughout a contract; the engineer must set up nearby stable stations to facilitate measurement.

13.6 STRUCTURES

Marks for plan and level control can be made on parts of the works by drawing pencil lines over waterproof crayon or paint on concrete, steel or brickwork.

Columns – prefabricated

Prefabricated columns are usually of steel (stanchions) and are secured by being bolted to their bases. The engineer should mark the column positions on the bases (two theodolites or theodolite and tape) and check that bolts or pockets are correctly positioned. The engineer must set a suitable thickness of packings (thin steel plates or 'shims') on each base so that a column, when lowered to the base, will take up its correct level.

Its position should be checked and the engineer may be required to plumb the columns by theodolite. This should be carried out in two directions at right angles (Fig. 13.6).

Precast concrete columns will usually sit in pockets formed in the bases. Mortar pads are used instead of steel shims to give the correct level. Otherwise the procedure is the same as for steel stanchions.

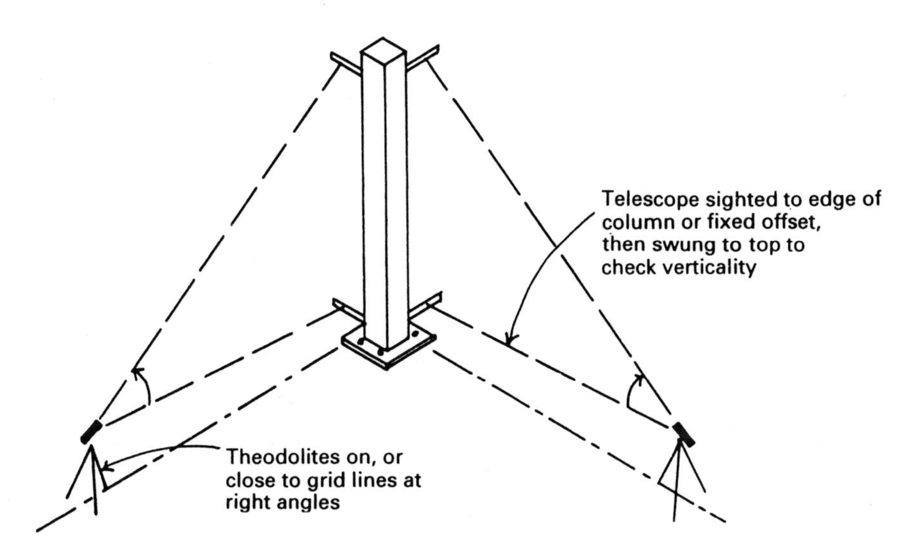

Fig. 13.6 Column plumbing.

Columns – in situ concrete

Before an in situ concrete column is cast, a kicker must be formed on the base. This is a 75 mm high upstand, formed to the shape of the column section during or after concreting the base. The engineer must set out and check its position.

The column formwork is butted up around the kicker and cramped into position. The engineer's job is then to check the formwork for verticality, using the two theodolite technique. The top level of concrete must be given, typically by using an inverted staff, and knocking a nail through the formwork at the correct level.

Walls

Walls in reinforced concrete are set out in the same way as in situ concrete columns. Brickwork is marked out on the footings; bricklayers should be capable of maintaining verticality and building the courses to the correct thickness; the engineer, if doubtful of this ability, should check.

Buildings

Buildings should initially be set out by pegging the corners and/or the intersections of grid lines. When these have been satisfactorily fixed (closing dimensions and diagonals should be measured), the building lines should be transferred to profiles and corner profiles set clear of construction (Fig. 13.7).

Once the ground floor slab has been concreted, building/grid lines can be marked on it. Profiles clear of the works may still be required to assist the transfer of alignment to upper storeys.

When columns or walls are of sufficient height, datum level marks can be set out. These are usually set 1.000 m above finished floor level, and will be used by site operatives. The marks can be established by levelling. A datum plane can also be swept out by a horizontally rotating laser mounted on a tripod.

A tripod or bracket mounted laser can rotate vertically to control alignment.

As work on a building proceeds, there will be a considerable amount of detail to be set out or checked by the engineer. Doors, windows,

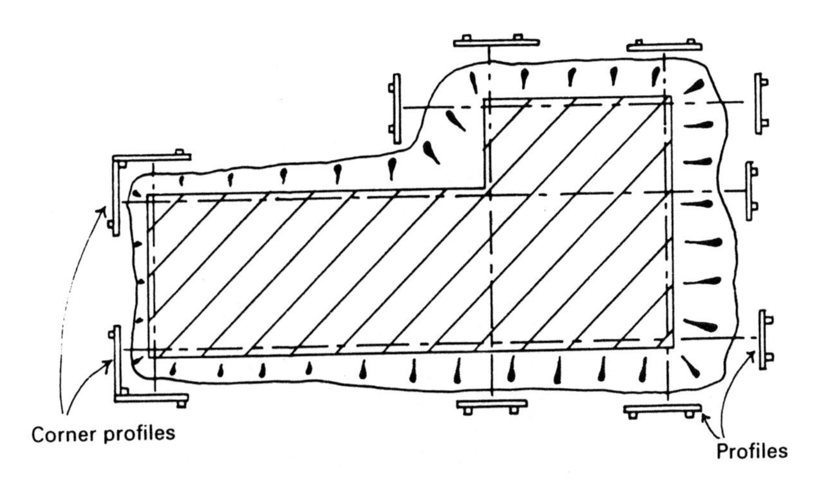

Fig. 13.7 Setting out a building.

cutouts for services, fixings and services themselves must be positioned. Taped measurements can be made from grid lines and from datum levels and planes; the engineer should check that no details are omitted.

Tall structures

Plan control: theodolite

The two theodolite method described for plumbing columns may be used. Aspects that the engineer should consider are:

(1) The theodolites must be placed on grid lines well clear of the structure (to keep elevation angles to minimum values).
(2) A mean of face left and right observations should be taken.

Plan control: automatic optical plummet

An automatic optical plummet is constructed in similar fashion to an automatic level, but with an objective lens pointing vertically upwards. It is provided with a downward plummet of the type fitted to theodolites.

When set up (and levelled) over a ground point, it can provide an accurate sight vertically upwards for plumbing tall structures (Fig. 13.8). In some models the telescope can be inverted to provide a precise downward sight.

The plummet is usually set close to a column, a short offset distance

Fig. 13.8 Wild ZNL automatic plummet.

from a grid line. The offset must be measured at the upper level to transfer the grid line upwards.

Checking of the upward and downward plummets is carried out by taking a sight, rotating the instrument horizontally through 180° and repeating the procedure. If the two sights do not coincide, adjustment is performed by resetting the cross-hairs.

Comments

A vertically rotating laser can be used provided that the engineer is satisfied that adequate accuracy will be achieved.

Where continuously climbing formwork is used (chimneys, lift shafts), the engineer must check that twisting of the formwork and working platform is not occurring.

Levelling

Levels can be transferred to upper storeys either by taping upwards (check verticality and tension) or by levelling up a stairwell.

13.7 BRIDGES

Bridges are usually either beam, arch, suspension or frame (often cantilever) types. There will always be considerable foundation and substructure works; the superstructure will depend on the type of bridge and the work may be prefabricated or formed in situ; many bridges involve a composite deck construction.

It is important that initial setting out is performed accurately. The bridge centreline and foundations must be positioned with considerable precision. Where a road or railway has to be crossed the location of the intersection point of the centrelines, and the establishment of the intersection angle, are key operations. Suitable reference points (off the line of an operational road or railway!) must be established, clearly marked and protected. Across watercourses, EDM should be used for distance measurements. Reciprocal levelling (by engineer's level or by theodolite/EDM) should be employed to overcome curvature and unpredictable refraction effects, and collimation or vertical angle indexing of instruments should be checked.

Foundations may involve piling and pile caps, strip footings (for wing walls), or concrete bases. In some cases, the pier or abutment is continued directly up from the base; in other cases, a kicker is formed for a smaller section pier. Steel columns will require baseplates setting correctly in plan and level. Foundations in a river are likely to require construction within a cofferdam, or caisson work. Control for such works should be by EDM set up on the banks. When work within the cofferdam, or caisson sinking and construction are sufficiently advanced, setting out can be controlled directly at the site of operation.

Work on bankseats, abutments, piers, columns and towers should be controlled as for structures. Much work is likely to be in in situ concrete, so that the position and verticality (or rake) of formwork must be checked. Continuously climbing formwork may be used for tall towers (for a suspension bridge) or high piers for a box section cantilever bridge in concrete. The site engineer must set out and check the positioning of springing points for arches, bearings for deck construction, and starter bars and kickers for the in situ concrete deck of a frame bridge.

The superstructure may well involve considerable temporary works, for example formwork, supported from the ground, for a concrete arch, frame or in situ deck. Apart from setting out the works, the site engineer may have to check that there has been no significant movement of formwork and falsework during concreting. Incremental (concrete) box section construction for cantilever bridges will require constant checking (instruments set up inside the box) and a check on span and bearing levels should be made before prefabricated beams are positioned.

The principles already covered for structural work should be followed. Checks on setting out should be regularly made, including ones on the control framework. Only after such checks can the engineer be confident that closure, on concreting the final span, or placing preformed beams, can be achieved.

13.8 PRECISE SETTING OUT OF MACHINERY – AUTOCOLLIMATION

Some process machinery (in paper making, for example) requires that components, whose absolute position is not critical, must be set perpendicular to a baseline and horizontally very accurately. The perpendicularity is achieved with a one second theodolite; to speed setting up the theodolite and to eliminate centring errors, autocollimation may be used.

Autocollimation makes use of a prism set up at one end of the baseline. The prism, with two reflecting surfaces, acts as a mirror in the horizontal plane and as a retroprism in the vertical plane. A theodolite with a lamp fitted near the eyepiece to project a beam forward through the cross-hairs and out of the objective lens is used.

Initially it must be set up precisely on, and sighting along, the baseline. The prism, in a tribrach on a tripod, is levelled and turned about its vertical axis until the engineer sees the cross-hairs and their image, as reflected by the prism, co-incidently.

To establish the theodolite subsequently, it is set approximately on the baseline, turned towards the prism and adjusted until the two cross-hair images are co-incident. The instrument is then sighting parallel to the baseline, and can be turned through 90° to set out the

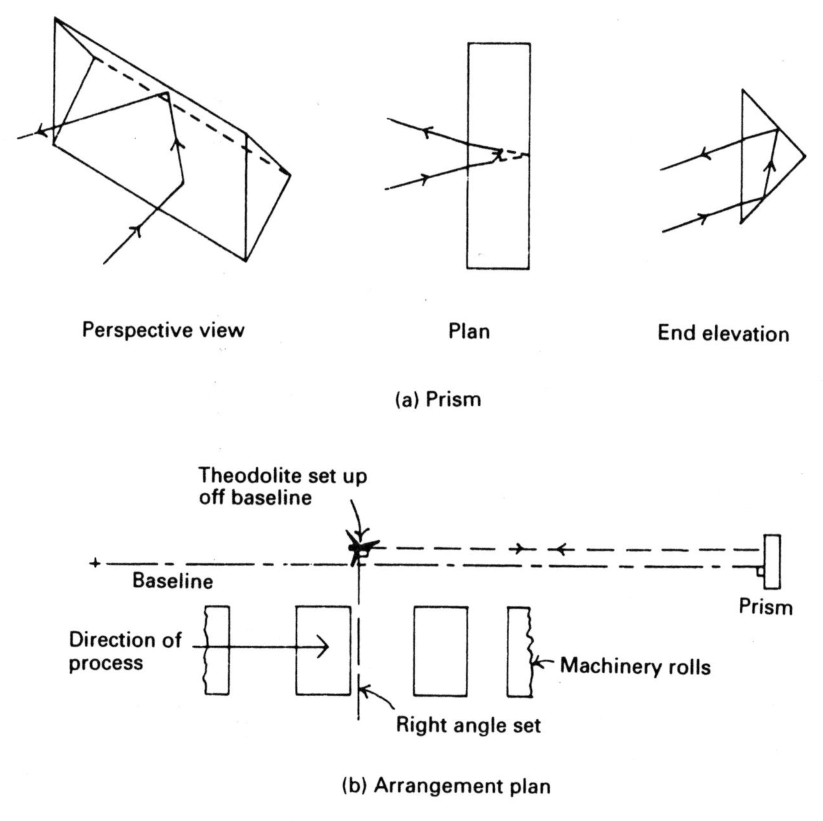

Fig. 13.9 Autocollimation.

machinery. A mean of both faces is necessary. Figure 13.9 illustrates the process.

13.9 COORDINATE MEASURING SYSTEMS

Coordinate measuring systems (also known as industrial measuring systems) have been developed to allow precise measurement of industrial components. In particular, complex surfaces can be surveyed and modelled in three dimensions. The method used is intersection, and precise electronic theodolites linked to a computer are employed. The theodolites are sighted on to one another at each end of a precisely measured baseline. Provided that a mark or target is available on the object to be surveyed, from horizontal angles the position of the remote object can be determined (Fig. 13.10).

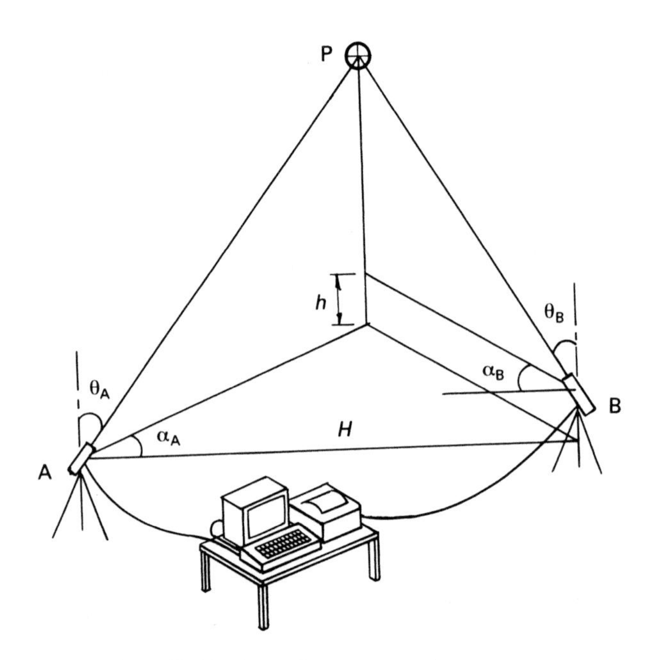

Fig. 13.10 Coordinate measuring system.

Frequently three-dimensional coordinates are required, and the difference in level of the theodolite trunnion axes and the vertical angles must be measured. Two vertical angles will produce one redundant measure so that a (3D) least squares process can be used to determine the probable coordinates.

Further redundancy can be brought in by using three (or more theodolites). A curved or multi-faceted piece of equipment can be surveyed by a group of surrounding theodolites.

Theodolites resolving to 1″ must be used, and the linked computer must be programmed to analyse results, carry out a least squares (variation of coordinates) adjustment, indicate whether measurements are of adequate precision and quote coordinates. It must be possible to measure the baseline and the difference in trunnion axes, although a more precise check can be applied by sighting with each theodolite the ends of a scale bar precisely measured to within a few microns. Where a number of theodolites are used, the bar can be used to 'register' each instrument (analysed on the computer) to save initial zeroing sights between the theodolites. Other clearly defined points are also used (their positions need not be known) in this 'bundle adjustment'.

A variety of targets for sighting are available; stick-on paper or card ones are often suitable – the thickness of any target should not be overlooked in precise measurement. Where the surface to be measured is inaccessible, a laser beam can be employed to provide a spot on which to sight. The simplest way of providing this is to use a laser eyepiece on one of the theodolites. A separate laster beam could be used, but beam dispersal may provide an insufficiently precise target.

An alternative method that can provide sub-millimetre accuracy uses a purpose built very precise total station. The signal is returned from a stick-on card target which acts as a retro-reflector. Calculations are handled by a cable connected keyboard terminal. Similar to a data logger, this accepts a program card, has a miniature printer, and can display computed coordinates of points surveyed.

Coordinate measuring systems are supplied as complete outfits comprising precise electronic instruments on movable mounting pillars, a computer, cables and all the necessary software.

14 Underground and Marine Works

14.1 UNDERGROUND WORKS – INTRODUCTION

In underground surveying very precise techniques are required. The basic 'working from the whole to the part' principle cannot be employed, and tunnels hundreds of metres in length have to be set out from baselines a few metres long. Simple visual checks on alignment are not possible and work has to be carried out in confined, dark, damp and potentially dangerous circumstances. There is often insufficient space for surveying to be performed while mining is in progress; the setting out may have to be done during evenings or weekends, or when work has been deliberately halted for the engineer to check and extend the line and level.

The consequences of faulty alignment can be considerable. At worst, work may have to be repeated; at best, inconvenience is caused; a heading, which has wandered in plan, will need two changes of direction to bring it back on line (Fig. 14.1). Deviations, other than by very small amounts, from correct vertical alignment may not be acceptable, as drain gradients may be seriously affected.

In the task of correlating underground work with surface surveys and controlling tunnelling there are four phases: surface survey, transfer underground, establishment of underground control, and control at the face. At each stage precise equipment must be used: one second theodolites and recently calibrated tapes or EDM.

14.2 SURFACE SURVEY

Headings will be driven from surface portals (roads, railways) or from bottoms of shafts (drains and intermediate sections of roads and railways). A proving survey locating these points must be carried out at ground level. A theodolite and EDM traverse should be

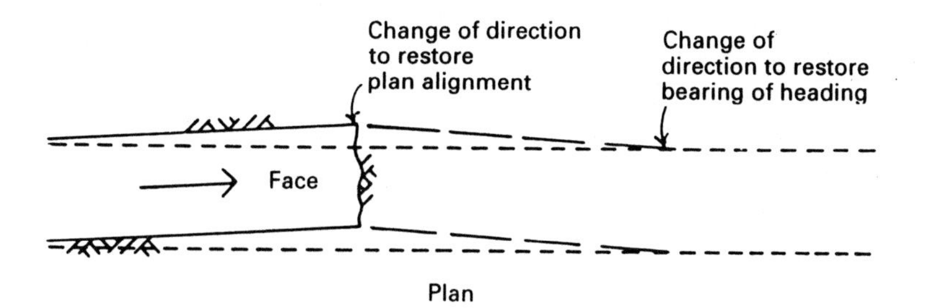

Fig. 14.1 Correction of poorly aligned tunnelling.

undertaken. Where there is open country, the line of the tunnel can be followed; in urban areas zig-zag courses may be necessary, and for tunnels under watercourses, long sights will be required. An 'out and back' loop traverse should be carried out so that the closing error, if replicated underground, will be within tolerance. Permanent stations must be established close to, but clear of, access points. Their positions should be such that checking is possible throughout the project. For all but the shortest tunnels, coordinates, local or national, should be used (for conversion from ground to grid coordinates see Chapter 5).

Shaft sinking

Access shafts are sunk vertically, lining rings being placed as excavation proceeds. The engineer must set out the position of each shaft, reference it, check verticality (usually with plumb wires) and indicate the bottom level.

14.3 TRANSFER UNDERGROUND

In plan

At each access point a baseline must be established. Where the heading is to run straight (drains, sewers) the baseline should be on the centreline. For works curved in plan (roads, railways) this may not be convenient, nor even possible where access shafts are offset from the tunnel line.

For surface portals, the baseline can be formed by two stations so that the theodolite can be set up over one to extend the line into the heading. The baseline should be the longest possible (for accuracy), should be free from disturbance and should avoid obstructions to sighting forwards.

At shafts a wire baseline should be provided. Plumb lines, made of 'piano wire' with heavy weights are fixed at the top to hang down the shafts forming the longest possible baseline. The wires may be set either in specific positions (along the centreline) or arbitrarily (but conveniently). To damp oscillations, the weights may be immersed in buckets of oil or water (Fig. 14.2(a)). They must be located by theodolite and steel tape; the Weisbach triangle (see below) is one method. The wires must be periodically checked for position and adjusted. Steps must be taken:

(1) To prevent disturbance of the wires from ventilation blasts (switch off the system).
(2) To prevent the wire kinking by winding it on large diameter reels.

Of level

Bench marks must be set up on stable surfaces close to heading portals. A closed level surface survey should be carried out. Levels

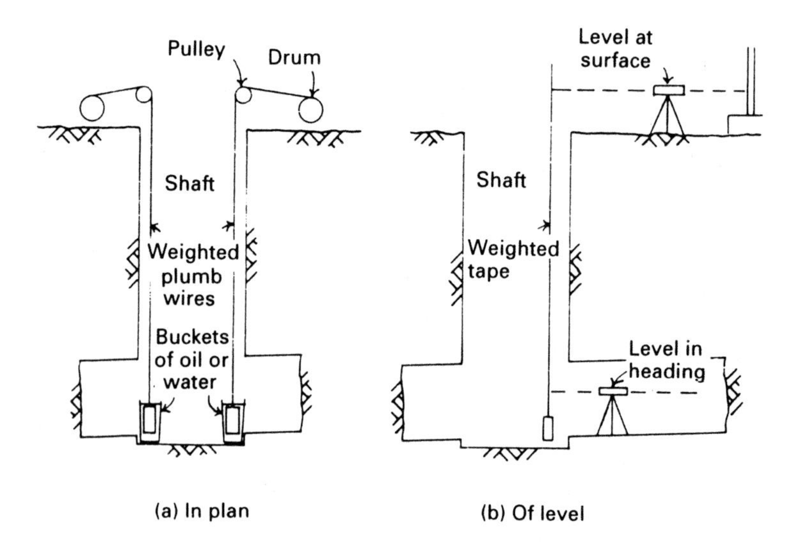

Fig. 14.2 Transfer of alignment.

can be transferred vertically down shafts by taping (check tension and verticality) and establishing bench marks at the feet of the shafts. Where the tape cannot be hung close to the shaft side, it may be necessary to suspend it in the middle of the shaft and take simultaneous readings at top and bottom with engineer's levels (Fig. 14.2(b)).

14.4 UNDERGROUND CONTROL

Plan control

The starting position of a heading, whether from a surface portal or a portal at the foot of a shaft, can be established by eye, as can the first few metres of the heading drive. It will then be necessary to transfer the setting out into the heading, establishing roof stations.

Headings for roads, railways and larger diameter sewers are lined as work progresses, so precise setting out is required at all stages. Only when a pilot tunnel is driven, subsequently to be opened out, or where small diameter pipes are to be laid in a square timbered heading, will it be possible to check a complete length before lining.

In a heading starting from a shaft, it will be necessary to set up the theodolite in a position related to the plumb wires. Two methods exist: co-planing and the Weisbach triangle.

Co-planing

By trial and error the theodolite is set on the line defined by the two wires. After one or two trials the tripod should be suitably placed and the theodolite can be aligned by the engineer releasing the centring clamp and sliding the instrument. If centring is below the footscrews, the instrument must be relevelled after movement. If the theodolite is between the wires, the telescope must be turned through 180° in azimuth (not over the top) between sightings. A mean of both faces should be taken; if there is a difference between face sightings, the instrument should be adjusted or returned to the suppliers at the earliest opportunity.

If the theodolite is beyond the wires, a single sighting will suffice. Within the range of focus adjustment it will be possible to see the far wire 'through' the near one.

Weisbach triangle

In this method the theodolite is set up beyond the wires and close to the line. Distances and angles are measured to determine the position of the theodolite relative to the wires so that further points underground can be set out.

The theodolite is set up at T so that the plumb wires at A and B may be observed (Fig. 14.3(a)). The distance TA should not be greater than the length of the baseline AB. Distances TA and AB are taped (AB should already be known) and angle $A\hat{T}B$ is measured.

By the sine rule:

$$\frac{\sin T\hat{B}A}{TA} = \frac{\sin A\hat{T}B}{AB}$$

For a small angle, its sine is proportional to its value, hence:

$$T\hat{B}A = \frac{TA}{AB} \times A\hat{T}B$$

Offset TX (if required) $= TA \sin (A\hat{T}B + T\hat{B}A)$.

Example 14.1

Figure 14.3(b) is a plan showing the connection of an underground

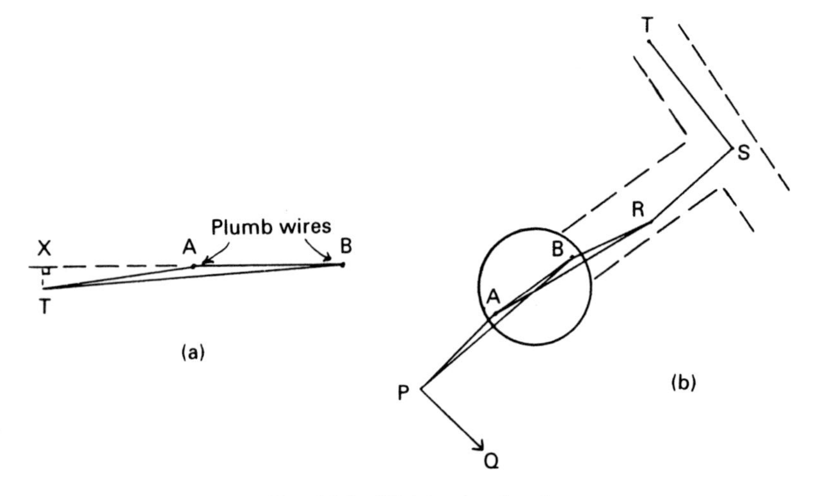

Fig. 14.3 Weisbach triangle.

survey to surface stations via piano wires at A and B in a vertical shaft. For the given data, calculate:

(1) The coordinates of S at the junction of a cross-passage with the main tunnel.
(2) The bearing ST.

Data:

Coordinates of P (411.365 m E; 296.470 m N)
Bearing PQ 132°17′42″

$$\text{Angle } A\hat{P}Q = 77°21'32'' \qquad \text{Length } PA = 3.113\,\text{m}$$
$$A\hat{P}B = 00°12'24'' \qquad AB = 3.487\,\text{m}$$
$$A\hat{R}B = 00°17'46'' \qquad BR = 2.904\,\text{m}$$
$$A\hat{R}S = 171°14'11'' \qquad RS = 5.235\,\text{m}$$
$$R\hat{S}T = 91°26'54''$$

Solution

$$P\hat{B}A = \frac{PA}{AB} \times A\hat{P}B = \frac{3.113}{3.487} \times 744'' = 664.2'' = 00°11'04.2''$$

Bearing of PA = 132°17′42″ − 77°21′32″ = 54°56′10″
Bearing of PB = 54°56′10″ + 00°12′24″ = 55°08′34″
Bearing of AB = 55°08′34″ + 00°11′04.2″ = 55°19′38.2″

$$B\hat{A}R = \frac{BR}{AB} \times A\hat{R}B = \frac{2.904}{3.487} \times 1066'' = 887.8'' = 00°14'47.8''$$

Bearing of AR = 55°19′38.2″ + 00°14′47.8″ = 55°34′26″
Bearing of BR = 55°34′26″ + 00°17′46″ = 55°52′12″
Bearing of RS = 55°34′26″ + 171°14′11″ − 180°00′00″ = 46°48′37″
Bearing of ST = 46°48′37″ + 91°26′54″ + 180°00′00″ = 318°15′31″

Table 14.1 shows the coordinate calculations.

Stations

Roof stations are usually employed underground, as floor stations are liable to disturbance and flooding. A mark can easily be made on the crown point of the lining. However, it is often necessary to hang plumb-bobs from roof stations; a plate or bracket fixed at the crown

Table 14.1

Line	Horiz Length	Bearing	Coord. Diff ΔE	ΔN	Station Coords E	N	Stn
					411.365	296.470	P
PA	3.113	54°56′10″	2.5480	1.7884			
					413.9130	298.2584	A
AB	3.487	55°19′38.2″	2.8678	1.9837			
					416.7808	300.2421	B
BR	2.904	55°52′12″	2.4038	1.6294			
					419.1846	301.8715	R
RS	5.235	46°48′37″	3.8168	3.5829			
					423.0014	305.4544	S
			Σ11.6364	8.9844			

Coordinates of S are (423.001 m E; 305.454 m N)
Bearing of ST = 318°15′31″

point will allow this. Use can often be made of lining segment bolt holes to secure station plates. In timbered headings a steel 'dog' (a U-shaped spike) can be hammered into a head tree and a hacksaw cut made at the required position.

A theodolite can be centred under a roof station by setting its plumbing thorn (on top of the telescope) under a plumb bob.

In small diameter headings, special short tripods may be necessary; alternatively, a stout piece of timber can be wedged horizontally across the lined tunnel at mid height and the theodolite set up on it, care being taken not to displace the instrument.

In large diameter headings, brackets can be fixed to the lining at the sides and theodolite stations established. For lines of sight parallel to the tunnel wall there is a risk of lateral refraction. Such a phenomenon occurred during construction of the Channel Tunnel, and zig-zag sights across the tunnel had to be taken.

Wriggle surveys

When a heading has been driven a considerable distance and lined, a *wriggle survey* can be carried out to ascertain how the 'as built' position differs, or wriggles, from the design centreline. Such a survey is of particular use in determining any realignment of the track for an underground railway to suit variations in tunnel position. An open traverse is conducted from portal to face. It should be closed by returning it to the portal, generally using different stations, although some of the original ones may be incorporated to provide periodic checks. When the heading has been thurled (broken through to the approaching heading or next shaft) a tied traverse between shafts can be carried out.

The gyrotheodolite

A gyrotheodolite can be used in tunnelling work. It consists of a gyroscope mounted on a theodolite so that orientation may be established underground without direct reference to the surface survey. Points to be considered by the engineer are:

(1) That position correlation with the surface will still be required.
(2) That the accuracy, typically 20″, may be no greater than that achievable by methods so far described.
(3) That conversion from true to grid bearings must be carried out (see Chapter 5).
(4) The practicality of setting it up in a confined heading.
(5) The cost of purchase or hire.

In long deep headings, the gyrotheodolite is likely to be preferable to conventional techniques which are likely to produce the cumulative errors associated with wire baselines in deep shafts and long drives.

More information can be found in the books by W. Schofield and F.A. Shepherd cited in Section 5.10.

Horizontal curves

It is clearly impossible to peg out a curve underground ahead of work. It will also be impracticable to set out individual lining rings with a theodolite. The engineer will have to set out stations defining chords, and calculate the offset of the centreline at the face for each lining ring (Fig. 14.4).

It will almost certainly be necessary to derive the offsets from rectangular coordinates. The engineer must allow for creep of the lining at bends, and should regularly check and extend the setting out. It will be necessary to check the positions of the lining rings as laid, in case modifications to the alignment are needed.

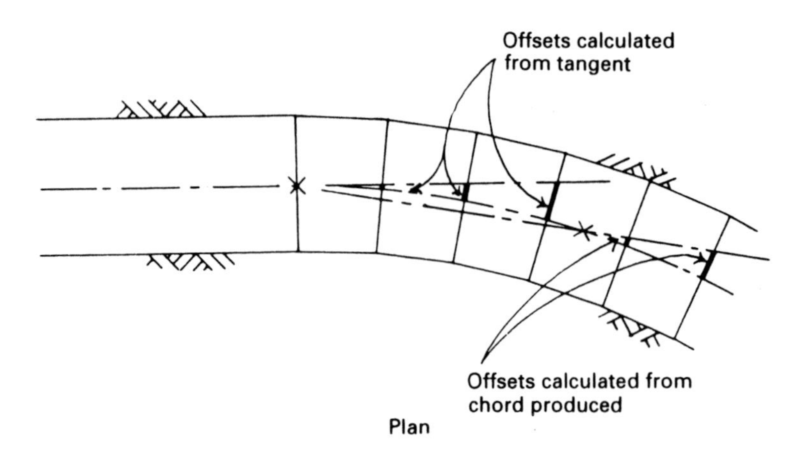

Fig. 14.4 Horizontal curve.

Level control

Levelling can be carried out underground more or less conventionally. A short staff will often be needed, as may a short tripod; alternatively, a baulk of timber can be used as described in 'Stations'. Bench marks must be carried forward along the heading, establishing them on the side (mid height) of the linings if possible. Roof bench marks (with the staff inverted) can also be set. All levelling should be returned to the point of origin.

14.5 CONTROL AT THE FACE

In plan

It will not be possible for the engineer to check each lining ring as it is fixed, so a method sufficiently simple and accurate for the miners to use must be employed. The usual method is to 'over-range' using plumb-bobs. Two bobs are suspended at crown points back up the heading and a further bob at the face is lined in. As far as excavation is concerned, alignment is a process of trial and adjustment, and the engineer must:

(1) Advance the stations promptly.
(2) Periodically check the alignment from portal to face.
(3) Occasionally check the miners' alignment at the face.

Of level

Overboning is used to level a heading at the face. In large headings, profiles can be set at the side of the works. In smaller headings, profile plates can be set in pairs flush on opposite sides of the linings; for boning, strings are tied across the plates. A traveller length of 1.000 m (above invert) is often used. Illumination of the strings by torches may be necessary.

Shield control

Where the ground at the face requires support, a shield is used. It is a cylindrical steel tube with a leading cutting edge. It is forced into the face by being jacked forwards off the lining rings already placed. Not only must the shield have its cutting edge correctly aligned, but it must be pointing in the required direction. The engineer must check line, level, lead, look-up and roll.

Line and level

Line and level are checked by methods already described. Where the diameter is large, a target, known as a 'fiddle', is fixed in the shield clear of the path of the muck tubs. The fiddle is a sheet of celluloid or similar material with graduated horizontal and vertical axes as in Fig. 14.5(a).

The position of the fiddle relative to sight lines provided by plumb-bobs, profiles, theodolite, level or laser must be calculated and checked by the engineer.

Lead

Lead refers to the direction (in plan) of pointing of the shield. It may be checked either by squaring off the centreline (or tangent) or by taping along each side of the heading from the portal.

On horizontal curves the engineer must calculate by how much the side of the shield on the inside of the curve should lead the outer side.

Look-up

Look-up refers to the gradient of the shield. It should be checked with

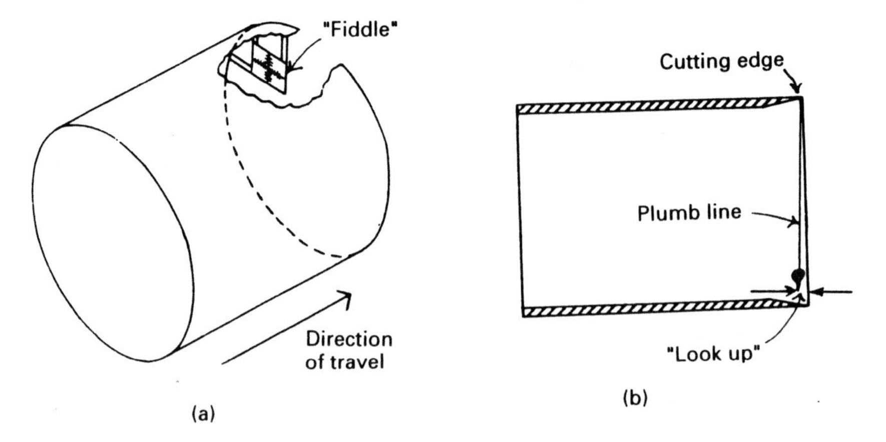

Fig. 14.5 Shield alignment.

a plumb-bob (Fig. 14.5(b)). On downhill drives, 'look-down' should occur.

Roll

Roll of the shield can also be checked with a plumb-bob. If a shield rolls accidentally, the fiddle, or other reference marks, will be out of position.

Occasionally roll is deliberately introduced. On horizontal curves, underground railway tracks are canted; if they are to be fixed to bolt holes in the lining rings, the latter must be rolled, requiring that the shield (which the lining rings follow) must also be rolled. The engineer must calculate and check the amount of roll.

Lasers

Lasers have found particular application in tunnelling, being used to control line and level. For a straight heading they should be fixed outside the portal to project a beam parallel to the centreline a short distance below the crown. They can also be set up inside large diameter headings on brackets fixed inside the lining rings. They must be aligned to accord with the underground control setting out. On curves a laser must be set to project a chord. The engineer must give offsets for distance and for level to the miners, and also frequently check the work.

The engineer must also be satisfied that the accuracy being achieved with laser control is adequate, particularly on long drives where beam misalignment can be critical and dispersal can occur. A laser suitable for tunnelling work, not a pipe-laying laser, should be used. A periodic check of the beam at the face should be carried out using a theodolite and level. The risk of lateral refraction mentioned earlier in this chapter should also be borne in mind.

Pipe jacking

Laser control is particularly suited to pipe jacking. The necessary checks at the face are similar to those for shield working, but the movement of the pipes, as they are jacked forward, prevents the establishment of fixed stations along the tunnel. The laser should be established at the tunnel entrance. A particular problem is finding a stable mounting as there are considerable ground pressures in the vicinity of the thrust pit. The engineer must carry out frequent checks on the alignment of the laser.

14.6 MARINE WORKS

Examples of marine works are dock and harbour engineering, coastal protection, bridge foundations in watercourses, submerged pipelines and tunnels and offshore platforms.

Many of the techniques described elsewhere in this book can be used or adapted to set out marine works. Instrument work should be carried out on land if possible, EDM having simplified measurement over water.

The engineer may also be responsible for monitoring water levels where tidal work is in progress, or where storm effects may be critical.

Sheet piling inshore

Used extensively and permanently in dock and harbour engineering, sheet piling is often installed temporarily for bridge foundations and for submerged pipelines. The basic principles described in Chapter 13 apply. A frame must be set up in the correct position. One way of achieving this is to drive isolated king piles from a barge or pontoon to support a stable frame in which the main piles can be precisely pitched.

The simplest way of positioning a barge or pontoon is to set a prism on it, and measure the distance and bearing from a shore station by theodolite and EDM. By trial and error the position can be fixed. Two theodolites from two shore stations can also be used, though synchronization may be a problem. Where a floating platform is to be in use for a considerable time, a system of shore targets giving collinearity with the inshore location may be established. Operatives on the platform can then check the position by eye.

Levels should be established by trigonometric means, usually from the vertical components from EDM measurements. Effects of curvature and refraction (Chapter 3) should be considered.

Permanent inshore works

When temporary works have been established, it should be possible to set up surveying equipment on a firm platform inshore. The tolerances for permanent construction (especially for bridgeworks) will be much smaller than for temporary works, and the engineer must take care in setting out and checking.

Offshore platforms

Offshore platforms for gas and oil recovery will entail long sights for setting out. Long range EDM equipment can be used on the shore or on the platform. For measurements from a platform, at least two shore stations must be used (linear resection). When the platform is still floating, it will be advisable to sight three stations and calculate a mean or most probable position. There will be considerable corrections for curvature and refraction in levelling to a distant platform.

Alternatively GPS may be used to locate offshore platforms and sounding vessels. The accuracy required for offshore positioning is often lower than that for works on dry land, and GPS has been suitable for offshore location for some years.

Coastal protection

Concrete structures (e.g. sea walls) should be set out in accordance with methods described in Section 13.6 (Structures). Stone breakwaters may be treated similarly to road embankments and set out by means of batter rails and profiles. The engineer may need to modify conventional setting-out methods to take account of water and tidal effects; on-shore reference stations should be established and EDM employed where possible.

14.7 FURTHER READING

Ingham, A.E. and Abbott, V.J. (1992) *Hydrography for the Surveyor and Engineer*. 3rd edn. Oxford: Blackwell Science.

Answers to Exercises

CHAPTER 2

2.1 (a)

Back sight	Inter sight	Fore sight	Rise	Fall	Height of collimation	Reduced level	Station
2.975					109.295	106.320	TBM
	2.700		0.275			106.595	A
	2.990			0.290		106.305	B
2.745		0.995	1.995		111.045	108.300	C
1.305		1.365	1.380		110.985	109.680	D
	0.785		0.520			110.200	E
	1.775			0.990		109.210	F
	1.980			0.205		109.005	G
1.805		2.815		0.835	109.975	108.170	H
	3.105			1.300		106.870	J
	3.010		0.095			106.965	K
		3.620		0.610		106.355	TBM
8.830		8.795	4.265	4.230		−106.320	
−8.795			−4.230				
0.035			0.035			0.035	

(b) Collimation error, failure to centre levelling bubbles, non-vertical staff, staff incorrectly extended, sinking level or staff, faulty reading, faulty booking.

2.2 (c) (i) No
(ii) 1.710 m

CHAPTER 3

3.1 $F\hat{I}G = 44°09'18''$ $G\hat{I}H = 46°56'44''$ $G\hat{I}J = 115°36'23.5''$

Reading of 71°17′10″ is inconsistent and should be ignored. It is quite possibly 71°10′17″ wrongly booked.

3.2 51.889 m, 14.001 m AOD (X), 13.363 m AOD (Y), 90°58′10″.

3.3 (a) 176.15 m AOD (b) 0.125 (c) 269.77 m AOD

CHAPTER 4

4.1 (a) AB = 58.406 m BC = 111.564 m (b) CA = 68.785 m

CHAPTER 5

5.1 X (687.081 m E; 432.560 m N) Level 127.000 m AOD
Y (683.086 m E; 432.767 m N) Level 127.000 m AOD

5.2 Q (598.22 m E; 354.37 m N) R (809.98 m E; 325.73 m N)
S (760.56 m E; 56.77 m N)

5.3 B (206.891 m E; 308.624 m N) C (309.668 m E; 267.825 m N)
D (438.377 m E; 281.054 m N)

5.4 P (261 825.6 m E; 376 430.5 m N)

5.5 5160.168 m (both methods)

5.6 (a) 1546.412 m (b) 226°41′01″
(c) 226°14′30″ (AB) and 46°13′42″ (BA) (d) 1547.075 m

CHAPTER 6

6.1 Line passes through marsh, through clump of trees, through a hedge. It crosses a road obliquely and runs up an embankment. The stations are not near features suitable for referencing. Well conditioned triangles have not been formed.

6.2 Plan distances: 82.965, 159.505, 126.972 (point 3 – both), 70.000, 48.500, 104.012 m.
Reduced levels: 242.460, 240.760, 233.235, 239.735, 235.050, 234.850, 232.200 m AOD.

(a) 1 in 80 (b) 6.500 m (c) (1213.708 m E; 1656.871 m N) (5)
(1273.338 m E; 1600.581 m N) (6),
1 in 72.25

(d) 7.410 m

6.3 See drawing 'Solution to Exercise 6.3'.

6.4 See drawing 'Solution to Exercise 6.4'.

CHAPTER 7

7.1 (a) 278.700 ± 0.017
(b) (1646.257 ± 0.007 m E; 1806.879 ± 0.006 m N)

7.2 $K\hat{L}M = 101°47'27''$ $M\hat{N}K = 50°45'07''$
$L\hat{M}K = 38°30'04''$ $N\hat{K}M = 50°17'31''$
$K\hat{M}N = 78°57'22''$ $M\hat{K}L = 39°42'29''$
$L\hat{M}N = 117°27'26''$

7.3 124.45 m (X), 129.84 m (Y)

7.4 (518.384 m E; 412.349 m N)

7.5 (467.375 m E; 374.827 m N)

CHAPTER 8

8.1 2085 m^2 (trapezoidal) 2089/2093 m^2 (Simpson)

8.2 214 170 m^3

8.3 6075 m^3 (trapezoidal) 5989 m^3 (prismoidal)

8.4 1 415 000 m^3 (trapezoidal) 1 407 000/1 404 000 m^3 (prismoidal) 240.8 m

8.5 5963 m^3 (trapezoidal) or 5951/5974 m^3 (triangular) 449 m^3

8.6 (a) see drawing 'Solution to Exercise 8.6' (b) £92 537.50 (c) £95 900; unlimited overhaul is recommended.

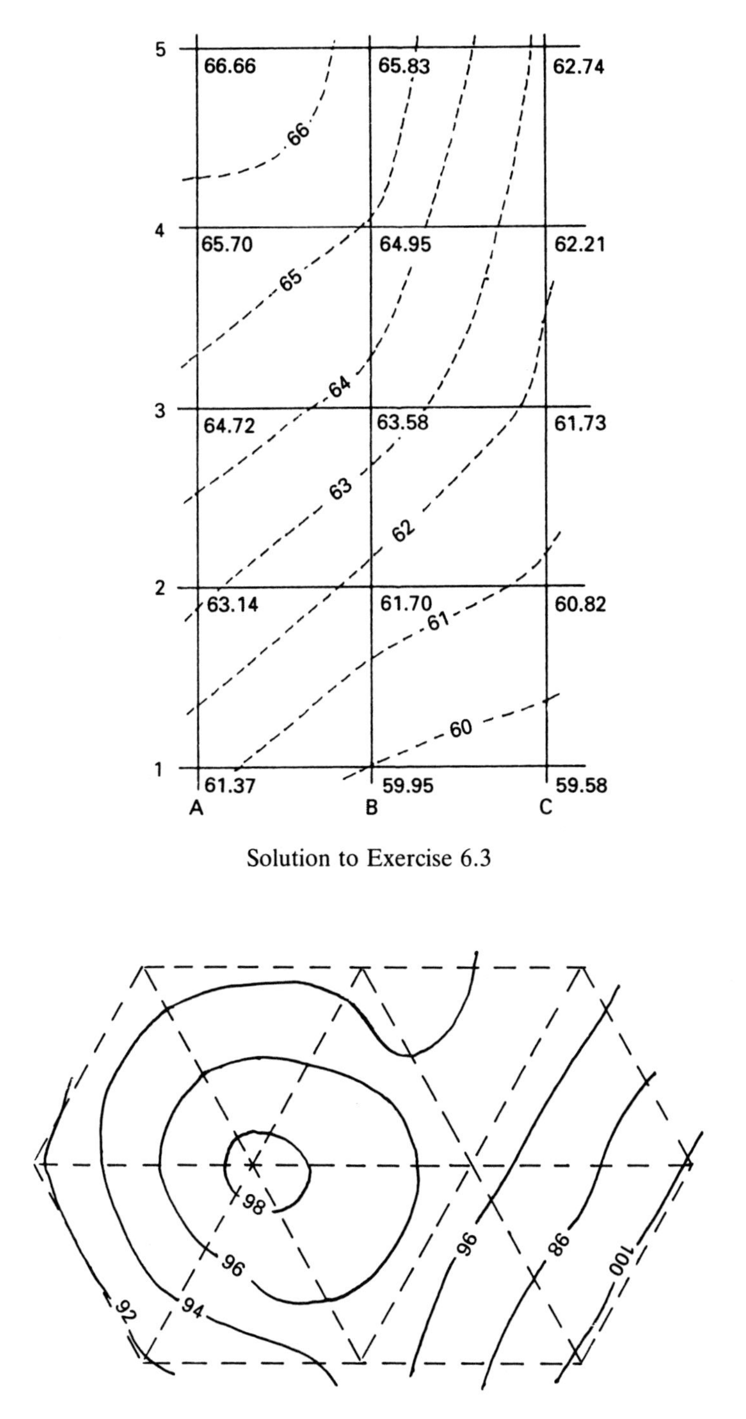

Solution to Exercise 6.3

Solution to Exercise 6.4

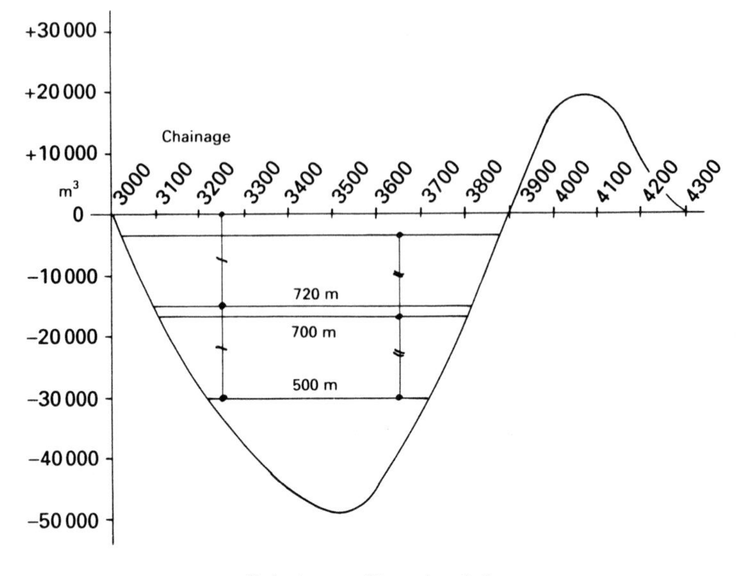

Solution to Exercise 8.6

CHAPTER 10

10.1

Chainage	Chord	Instrument angle	2nd inst angle
2342.88		00°00′00″	
2360.00	17.12	00°57′42″	
2380.00	20.00	02°05′06″	
2400.00	20.00	03°12′31″	
2420.00	20.00	04°19′55″	
2440.00	20.00	05°27′20″	
2460.00	20.00	06°34′44″	
2480.00	20.00	07°42′09″	
2500.00	20.00	08°49′33″	
2520.00	20.00	09°56′58″	
2540.00	20.00	11°04′22″	
2560.00	20.00	12°11′47″	
2580.00	20.00	13°19′11″	
2600.00	20.00	14°26′36″	

continued

Chainage	Chord	Instrument angle	2nd inst angle
2620.00	20.00	15°34′00″	
2640.00	20.00	16°41′25″	
2660.00	20.00	17°48′49″	356°03′49″
2680.00	20.00	18°56′14″	357°11′14″
2700.00	20.00	20°03′38″	358°18′38″
2720.00	20.00	21°11′03″	359°26′03″
2730.08	10.08	21°45′01″	00°00′00″

10.2 (a) 980.000 m (b) 14°00′00″, 244.341 m

(c)

Chainage	Chord	Instrument angle
3571.082		00°00′00″
3600.000	28.918	359°09′17″
3650.000	50.000	357°41′35″
3700.000	50.000	356°13′53″
3750.000	50.000	354°46′11″
3800.000	50.000	353°18′29″
3850.000	50.000	351°50′47″
3900.000	50.000	350°23′05″
3950.000	50.000	348°55′23″
4000.000	50.000	347°27′41″
4050.000	50.000	345°59′59″

10.3 (a) 130.000 m (Theoretical length of 82.305 m increased to 129.877 m to conform with permitted kerb gradient difference, then rounded up)
(b) 943.380 m
(c) 3.40%
(d) 0.5%
(e)

Transition Chainage	Long chord	Instrument angle
2770.000		00°00′00″
2780.000	10.000	00°00′18″
2800.000	30.000	00°02′39″
2820.000	50.000	00°07′21″
2840.000	70.000	00°14′24″
2860.000	90.000	00°23′48″
2880.000	110.000	00°35′33″
2900.000	130.000	00°49′39″

Circular: sight T_1 reading 178°20′41″		
Chainage	Short chord	Instrument angle
2900.000		00°00′00″
2920.000	20.000	00°22′55″
2940.000	20.000	00°45′50″
2960.000	20.000	01°08′45″
2980.000	20.000	01°31′40″

10.4 (b)

Distance	Level (m)
00	61.350
20	62.219
40	63.025
60	63.769
80	64.450
100	65.069
120	65.625
140	66.119
160	66.550
180	66.919
200	67.225
220	67.469
240	67.650

(c) 173.620 m

10.5 (a) 180.000 m

(b)

Chainage	Level (m)
2040	155.600
2060	156.267
2080	156.867
2100	157.400
2120	157.867
2140	158.267
2160	158.600
2180	158.867
2200	159.067
2220	159.200

(c) 112.25 m. Curve length is between ‘desirable’ and ‘one step below desirable’

Index

accuracy, 2, 7, 190, 192
in angle measurement, 60
in EDM, 79, 165
in levelling, 24
in setting out, 286–8
in stadia tacheometry, 171
in taping, 68
adjustment of observations, 106–15, 188–237
aerial photogrammetry, 175
Airy spheroid, 127
alidade, 40
altitude
bubble, 40, 51
correction, 68, 77
angle closure check, 99
angular/linear relationship, 287
angular measurement
accuracy, 60
in control surveys, 88
horizontal, 45–50
in traversing, 97
vertical, 50–2
answers to exercises, 402–408
apex length, 300
arbitrary north, 91
arc–chord correction, 134
architect, 279
area and volume measurement, 238–75
area measurements, 2, 238–49
of irregular shapes, 240
by planimeter, 245
of regular shapes, 239
of road cross-sections, 245
Simpson's method for, 242
trapezoidal method for, 241
within traverses, 239
atmospheric correction, 76
autocollimation, 384
automatic level, 14
automatic optical plummet, 381
back
angle, 317, 318, 323
bearing, 90
sight, 18, 19
base, road, setting out, 355, 358
bases, setting out, 375–6
baseline
adjustment, worked example of, 202, 204
corrections to, 66–9, 76–8
batter rails (rules), 291, 350–54
bearings, 89–92
calculations of, 100
bench marks, 1, 17, 88
Ordnance (OBM), 1, 10
temporary (TBM), 10, 291, 295
boning in, 278
drains, 363–7
roads, 354–8
boning rod *see* traveller
booking, 6
EDM surveys, 163–6
levelling, 18–23
linear surveys, 156
tacheometric surveys, 170
borrow, 262
Bowditch's adjustment, 106–14
braced quadrilateral, 120
break lines, 173
bridges, setting out, 377, 383
buildings, setting out, 380
bulking, 270

cant, 312
central meridian, 128
centre point polygon, 120
centring errors, 60
chain
lines, 155
survey, 155
chainage, 156, 301

chainmen, 285
change point, 17
circle
  reading (theodolite), 45–8
  setting, 49
client, 279
closed traverses, 3, 95
closing errors
  levelling, 23, 24
  traversing, 99, 106
closure of levelling, 21, 24
clothoid, 315, 316, 318
coastal protection (setting out), 401
coefficient
  of friction, 315
  linear expansion, 68, 69
  of refraction, 30, 53
cofferdams *see* sheet piles
collimation
  error
    EDM, 75
    levels, 34
  method, 22
  test levels, 33
column bases, setting out, 375–6
column setting out
  in situ, 380
  prefabricated, 379
comfort criterion (vertical curves), 331, 337
composite curves, 312, 322–30
  worked example, 324
compound curves, 331
compromise tension, 68
computer
  design (curves), 331
  use of, 5, 178–83
condition method, 203
connecting traverse, 95
connection to National Grid, 127, 130–35
consumables, 285
contouring, 28, 172–7
contractor, 280
control
  networks, 119, 121–7
  setting out, 276, 295
  stations for setting out, 291, 295
  surveys, 2, 86–152
    adjustment of, 202–37
    angular measurement for, 88
    bench marks for, 88
    levelling in, 88, 135
    linear measurement for, 88
    planning of, 87
    station selection for, 87, 96, 122
    underground, 391–6
conventional symbols, 159, 160
convergence of meridians, 132
converging plumb lines (normals), 124
coordinate
  differences, 91
  measuring systems, 117, 385
  systems, 89
coordinated curve design, 345
co-planning, 391
corner profiles, 291, 382
covariance, 222
corrections
  in EDM, 76
  in taping, 66
crest
  curves, 331
  sight distance, 331, 338
crossfall (normal), 315
cross-hairs, 12, 13, 38, 43, 44
  test (theodolite) and adjustment, 56
cross-sections, 26
  of roads, 245–9
cubic
  parabola, 318
  spiral, 317
curvature effect
  in levelling, 29
  in theodolite use, 53
curve(s)
  calculations, 299–345
  circular horizontal, 299–312, 350
  composite, 312, 322–30
  compound, 330
  large radius, 301–308
  obstructions, 307, 330
  radius, 300, 310, 313
  setting out, 301–10, 306–30, 348
  small radius, 309
  through a given point
    horizontal, 310
    vertical, 341
  transition, 312–30
  vertical, 331–43
curved solids, volumes of, 260
cutting, setting out, 349

data
  loggers, 5, 81, 161, 162, 178–81
  memory cards, 83, 179

storage, 178–81
datum levels, 380
deck construction, setting out, 377, 378, 384
drains, 368
foundations, 376
deflection angles
for horizontal circular curves, 302
for transition curves, 317
degree of curvature, 300
Department of Transport Standard, 299, 313
depression angle, 37
detail surveys, 2, 3, 153–87
electromagnetic (EDM) method, 159–68
linear method, 155–9
optical tacheometry method, 168–71
differential GPS, 143
digital
ground (terrain) model, 178
level, 16
dipping, 360
distance measurement *see* linear measurement
double ended trig heighting, 54
double crossfall, 356, 360
drain-off chambers, 372
drains and pipelines, 361–72

earthworks
setting out, 349–58
volumes, 249–70
eastings, 89
eccentricity of circle, 57
electromagnetic distance measurement (EDM), 70–84
accuracy, 79
applications, 79
atmospheric correction, 76
battery problems, 74
calibration, 79, 213
corrections, 76
in curve ranging, 307
description and principle, 70
developments, 80
errors, 95–9
operation, 72
prism constant, 79, 212
problems, 74
electromagnetic (EDM) detailing, 159–68
electronic theodolites, 39, 40
elevation angle, 37
ellipsoid, 127
ellipsoidal height, 129
embankment setting out, 353
employer, 279
engineer, 279
engineering surveying, 1
engineer's level, 10, 11
error ellipses, 223
worked example, 225
errors (and adjustment), 188–237
accumulation of, 2, 286
in angular measurement, 55–60
in EDM measurement, 75–9
in levelling, 23
random, 2, 189
systematic, 2, 188
Euler spiral, 315
eyepiece
focusing, 13, 15, 43, 44, 46
laser, 293

face, 44
control, 396
fieldbooks, 6, 18, 156, 163, 166, 170
figure checks, 120
focusing, 12, 13, 15, 38, 43, 44, 46
footings setting out, 375–6
foresight, 17, 19
formation setting out, 354
formwork, setting out, 377, 378, 380
forward chainage, 301
foundations, setting out, 373–7
freehaul distance, 263

geodetic
surveying, 1
surveys, 127–35
geoid, 128
Global Positioning System (GPS), 3, 5, 11, 86, 88, 135, 136–49
ambiguity resolution, 139
atmospheric effects, 142
detailing, 171
differential GPS, 143
dilution of precision, 140
equipment, 137, 140
errors, 142
kinematic, 147
multipath errors, 143
pseudo-kinematic, 145
pseudo range, 137
rapid static, 145

selective availability, 143
static, 144
stop and go, 147
transmitted signal, 137
vertical control, 144
good practice, 6–8
grazing rays, 18, 23, 46, 122
grid/ground conversion, 127–35
grid
levelling, 24
north, 132
origin, 89, 130
gross errors in traverses, 115
gully connections, 362, 371
gullies, setting out, 359
gyrotheodolite, 395

halving and quartering, 310
haul, 263
distance, 263
headlight distances, 331, 340
height control, 135
height of collimation (HPC) method, 22
highest point on crest curve, 336
Highway Design Standard, 299, 313, 346
highway transition tables, 316, 319, 320, 321
horizontal
angle measurement, 45–50
circle eccentricity test, 57
circular curves, 299–312
collimation test and adjustment, 56
curves, 299–331
circular, 299–312
composite, 324–31
compound, 331
setting out, 301–10, 316–30, 348
transition, 312–31
underground, 395
transition curves, 312–31

included angles, 98
indexing vertical circle, 51
initial
bearing, 91
setting out (roads), 347
survey (roads), 346
inshore works, setting out, 399
in situ columns, setting out, 380
in situ slab, setting out, 379
ICE Conditions of Contract, 279
industrial measuring systems, 117, 385
instrument checks and adjustments, 2, 284
for EDM equipment, 77
for levels, 33
for theodolites, 55–9
intermediate sight, 19
intersection, 116
angle, 299
point, 299
inverted staff readings, 18, 20
irregular
areas, 240–49
solids (volumes), 249–62

K values, 334, 339, 341
kerb setting out, 359
keyboard data logger, 180
kinematic GPS, 147

land surveying, 1
lasers, 292–5
eyepiece attachments, 293
pipelaying, 293, 363, 367
rotating, 293
underground, 293, 398
latitude, 127
least squares adjustment, 115, 116, 119, 126, 203–34
left-hand curves, 303
level (engineer's), 10
automatic, 14
digital, 16
permanent adjustments of, 33
quickset, 13
tilting, 14
level surface, 10
levelling, 5, 10–36, 88
accuracy, 24
adjustment, worked example of, 207
arithmetic checks, 20
booking, 18
closure, 21, 24
collimation method of, 22
curvature effect on, 29
grids, 24
instrument tests, 33
precise method of, 28
principle, 10
procedure, 17–23
reducing, 18–23
reciprocal method of, 31
refraction effect on, 18, 23, 30

rise and fall method of, 20
staff, 10, 16, 18, 19
underground, 396–7
linear measurement, 64–85, 88
corrections
EDM, 76
tape, 66
electromagnetic, 70–84
taped, 64–9
linear surveying, 155–9
booking, 156
fieldwork, 156
obstructions, 158
offsets, 156
plotting, 159
slope correction, 157
station selection, 156
tie lines, 156
link traverse *see* tied traverse
local (magnetic) attraction, 91
local scale factors, 130
longitude, 128
longitudinal sections, 26
loop traverse, 95, 108
lowest point on sag curve, 336

machinery setting out, 384
magnetic
declination, 91
north, 91
manhole setting out, 361, 370
marine works, 399–401
markers (for setting out), 289–91, 295
mass-haul calculations, 262–70
worked examples, 263, 267
matrix solution (least squares), 221
mean, 192
meridian, 89, 127, 132
micrometer operation (theodolite), 47
mistakes, 2, 188
most probable value (MPV), 194

National Grid, 1
connection to, 127–35
networks, 119, 121–7
Newlyn datum, 1, 10
normal
distribution, 192
equations, 204, 206, 207, 213
north, 89, 91, 132, 159
northings, 89

observation equations, 204, 205
obstructions
to curve ranging, 307, 330
in linear surveys, 158
offsets, 156
offshore platforms, 400
open traverse, 95
optical
plummet (automatic), 381
plummet test and adjustment, 58
tacheometry, 168–71
theodolite, 40
Ordnance datum, 1
Ordnance Survey, 1
bench marks, 10
orienting bearings, 91
origin of grid, 89, 130
orthometric height, 129
OSGB36, 127
overhaul, 263
overtaking sight distance, 338

parallax, 14, 15
parallel plate micrometer, 28
passing a curve through a given point
horizontally, 310
vertically, 341
pavement setting out, 354–60
permanent adjustments
level, 33
theodolite, 55–9
permanent inshore works, 400
photogrammetry, 117, 177
pile(s)
caps, setting out, 375–6
load bearing, setting out, 373
sheet, setting out, 378, 399
pipe jacking, 399
pipelaying lasers, 295, 363, 367
pipelines, 361–72
plane surveys, 89
planimeter, 245
plate bubble test and adjustment, 55
plotting
by automated means, 5, 163–5, 178, 181, 183
linear surveys, 159
surveys, 5, 89
tacheometric surveys, 170
plumb wires, 390
plumbing structures, 379, 381
polar/rectangular coordinate conversion, 91
precise levelling, 28

precision, 2, 6, 7, 154, 192, 222
prism
  constant, 77, 212
  EDM, 70
prismoid, volume of, 249
prismodial rule for volumes, 251
profiles, 279, 291
  corner, 291
  drain, 363–7
  road, 354–8
promoter, 279
proving survey, 347, 388
provisional
  coordinates, 215
  values, 204
pseudo kinematic GPS, 145

quadrant, 90
quadrantal bearing, 90
quickset level, 13

radius of curvature, 300, 310, 313
railway alignment, 299, 310
random errors, 2, 189
Rankine's method (curves), 301
rapid static GPS, 145
reciprocal
  levelling, 31
  trigonometric heighting, 54
reconnaissance, 6
record drawings, 298
rectangular/polar coordinate conversion, 93
reduced
  bearing, 90
  levels, 10, 20–22
reducing levels, 18–22
redundant measures, 204
reference object, 45
referencing, 87, 296, 347, 362, 389
refraction, 18, 23, 30, 53, 77
  coefficient for light and radio waves, 126
  coefficient of, 30, 53
  lateral, 394, 399
regular area calculations, 239
rejection criteria, 195
related quantities, 202
resection, 117, 218
  worked example, 118
reservoir volumes, 252
residual, 203, 204, 206, 222, 234
right-hand curve, 303
rise and fall method, 20
road
  cross-sections, 245
  earthworks, 349–58
  profiles, 354–8
roadworks
  curve calculations, 299–345
  setting out, 301–10, 316–31, 346–60
rounding off, 8
round of angles, 49
RIBA Conditions of Contract, 279

sag
  correction, 67
  curves, 332
  headlight distances, 331, 340
scale, 154, 159
  factor, 130
setting out, 3, 6, 276–98
  accuracy in, 286
  angles, 49
  bases, 375–6
  batter rails (rules), 291, 350–54
  bridges, 377, 378, 383–4
  buildings, 380
  coastal protection, 401
  column bases, 375–6
  columns, 379, 380
  control, 276, 295
  curves by EDM (off-line), 307
  drains and pipelines, 361–72
  earthworks, 349–58
  equipment for, 283
  footings, 375–6
  foundations, 373–7
  horizontal curves, 301–10, 316–31, 348
  levels, 291
  machinery, 384
  marine works, 399–401
  markers for, 289
  organization and planning, 283
  pile caps, 375–6
  piles (loadbearing), 373
  profiles, 279, 291, 292, 354–8, 363–7, 380
  road centrelines, 301
  roadworks, 301–10, 316–21, 346–60
  sheet piles, 378, 399
  sight rails (profiles), 278, 292, 354–8, 363–7
  sketch of, 296, 367
  slope stakes, 350
  structures, 379–83

temporary works, 377–8
transition curves, 316–31
underground works, 388–99
using polar coordinates (EDM), 80, 307
walls, 380
sewers *see* drains and pipelines
shaft sinking, 389
sheet piles, setting out, 378, 399
shield control, 397
shift, 318, 322
short chords, 301
shrinkage, 270
side condition checks, 121
sight
distances (vertical curves), 331, 338–41
rails, 278, 292, 354–8, 363–7
Simpson's rule
for areas, 242
for volumes, 251
site engineer, 280
site surveys, 2
sketch of setting out, 296, 367
slope
corrections (reductions), 66, 157
stakes, 350
spherical excess, 120, 126
spheroidal
coordinates, 127
correction, 77, 126
spire test and adjustment, 56
spirit levelling, 10
stadia tacheometry, 168–71
staff, levelling, 10, 18
standard deviation, 192
standard error
for derived quantities, 197
in east/north directions, 223
maximum/minimum, 225
of mean, 193
for single observation, 193
standardisation correction, 66
static GPS, 144
station
checks and adjustments, 120
coordinates, 104
selection, 87, 96, 120, 156, 295, 393
steel pins, 292, 358
stop and go GPS, 147
stopping sight distance, 339
strings, 163, 173
structures (setting out), 379–83
sub chords, 303
summit (crest) sight distances, 331, 338
summit curves, 332
superelevation, 312, 315
symbols, conventional, 159, 160
systematic errors, 2, 7, 188

tacheometry
electromagnetic, 159–68
optical, 168–71
tall structures, setting out, 381
tangent
length, 300
point, 299
tangential angle method, 301
taping, 64–9
accuracy, 68
corrections, 66
procedure, 64
targets, 46, 88, 97
temperature correction, 68
temporary
bench marks, 10, 291, 295
works (setting out), 377–8
tension correction, 67
terrestrial photogrammetry, 117, 177
theodolite, 37–63
alidade, 40
and tape method (curve ranging), 301
angle measurement
horizontal, 45–50
vertical, 50
construction, 37
focusing, 43, 44, 45
heighting, 53
instrumental errors, 59
micrometer operation, 47
permanent adjustments, 55–9
setting up, 41
tribrach, 41
three tripod equipment, 97
tied traverse, 95
Tienstra formula, 119
tilting level, 12
tools and tackle (setting out), 284
topographic survey, 1
topsoil strip, 349
total
deflection angle, 303
station, 70
transfer underground of setting out, 389
transition
curves, 312–30
length and shape of, 313

setting out of, 316–30
tables, 316, 318–21
transverse Mercator projection, 130
trapezoidal rule
for areas, 241
for volumes, 251
traveller, 278, 350, 353, 354, 356, 358, 363, 368, 372
traversing, 95–116
adjustments, 99, 100–104, 106–15, 215–8, 227–31
Bowditch's method, 106–14
closure checks, 99
fieldwork, 97
gross errors in, 115
least squares adjustment, 115
station selection, 96
worked examples, 108–14, 227–31
triangulateration, 122
triangulation, 120
adjustment by least squares, 209
points, 1, 121
tribrach, 41, 97
trigonometric heighting, 53
trig points, 121
trilateration, 121
tripod, 13, 14, 41
true
north, 91, 132
spiral, 315
trunnion axis test and adjustment, 56
t–T correction, 134
tunnelling
laser, 293, 398
see underground surveying and setting out
two peg test and adjustment, 33
two theodolite method (curve ranging), 305

underground surveying and setting out, 388–99
unit variance, 222

variable stadia tacheometry, 710
variance, 194, 195, 222
variance–covariance matrix, 222
variation
of coordinates, 116, 126, 215–32
method, 203–15
program, 233
vertical
angle measurement, 50
circle index test and adjustment, 57
curves, 331–43
comfort criterion for, 331, 337
design of, 333–4
highest/lowest point of, 336
length of, 333
sight distances along, 331, 338
through given points, 341
volume measurement, 5, 238, 249–70
of curved solids, 260
as fill changes to cut, 253
of irregular solids, 248
from level grids, 258
by mass-haul methods, 262–70
prismoidal (Simpson's) method for, 251
of regular solids, 248
of reservoirs, 252
trapezoidal method for, 251

walls, setting out, 380
wash-out chambers, 372
waste, 262
weighting, 195, 218
Weisbach triangle, 392
WGS84, 127
whole circle bearings, 89
working from the whole to the part, 2
wriggle survey, 394

zenith angle, 37

Printed in the United Kingdom
by Lightning Source UK Ltd.
133062UK00001B/79-84/A